As

Dick Aufmann

Joanne Lockwood

We have taught math for many years. During that time, we have had students ask us a number of questions about mathematics and this course. Here you find some of the questions we have been asked most often, starting with the big one.

Why do I have to take this course?   You may have heard that *"Math is everywhere."* That is probably a slight exaggeration but math does find its way into many disciplines. There are obvious places like engineering, science, and medicine. There are other disciplines such as business, social science, and political science where math may be less obvious but still essential. If you are going to be an artist, writer, or musician, the direct connection to math may be even less obvious. Even so, as art historians who have studied the Mona Lisa have shown, there is a connection to math. But, suppose you find these reasons not all that compelling. **There is still a reason to learn basic math skills: You will be a better consumer and able to make better financial choices for you and your family.** For instance, is it better to buy a car or lease a car? Math can provide an answer.

I find math difficult. Why is that?   It is true that some people, even very smart people, find math difficult. Some of this can be traced to previous math experiences. If your basic skills are lacking, it is more difficult to understand the math in a new math course. Some of the difficulty can be attributed to the ideas and concepts in math. They can be quite challenging to learn. Nonetheless, most of us can learn and understand the ideas in the math courses that are required for graduation. **If you want math to be less difficult, practice. When you have finished practicing, practice some more.** Ask an athlete, actor, singer, dancer, artist, doctor, skateboarder, or (name a profession) what it takes to become successful and the one common characteristic they all share is that they practiced—a lot.

Why is math important?   As we mentioned earlier, math is found in many fields of study. There are, however, other reasons to take a math course. Primary among these reasons is to become a better problem solver. Math can help you learn critical thinking skills. It can help you develop a logical plan to solve a problem. Math can help you see relationships between ideas and to identify patterns. **When employers are asked what they look for in a new employee, being a problem solver is one of the highest ranked criteria.**

What do I need to do to pass this course?   The most important thing you must do is to know and understand the requirements outlined by your instructor. These requirements are usually given to you in a syllabus. Once you know what is required, you can chart a course of action. Set time aside to study and do homework. If possible, choose your classes so that you have a free hour after your math class. Use this time to review your lecture notes, rework examples given by the instructor, and to begin your homework. All of us eventually need help, so know where you can get assistance with this class. This means knowing your instructor's office hours, know the hours of the math help center, and how to access available online resources. And finally, do not get behind. **Try to do some math EVERY day, even if it is for only 20 minutes.**

Essential Mathematics
with Applications

Essential Mathematics
with Applications

EIGHTH EDITION

Richard N. Aufmann
Palomar College

Joanne S. Lockwood
Nashua Community College

BROOKS/COLE
CENGAGE Learning™

Australia • Brazil • Japan • Korea • Mexico • Singapore • Spain • United Kingdom • United States

BROOKS/COLE
CENGAGE Learning™

Essential Mathematics with Applications, Eighth Edition
Richard N. Aufmann and Joanne S. Lockwood

Acquisitions Editor: Marc Bove

Developmental Editor: Erin Brown

Assistant Editor: Shaun Williams

Editorial Assistant: Kyle O'Loughlin

Media Editor: Heleny Wong

Marketing Manager: Gordon Lee

Marketing Assistant: Erica O'Connell

Marketing Communications Manager: Katy Malatesta

Content Project Manager: Cheryll Linthicum

Creative Director: Rob Hugel

Art Director: Vernon Boes

Print Buyer: Becky Cross

Rights Acquisitions Account Manager, Text: Roberta Broyer

Rights Acquisitions Account Manager, Image: Don Schlotman

Production Service: Graphic World Inc.

Text Designer: The Davis Group

Photo Researcher: Jennifer Lim

Copy Editor: Jean Bermingham

Illustrator: Graphic World Inc.

Cover Designer: Irene Morris

Cover Image: Vito Palmisano/ Photographer's Choice/Getty Images

Compositor: Graphic World Inc.

For product information and technology assistance, contact us at **Cengage Learning Customer & Sales Support, 1-800-354-9706**

For permission to use material from this text or product, submit all requests online at **www.cengage.com/permissions**
Further permissions questions can be e-mailed to **permissionrequest@cengage.com**

Library of Congress Control Number: 200998193

ISBN-13: 978-1-4390-4697-5

ISBN-10: 1-4390-4697-2

Brooks/Cole
20 Davis Drive
Belmont, CA 94002-3098
USA

Cengage Learning is a leading provider of customized learning solutions with office locations around the globe, including Singapore, the United Kingdom, Australia, Mexico, Brazil, and Japan. Locate your local office at **www.cengage.com/global.**

Cengage Learning products are represented in Canada by Nelson Education, Ltd.

To learn more about Brooks/Cole, visit **www.cengage.com/brookscole**
Purchase any of our products at your local college store or at our preferred online store **www.ichapters.com**

Printed in the United States of America
2 3 4 5 6 7 13 12 11 10

Contents

CHAPTER 4

Ratio and Proportion
173

CHAPTER 5

Percents
201

CHAPTER 6

Applications for Business and Consumers   233

Preface

The goal in any textbook revision is to improve upon the previous edition, taking advantage of new information and new technologies, where applicable, in order to make the book more current and appealing to students and instructors. While change goes hand-in-hand with revision, a revision must be handled carefully, without compromise to valued features and pedagogy. In the eighth edition of *Essential Mathematics with Applications,* we endeavored to meet these goals.

As in previous editions, the focus remains on the **Aufmann Interactive Method (AIM).** Students are encouraged to be active participants in the classroom and in their own studies as they work through the How To examples and the paired Examples and You Try It problems. The role of "active participant" is crucial to success. Providing students with worked examples, and then affording them the opportunity to immediately work similar problems, helps them build their confidence and eventually master the concepts.

To this point, simplicity plays a key factor in the organization of this edition, as in all other editions. All lessons, exercise sets, tests, and supplements are organized around a carefully constructed hierarchy of objectives. This "objective-based" approach not only serves the needs of students, in terms of helping them to clearly organize their thoughts around the content, but instructors as well, as they work to design syllabi, lesson plans, and other administrative documents.

In order to enhance the AIM and the organization of the text around objectives, we have introduced a new design. We believe students and instructors will find the page even easier to follow. Along with this change, we have introduced several new features and modifications that we believe will increase student interest and renew the appeal of presenting the content to students in the classroom, be it live or virtual.

Changes to the Eighth Edition

With the eighth edition, previous users will recognize many of the features that they have come to trust. Yet, they will notice some new additions and changes:

- Enhanced WebAssign® now accompanies the text
- Revised exercise sets with new applications
- New **In the News** applications
- New **Think About It** exercises
- Revised Chapter Review Exercises and Chapter Tests
- End-of-chapter materials now include Concept Reviews
- Revised Chapter Openers, now with Prep Tests

Take AIM and Succeed!

Essential Mathematics with Applications is organized around a carefully constructed hierarchy of **OBJECTIVES**. This "objective-based" approach provides an integrated learning environment that allows students and professors to find resources such as assessment (both within the text and online), videos, tutorials, and additional exercises.

Each Chapter Opener outlines the **OBJECTIVES** that appear in each section. The list of objectives serves as a resource to guide you in your study and review of the topics.

ARE YOU READY? outlines what you need to know to be successful in the coming chapter.

Complete each **PREP TEST** to determine which topics you may need to study more carefully, versus those you may only need to skim over to review.

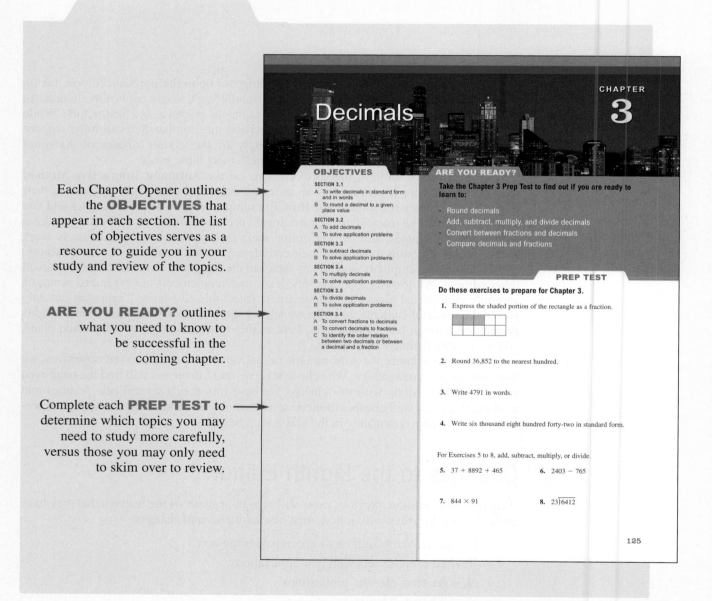

Decimals

CHAPTER 3

OBJECTIVES

SECTION 3.1
A To write decimals in standard form and in words
B To round a decimal to a given place value

SECTION 3.2
A To add decimals
B To solve application problems

SECTION 3.3
A To subtract decimals
B To solve application problems

SECTION 3.4
A To multiply decimals
B To solve application problems

SECTION 3.5
A To divide decimals
B To solve application problems

SECTION 3.6
A To convert fractions to decimals
B To convert decimals to fractions
C To identify the order relation between two decimals or between a decimal and a fraction

ARE YOU READY?

Take the Chapter 3 Prep Test to find out if you are ready to learn to:

- Round decimals
- Add, subtract, multiply, and divide decimals
- Convert between fractions and decimals
- Compare decimals and fractions

PREP TEST

Do these exercises to prepare for Chapter 3.

1. Express the shaded portion of the rectangle as a fraction.

2. Round 36,852 to the nearest hundred.

3. Write 4791 in words.

4. Write six thousand eight hundred forty-two in standard form.

For Exercises 5 to 8, add, subtract, multiply, or divide.

5. $37 + 8892 + 465$ 6. $2403 - 765$

7. 844×91 8. $23\overline{)6412}$

125

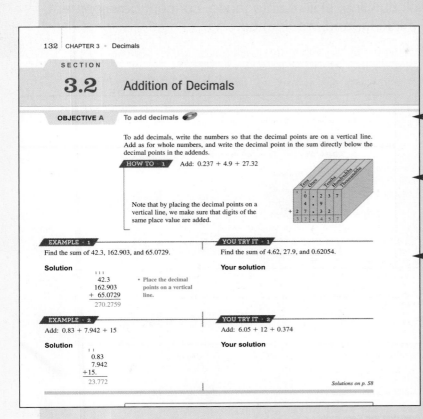

SECTION

3.2 Addition of Decimals

OBJECTIVE A To add decimals

To add decimals, write the numbers so that the decimal points are on a vertical line. Add as for whole numbers, and write the decimal point in the sum directly below the decimal points in the addends.

HOW TO 1 Add: 0.237 + 4.9 + 27.32

Note that by placing the decimal points on a vertical line, we make sure that digits of the same place value are added.

EXAMPLE · 1

Find the sum of 42.3, 162.903, and 65.0729.

Solution

$$
\begin{array}{r}
1\,1\,1\\
42.3\\
162.903\\
+65.0729\\
\hline
270.2759
\end{array}
$$

• Place the decimal points on a vertical line.

YOU TRY IT · 1

Find the sum of 4.62, 27.9, and 0.62054.

Your solution

EXAMPLE · 2

Add: 0.83 + 7.942 + 15

Solution

$$
\begin{array}{r}
1\,1\\
0.83\\
7.942\\
+15.\\
\hline
23.772
\end{array}
$$

YOU TRY IT · 2

Add: 6.05 + 12 + 0.374

Your solution

Solutions on p. S8

In each section, **OBJECTIVE STATEMENTS** introduce each new topic of discussion.

In each section, the **HOW TO'S** provide detailed explanations of problems related to the corresponding objectives.

The **EXAMPLE/YOU TRY IT** matched pairs are designed to actively involve you in learning the techniques presented. The You Try Its are based on the Examples. They appear side-by-side so you can easily refer to the steps in the Examples as you work through the You Try Its.

Complete, **WORKED-OUT SOLUTIONS** to the You Try It problems are found in an appendix at the back of the text. Compare your solutions to the solutions in the appendix to obtain immediate feedback and reinforcement of the concept(s) you are studying.

SOLUTIONS TO CHAPTER 3 "YOU TRY IT"

SECTION 3.1

You Try It 1 The digit 4 is in the thousandths place.

You Try It 2 $\frac{501}{1000} = 0.501$
(five hundred one thousandths)

You Try It 3 $0.67 = \frac{67}{100}$ (sixty-seven hundredths)

You Try It 4 Fifty-five and six thousand eighty-three ten-thousandths

You Try It 5 806.00491 • 1 is in the hundred-thousandths place.

You Try It 6 ⌐ Given place value
3.675849
└ 4 < 5
3.675849 rounded to the nearest ten-thousandth is 3.6758.

You Try It 7 ⌐ Given place value
48.907
└ 0 < 5
48.907 rounded to the nearest tenth is 48.9.

65 to 74, and 75 and over.

Solution
$$
\begin{array}{r}
4.48\\
4.31\\
5.41\\
+3.80\\
\hline
18.00
\end{array}
$$

18 million Americans ages 45 and older are hearing-impaired.

You Try It 4

Strategy To find the total income, add the four commissions (985.80, 791.46, 829.75, and 635.42) to the salary (875).

Solution 875 + 985.80 + 791.46 + 829.75 + 635.42 = 4117.43

Anita's total income was $4117.43.

SECTION 3.3

You Try It 1
$$
\begin{array}{r}
11\ 9\\
6\ \cancel{7}\ \cancel{10}13\\
7\,2.0\,3\,9\\
-8.4\,7\\
\hline
6\,3.5\,6\,9
\end{array}
$$
Check:
$$
\begin{array}{r}
1\ 11\\
8.47\\
+63.569\\
\hline
72.039
\end{array}
$$

Essential Mathematics with Applications contains **A WIDE VARIETY OF EXERCISES** that promote skill building, skill maintenance, concept development, critical thinking, and problem solving.

THINK ABOUT IT exercises promote conceptual understanding. Completing these exercises will deepen your understanding of the concepts being addressed.

Working through the application exercises that contain **REAL DATA** will help prepare you to answer questions and/or solve problems based on your own experiences, using facts or information you gather.

Completing the **WRITING** exercises will help you to improve your communication skills, while increasing your understanding of mathematical concepts.

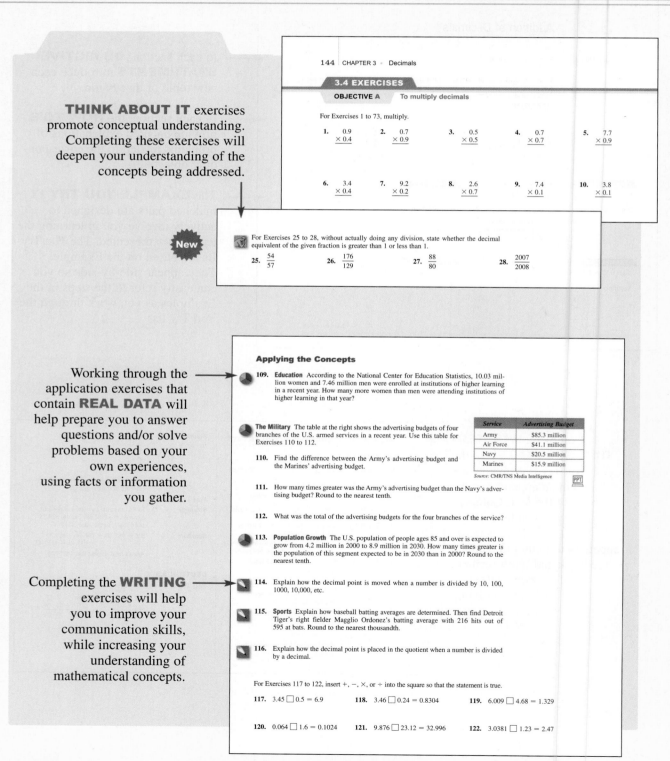

144   CHAPTER 3 · Decimals

3.4 EXERCISES

OBJECTIVE A   To multiply decimals

For Exercises 1 to 73, multiply.

1.	0.9 × 0.4	2.	0.7 × 0.9	3.	0.5 × 0.5	4.	0.7 × 0.7	5.	7.7 × 0.9
6.	3.4 × 0.4	7.	9.2 × 0.2	8.	2.6 × 0.7	9.	7.4 × 0.1	10.	3.8 × 0.1

New

For Exercises 25 to 28, without actually doing any division, state whether the decimal equivalent of the given fraction is greater than 1 or less than 1.

25. $\frac{54}{57}$ 26. $\frac{176}{129}$ 27. $\frac{88}{80}$ 28. $\frac{2007}{2008}$

Applying the Concepts

109. **Education** According to the National Center for Education Statistics, 10.03 million women and 7.46 million men were enrolled at institutions of higher learning in a recent year. How many more women than men were attending institutions of higher learning in that year?

The Military The table at the right shows the advertising budgets of four branches of the U.S. armed services in a recent year. Use this table for Exercises 110 to 112.

Service	Advertising Budget
Army	$85.3 million
Air Force	$41.1 million
Navy	$20.5 million
Marines	$15.9 million

Source: CMR/TNS Media Intelligence   PPT

110. Find the difference between the Army's advertising budget and the Marines' advertising budget.

111. How many times greater was the Army's advertising budget than the Navy's advertising budget? Round to the nearest tenth.

112. What was the total of the advertising budgets for the four branches of the service?

113. **Population Growth** The U.S. population of people ages 85 and over is expected to grow from 4.2 million in 2000 to 8.9 million in 2030. How many times greater is the population of this segment expected to be in 2030 than in 2000? Round to the nearest tenth.

114. Explain how the decimal point is moved when a number is divided by 10, 100, 1000, 10,000, etc.

115. **Sports** Explain how baseball batting averages are determined. Then find Detroit Tiger's right fielder Magglio Ordonez's batting average with 216 hits out of 595 at bats. Round to the nearest thousandth.

116. Explain how the decimal point is placed in the quotient when a number is divided by a decimal.

For Exercises 117 to 122, insert $+$, $-$, $\times$, or $\div$ into the square so that the statement is true.

117. 3.45 □ 0.5 = 6.9   **118.** 3.46 □ 0.24 = 0.8304   **119.** 6.009 □ 4.68 = 1.329

120. 0.064 □ 1.6 = 0.1024   **121.** 9.876 □ 23.12 = 32.996   **122.** 3.0381 □ 1.23 = 2.47

23. Girl Scout Cookies Using the information in the news clipping at the right, calculate the cash generated annually **a.** from sales of Thin Mints and **b.** from sales of Trefoil shortbread cookies.

In the News

Thin Mints Biggest Seller

Every year, sales from all the Girl Scout cookies sold by about 2.7 million girls total $700 million. The most popular cookie is Thin Mints, which earn 25% of total sales, while sales of the Trefoil shortbread cookies represent only 9% of total sales.

Source: Southwest Airlines Spirit Magazine 2007

24. Charities The American Red Cross spent $185,048,179 for administrative expenses. This amount was 3.16% of its total revenue. Find the American Red Cross's total revenue. Round to the nearest hundred million.

25. Poultry In a recent year, North Carolina produced 1,300,000,000 pounds of turkey. This was 18.6% of the U.S. total in that year. Calculate the U.S. total turkey production for that year. Round to the nearest billion.

26. Mining During 1 year, approximately 2,240,000 ounces of gold went into the manufacturing of electronic equipment in the United States. This is 16% of all the gold mined in the United States that year. How many ounces of gold were mined in the United States that year?

27. Education See the news clipping at the right. What percent of the baby boomers living in the United States have some college experience but have not earned a college degree? Round to the nearest tenth of a percent.

In the News

Over Half of Baby Boomers Have College Experience

Of the 78 million baby boomers living in the United States, 45 million have some college experience but no college degree. Twenty million baby boomers have one

---

New

IN THE NEWS application exercises help you master the utility of mathematics in our everyday world. They are based on information found in popular media sources, including newspapers and magazines, and the Web.

APPLYING THE CONCEPTS exercises may involve further exploration of topics, or they may involve analysis. They may also integrate concepts introduced earlier in the text. **Optional** scientific calculator exercises are included, denoted by .

Applying the Concepts

77. Air Pollution An emissions test for cars requires that of the total engine exhaust, less than 1 part per thousand $\left(\frac{1}{1000} = 0.001\right)$ be hydrocarbon emissions. Using this figure, determine which of the cars in the table at the right would fail the emissions test.

Car	Total Engine Exhaust	Hydrocarbon Emission
1	367,921	360
2	401,346	420
3	298,773	210
4	330,045	320
5	432,989	450

---

164   CHAPTER 3 • Decimals

PROJECTS AND GROUP ACTIVITIES

Fractions as Terminating or Repeating Decimals

✓ **Take Note**
If the denominator of a fraction in simplest form is 20, then it can be written as a terminating decimal because 20 = 2 · 2 · 5 (only prime factors of 2 and 5). If the denominator of a fraction in simplest form is 6, it represents a repeating decimal because it contains the prime factor 3 (a number other than 2 or 5).

The fraction $\frac{3}{4}$ is equivalent to 0.75. The decimal 0.75 is a **terminating decimal** because there is a remainder of zero when 3 is divided by 4. The fraction $\frac{1}{3}$ is equivalent to 0.333 . . . . The three dots mean the pattern continues on and on. 0.333 . . . is a **repeating decimal.** To determine whether a fraction can be written as a terminating decimal, first write the fraction in simplest form. Then look at the denominator of the fraction. If it contains prime factors of only 2s and/or 5s, then it can be expressed as a terminating decimal. If it contains prime factors other than 2s or 5s, it represents a repeating decimal.

1. Assume that each of the following numbers is the denominator of a fraction written in simplest form. Does the fraction represent a terminating or repeating decimal?
a. 4 **b.** 5 **c.** 7 **d.** 9 **e.** 10 **f.** 12 **g.** 15
h. 16 **i.** 18 **j.** 21 **k.** 24 **l.** 25 **m.** 28 **n.** 40

2. Write two other numbers that, as denominators of fractions in simplest form, represent terminating decimals, and write two other numbers that, as denominators of fractions in simplest form, represent repeating decimals.

PROJECTS AND GROUP ACTIVITIES appear at the end of each chapter. Your instructor may assign these to you individually, or you may be asked to work through the activity in groups.

Essential Mathematics with Applications addresses
students' broad range of study styles
by offering **A WIDE VARIETY OF TOOLS FOR REVIEW**.

At the end of each chapter you will find a **SUMMARY** with **KEY WORDS** and **ESSENTIAL RULES AND PROCEDURES**. Each entry includes an example of the summarized concept, an objective reference, and a page reference to show where each concept was introduced.

CHAPTER 3

SUMMARY

KEY WORDS	EXAMPLES
A number written in *decimal notation* has three parts: a *whole-number part*, a *decimal point*, and a *decimal part*. The decimal part of a number represents a number less than 1. A number written in decimal notation is often simply called a *decimal*. [3.1A, p. 126]	For the decimal 31.25, 31 is the whole-number part and 25 is the decimal part.

ESSENTIAL RULES AND PROCEDURES	EXAMPLES
To write a decimal in words, write the decimal part as if it were a whole number. Then name the place value of the last digit. The decimal point is read as "and." [3.1A, p. 126]	The decimal 12.875 is written in words as twelve and eight hundred seventy-five thousandths.
To write a decimal in standard form when it is written in words, write the whole-number part, replace the word *and* with a decimal point, and write the decimal part so that the last digit is in the given place-value position. [3.1A, p. 127]	The decimal forty-nine and sixty-three thousandths is written in standard form as 49.063.

166   CHAPTER 3 • Decimals

CHAPTER 3

CONCEPT REVIEW

Test your knowledge of the concepts presented in this chapter. Answer each question. Then check your answers against the ones provided in the Answer Section.

1. How do you round a decimal to the nearest tenth?

2. How do you write the decimal 0.37 as a fraction?

3. How do you write the fraction $\frac{173}{10,000}$ as a decimal?

CONCEPT REVIEWS actively engage you as you study and review the contents of a chapter. The **ANSWERS** to the questions are found in an appendix at the back of the text. After each answer, look for an objective reference that indicates where the concept was introduced.

By completing the chapter **REVIEW EXERCISES**, you can practice working problems that appear in an order that is different from the order they were presented in the chapter. The **ANSWERS** to these exercises include references to the section objectives upon which they are based. This will help you to quickly identify where to go to review the concepts if needed.

CHAPTER 3

REVIEW EXERCISES

1. Find the quotient of 3.6515 and 0.067.

2. Find the sum of 369.41, 88.3, 9.774, and 366.474.

3. Place the correct symbol, < or >, between the two numbers.
   0.055   0.1

4. Write 22.0092 in words.

5. Round 0.05678235 to the nearest hundred-thousandth.

6. Convert $2\frac{1}{3}$ to a decimal. Round to the nearest hundredth.

7. Convert 0.375 to a fraction.

8. Add: 3.42 + 0.794 + 32.5

Each chapter **TEST** is designed to simulate a possible test of the concepts covered in the chapter. The **ANSWERS** include references to section objectives. References to How Tos, worked Examples, and You Try Its, that provide solutions to similar problems, are also included.

TEST

1. Place the correct symbol, < or >, between the two numbers.
0.66   0.666

2. Subtract: 13.027
              − 8.94

3. Write 45.0302 in words.

4. Convert $\frac{9}{13}$ to a decimal. Round to the nearest thousandth.

5. Convert 0.825 to a fraction.

6. Round 0.07395 to the nearest ten-thousandth.

7. Find 0.0569 divided by 0.037. Round to the nearest thousandth.

8. Find 9.23674 less than 37.003.

9. Round 7.0954625 to the nearest thousandth.

10. Divide: $0.006\overline{)1.392}$

CUMULATIVE REVIEW EXERCISES

1. Divide: $89\overline{)20,932}$

2. Simplify: $2^3 \cdot 4^2$

3. Simplify: $2^2 - (7 - 3) \div 2 + 1$

4. Find the LCM of 9, 12, and 24.

5. Write $\frac{22}{5}$ as a mixed number.

6. Write $4\frac{5}{8}$ as an improper fraction.

7. Write an equivalent fraction with the given denominator.
$\frac{5}{12} = \frac{}{60}$

8. Add: $\frac{3}{8} + \frac{5}{12} + \frac{9}{16}$

9. What is $5\frac{7}{12}$ increased by $3\frac{7}{18}$?

10. Subtract: $9\frac{5}{9} - 3\frac{11}{12}$

11. Multiply: $\frac{9}{16} \times \frac{4}{27}$

12. Find the product of $2\frac{1}{8}$ and $4\frac{5}{17}$.

13. Divide: $\frac{11}{12} \div \frac{3}{4}$

14. What is $2\frac{3}{8}$ divided by $2\frac{1}{2}$?

CUMULATIVE REVIEW EXERCISES, which appear at the end of each chapter (beginning with Chapter 2), help you maintain skills you previously learned. The **ANSWERS** include references to the section objectives upon which the exercises are based.

Other Key Features

MARGINS Within the margins, students can find the following.

 Take Note boxes alert students to concepts that require special attention.

 Point of Interest boxes, which may be historical in nature or be of general interest, relate to topics under discussion.

 Integrated Technology boxes, which are offered as optional instruction in the proper use of the scientific calculator, appear for selected topics under discussion.

 Tips for Success boxes outline good study habits.

ESTIMATION

Estimating the Sum of Two or More Decimals

Calculate 23.037 + 16.7892. Then use estimation to determine whether the sum is reasonable.

Add to find the exact sum.   23.037 ⊞ 16.7892 ⊟ 39.8262

To estimate the sum, round each number to
the same place value. Here we have
rounded to the nearest whole number. Then
add. The estimated answer is 40, which is
very close to the exact sum, 39.8262.

$$\begin{array}{r} 23.037 \approx \quad 23 \\ +16.7892 \approx +17 \\ \hline 40 \end{array}$$

ESTIMATION Throughout the textbook, Estimation boxes appear, where appropriate. Tied to relevant content, the Estimation boxes demonstrate how estimation may be used to check answers for reasonableness.

PROBLEM-SOLVING STRATEGIES The text features a carefully developed approach to problem solving that encourages students to develop a Strategy for a problem and then to create a Solution based on the Strategy.

EXAMPLE · 3

Determine the number of Americans under the age of 45 who are hearing-impaired.

Strategy
To determine the number, add the numbers of hearing impaired ages 0 to 17, 18 to 34, and 35 to 44.

Solution
$$\begin{array}{r} 1.37 \\ 2.77 \\ +4.07 \\ \hline 8.21 \end{array}$$

8.21 million Americans under the age of 45 are hearing-impaired.

YOU TRY IT · 3

Determine the number of Americans ages 45 and older who are hearing-impaired.

Your strategy

Your solution

EXAMPLE · 4

Dan Burhoe earned a salary of $210.48 for working 3 days this week as a food server. He also received $82.75, $75.80, and $99.25 in tips during the 3 days. Find his total income for the 3 days of work.

YOU TRY IT · 4

Anita Khavari, an insurance executive, earns a salary of $875 every 4 weeks. During the past 4-week period, she received commissions of $985.80, $791.46, $829.75, and $635.42. Find her total income for the past 4-week period.

FOCUS ON PROBLEM SOLVING At the end of each chapter, the Focus on Problem Solving fosters further discovery of new problem-solving strategies, such as applying solutions to other problems, working backwards, inductive reasoning, and trial and error.

FOCUS ON PROBLEM SOLVING

Relevant Information

Problems in mathematics or real life involve a question or a need and information or circumstances related to that question or need. Solving problems in the sciences usually involves a question, an observation, and measurements of some kind.

One of the challenges of problem solving in the sciences is to separate the information that is relevant to the problem from other information. Following is an example from the physical sciences in which some relevant information was omitted.

Hooke's Law states that the distance that a weight will stretch a spring is directly proportional to the weight on the spring. That is, $d = kF$, where d is the distance the spring is stretched and F is the force. In an experiment to verify this law, some physics students were continually getting inconsistent results. Finally, the instructor discovered that the heat produced when the lights were turned on was affecting the experiment. In this case, relevant information was omitted—namely, that the temperature of the spring can affect the distance it will stretch.

A lawyer drove 8 miles to the train station. After a 35-minute ride of 18 miles, the lawyer walked 10 minutes to the office. Find the total time it took the lawyer to get to work.

From this situation, answer the following before reading on.

a. What is asked for?

b. Is there enough information to answer the question?

c. Is information given that is not needed?

General Revisions

- Chapter Openers now include Prep Tests for students to test their knowledge of prerequisite skills for the new chapter.

- Each exercise set has been thoroughly reviewed to ensure that the pace and scope of the exercises adequately cover the concepts introduced in the section.

- The variety of word problems has increased. This will appeal to instructors who teach to a range of student abilities and want to address different learning styles.

- Think About It exercises, which are conceptual in nature, have been added. They are meant to assess and strengthen a student's understanding of the material presented in an objective.

- In the News exercises have been added and are based on a media source such as a newspaper, a magazine, or the Web. The exercises demonstrate the pervasiveness and utility of mathematics in a contemporary setting.

- Concept Reviews now appear in the end-of-chapter materials to help students more actively study and review the contents of the chapter.

- The Chapter Review Exercises and Chapter Tests have been adjusted to ensure that there are questions that assess the key ideas in the chapter.

- The design has been significantly modified to make the text even easier for students to follow.

Acknowledgments

The authors would like to thank the people who have reviewed this manuscript and provided many valuable suggestions.

> Dorothy Fujimura, *CSU East Bay*
> Rinav Mehta, *Seattle Central Community College*
> Joseph Phillips, *Warren County Community College*
> Yan Tian, *Palomar College*

The authors would also like to thank the people who reviewed the seventh edition.

> Dorothy A. Brown, *Camden County College, NJ*
> Kim Doyle, *Monroe Community College, NY*
> Said Fariabi, *San Antonio College, TX*
> Kimberly A. Gregor, *Delaware Technical and Community College, DE*
> Allen Grommet, *East Arkansas Community College, AR*
> Anne Haney
> Rose M. Kaniper, *Burlington County College, NJ*
> Mary Ann Klicka, *Bucks County Community College, PA*
> Helen Medley, *Kent State University, OH*
> Steve Meidinger, *Merced College, CA*
> James R. Perry, *Owens Community College, OH*
> Gowribalan Vamadeva, *University of Cincinnati, OH*
> Susan Wessner, *Tallahassee Community College, FL*

Special thanks go to Jean Bermingham for copyediting the manuscript and proofreading pages, to Carrie Green for preparing the solutions manuals, and to Lauri Semarne for her work in ensuring the accuracy of the text. We would also like to thank the many people at Cengage Learning who worked to guide the manuscript from development through production.

Instructor Resources

Print Ancillaries

Complete Solutions Manual (0-538-49542-1)
Carrie Green

The Complete Solutions Manual provides worked-out solutions to all of the problems in the text.

Instructor's Resource Binder (0-538-49775-0)
Maria H. Andersen, *Muskegon Community College*

New The Instructor's Resource Binder contains uniquely designed Teaching Guides, which include instruction tips, examples, activities, worksheets, overheads, and assessments, with answers to accompany them.

Appendix to accompany Instructor's Resource Binder (0-538-49775-0)
Richard N. Aufmann, *Palomar College*
Joanne S. Lockwood, *Nashua Community College*

New! The Appendix to accompany the Instructor's Resource Binder contains teacher resources that are tied directly to *Basic College Mathematics: An Applied Approach,* 9e. Organized by objective, the Appendix contains additional questions and short, in-class activities. The Appendix also includes answers to Writing Exercises, Focus on Problem Solving, and Projects and Group Activities found in the text.

Electronic Ancillaries

Enhanced WebAssign **ENHANCED** **WebAssign**

New Used by over one million students at more than 1,100 institutions, WebAssign allows you to assign, collect, grade, and record homework assignments via the Web. This proven and reliable homework system includes thousands of algorithmically generated homework problems, links to relevant textbook sections, video examples, problem-specific tutorials, and more.

Solution Builder (0-840-03659-0)

This online solutions manual allows instructors to create customizable solutions that they can print out to distribute or post as needed. This is a convenient and expedient way to deliver solutions to specific homework sets.

PowerLecture with Diploma®
(0-538-49667-3)

New This CD-ROM provides the instructor with dynamic media tools for teaching. Create, deliver, and customize tests (both print and online) in minutes with Diploma's Computerized Testing featuring algorithmic equations. Easily build solution sets for homework or exams using Solution Builder's online solutions manual. Quickly and easily update your syllabus with the new Syllabus Creator, which was created by the authors and contains the new edition's table of contents. Practice Sheets, First Day of Class PowerPoint® lecture slides, art and figures from the book, and a test bank in electronic format are also included on this CD-ROM.

Text Specific DVDs (0-538-73632-1)

Hosted by Dana Mosely and captioned for the hearing-impaired, these DVDs cover all sections in the text. Ideal for promoting individual study and review, these comprehensive DVDs also support students in online courses or those who may have missed a lecture.

Student Resources

Print Ancillaries

Student Solutions Manual (0-538-49352-6)
Carrie Green

The Student Solutions Manual provides worked-out solutions to the odd-numbered problems in the textbook.

Student Workbook (0-538-49399-2)
Maria H. Andersen, *Muskegon Community College*

New Get a head-start! The Student Workbook contains assessments, activities, and worksheets from the Instructor's Resource Binder. Use them for additional practice to help you master the content.

Electronic Ancillaries

Enhanced WebAssign **ENHANCED** **WebAssign**

New If you are looking for extra practice or additional support, Enhanced WebAssign offers practice problems, videos, and tutorials that are tied directly to the problems found in the textbook.

Text Specific DVDs (0-538-73632-1)

Hosted by Dana Mosley, an experienced mathematics instructor, the DVDs will help you to get a better handle on topics found in the textbook. A comprehensive set of DVDs for the entire course is available to order.

AIM for Success: Getting Started

Welcome to *Essential Mathematics with Applications!* Students come to this course with varied backgrounds and different experiences in learning math. We are committed to your success in learning mathematics and have developed many tools and resources to support you along the way. Want to excel in this course? Read on to learn the skills you'll need and how best to use this book to get the results you want.

Motivate Yourself

You'll find many real-life problems in this book, relating to sports, money, cars, music, and more. We hope that these topics will help you understand how you will use mathematics in your real life. However, to learn all of the necessary skills and how you can apply them to your life outside this course, you need to stay motivated.

✓ Take Note

Motivation alone won't lead to success. For example, suppose a person who cannot swim is rowed out to the middle of a lake and thrown overboard. That person has a lot of motivation to swim, but will most likely drown without some help. You'll need motivation and learning in order to succeed.

> **THINK ABOUT WHY YOU WANT TO SUCCEED IN THIS COURSE. LIST THE REASONS HERE (NOT IN YOUR HEAD . . . ON THE PAPER!):**
>
> _____
>
> _____

We also know that this course may be a requirement for you to graduate or complete your major. That's OK. If you have a goal for the future, such as becoming a nurse or a teacher, you will need to succeed in mathematics first. Picture yourself where you want to be, and use this image to stay on track.

Photodisc

Make the Commitment

Stay committed to success! With practice, you will improve your math skills. Skeptical? Think about when you first learned to ride a bike or drive a car. You probably felt self-conscious and worried that you might fail. But with time and practice, it became second nature to you.

You will also need to put in the time and practice to do well in mathematics. Think of us as your "driving" instructors. We'll lead you along the path to success, but we need you to stay focused and energized along the way.

> **LIST A SITUATION IN WHICH YOU ACCOMPLISHED YOUR GOAL BY SPENDING TIME PRACTICING AND PERFECTING YOUR SKILLS (SUCH AS LEARNING TO PLAY THE PIANO OR PLAYING BASKETBALL):**
>
> _____
>
> _____
>
> _____
>
> _____

Photodisc

If you spend time learning and practicing the skills in this book, you will also succeed in math.

Think You Can't Do Math? Think Again!

You can do math! When you first learned the skills you just listed, you may have not done them well. With practice, you got better. With practice, you will be better at math. Stay focused, motivated, and committed to success.

It is difficult for us to emphasize how important it is to overcome the "I Can't Do Math Syndrome." If you listen to interviews of very successful athletes after a particularly bad performance, you will note that they focus on the positive aspect of what they did, not the negative. Sports psychologists encourage athletes to always be positive—to have a "Can Do" attitude. Develop this attitude toward math and you will succeed.

Photodisc.

Skills for Success

GET THE BIG PICTURE If this were an English class, we wouldn't encourage you to look ahead in the book. But this is mathematics—go right ahead! Take a few minutes to read the table of contents. Then, look through the entire book. Move quickly: scan titles, look at pictures, notice diagrams.

Getting this big picture view will help you see where this course is going. To reach your goal, it's important to get an idea of the steps you will need to take along the way.

As you look through the book, find topics that interest you. What's your preference? Horse racing? Sailing? TV? Amusement parks? Find the Index of Applications at the back of the book and pull out three subjects that interest you. Then, flip to the pages in the book where the topics are featured and read the exercises or problems where they appear.

Photodisc.

WRITE THE TOPIC HERE:	WRITE THE CORRESPONDING EXERCISE/PROBLEM HERE:
_____	_____
_____	_____
_____	_____

You'll find it's easier to work at learning the material if you are interested in how it can be used in your everyday life.

Use the following activities to think about more ways you might use mathematics in your daily life. Flip open your book to the following exercises to answer the questions.

- (see p. 83, #82) I just started a new job and will be paid hourly, but my hours change every week. I need to use mathematics to . . .

_____

_____

_____

_____

- (see p. 228, #24) I'd like to buy a new video camera, but it's very expensive. I need math to . . .

_____

_____

_____

- (see p. 147, #106) I want to rent a car, but I have to find the company that offers the best overall price. I need mathematics to . . .

_____

_____

_____

You know that the activities you just completed are from daily life, but do you notice anything else they have in common? That's right—they are **word problems.** Try not to be intimidated by word problems. You just need a strategy. It's true that word problems can be challenging because we need to use multiple steps to solve them:

- Read the problem.
- Determine the quantity we must find.
- Think of a method to find it.
- Solve the problem.
- Check the answer.

In short, we must come up with a **strategy** and then use that strategy to find the **solution.**

We'll teach you about strategies for tackling word problems that will make you feel more confident in branching out to these problems from daily life. After all, even though no one will ever come up to you on the street and ask you to solve a multiplication problem, you will need to use math every day to balance your checkbook, evaluate credit card offers, etc.

Take a look at the following example. You'll see that solving a word problem includes finding a *strategy* and using that strategy to find a *solution.* If you find yourself struggling with a word problem, try writing down the information you know about the problem. Be as specific as you can. Write out a phrase or a sentence that states what you are trying to find. Ask yourself whether there is a formula that expresses the known and unknown quantities. Then, try again!

Photodisc

EXAMPLE · 7	YOU TRY IT · 7
It costs \$.036 an hour to operate an electric motor. How much does it cost to operate the motor for 120 hours?	The cost of electricity to run a freezer for 1 hour is \$.035. This month the freezer has run for 210 hours. Find the total cost of running the freezer this month.
Strategy	**Your strategy**
To find the cost of running the motor for 120 hours, multiply the hourly cost (0.036) by the number of hours the motor is run (120).	
Solution	**Your solution**

$$
\begin{array}{r}
0.036 \\
\times \quad 120 \\
\hline
720 \\
36 \quad \\
\hline
4.320
\end{array}
$$

The cost of running the motor for 120 hours is \$4.32.

Page 143

✓ **Take Note**

Take a look at your syllabus to see if your instructor has an **attendance policy** that is part of your overall grade in the course.

The attendance policy will tell you:

- How many classes you can miss without a penalty
- What to do if you miss an exam or quiz
- If you can get the lecture notes from the professor if you miss a class

✓ **Take Note**

When planning your schedule, give some thought to how much time you realistically have available each week. For example, if you work 40 hours a week, take 15 units, spend the recommended study time given at the right, and sleep 8 hours a day, you will use over 80% of the available hours in a week. That leaves less than 20% of the hours in a week for family, friends, eating, recreation, and other activities.

Visit http://college. cengage.com/masterstudent/ shared/content/time_chart/ chart.html and use the Interactive Time Chart to see how you're spending your time—you may be surprised.

GET THE BASICS   On the first day of class, your instructor will hand out a **syllabus** listing the requirements of your course. Think of this syllabus as your personal roadmap to success. It shows you the destinations (topics you need to learn) and the dates you need to arrive at those destinations (by when you need to learn the topics). Learning mathematics is a journey. But, to get the most out of this course, you'll need to know what the important stops are and what skills you'll need to learn for your arrival at those stops.

You've quickly scanned the table of contents, but now we want you to take a closer look. Flip open to the table of contents and look at it next to your syllabus. Identify when your major exams are and what material you'll need to learn by those dates. For example, if you know you have an exam in the second month of the semester, how many chapters of this text will you need to learn by then? What homework do you have to do during this time? Managing this important information will help keep you on track for success.

MANAGE YOUR TIME   We know how busy you are outside of school. Do you have a full-time or a part-time job? Do you have children? Visit your family often? Play basketball or write for the school newspaper? It can be stressful to balance all of the important activities and responsibilities in your life. Making a **time management plan** will help you create a schedule that gives you enough time for everything you need to do.

Photodisc

Let's get started! Create a weekly schedule.

First, list all of your responsibilities that take up certain set hours during the week. Be sure to include:

- each class you are taking
- time you spend at work
- any other commitments (child care, tutoring, volunteering, etc.)

Then, list all of your responsibilities that are more flexible. Remember to make time for:

- **STUDYING** You'll need to study to succeed, but luckily you get to choose what times work best for you. Keep in mind:
  - Most instructors ask students to spend twice as much time studying as they do in class (3 hours of class = 6 hours of study).
  - Try studying in chunks. We've found it works better to study an hour each day, rather than studying for 6 hours on one day.
  - Studying can be even more helpful if you're able to do it right after your class meets, when the material is fresh in your mind.
- **MEALS** Eating well gives you energy and stamina for attending classes and studying.
- **ENTERTAINMENT** It's impossible to stay focused on your responsibilities 100% of the time. Giving yourself a break for entertainment will reduce your stress and help keep you on track.
- **EXERCISE** Exercise contributes to overall health. You'll find you're at your most productive when you have both a healthy mind and a healthy body.

Here is a sample of what part of your schedule might look like:

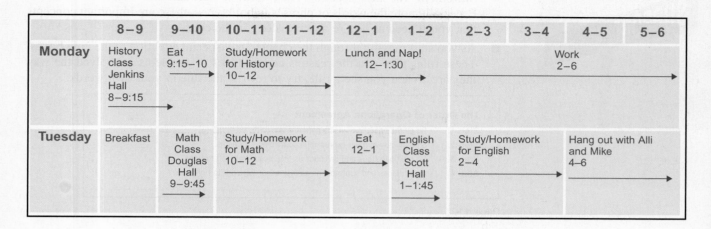

	8–9	9–10	10–11	11–12	12–1	1–2	2–3	3–4	4–5	5–6
Monday	History class Jenkins Hall 8–9:15	Eat 9:15–10	Study/Homework for History 10–12		Lunch and Nap! 12–1:30		Work 2–6			
Tuesday	Breakfast	Math Class Douglas Hall 9–9:45	Study/Homework for Math 10–12		Eat 12–1	English Class Scott Hall 1–1:45	Study/Homework for English 2–4		Hang out with Alli and Mike 4–6	

Features for Success in This Text

ORGANIZATION   Let's look again at the Table of Contents. There are six chapters in this book. You'll see that every chapter is divided into **sections,** and each section contains a number of **learning objectives.** Each learning objective is labeled with a letter from A to D. Knowing how this book is organized will help you locate important topics and concepts as you're studying.

PREPARATION   Ready to start a new chapter? Take a few minutes to be sure you're ready, using some of the tools in this book.

- ■ **CUMULATIVE REVIEW EXERCISES:**   You'll find these exercises after every chapter, starting with Chapter 2. The questions in the Cumulative Review Exercises are taken from the previous chapters. For example, the Cumulative Review for Chapter 3 will test all of the skills you have learned in Chapters 1, 2, and 3. Use this to refresh yourself before moving on to the next chapter, or to test what you know before a big exam.

Here's an example of how to use the Cumulative Review:

- • Turn to page 171 and look at the questions for the Chapter 3 Cumulative Review, which are taken from the current chapter and the previous chapters.
- • We have the answers to all of the Cumulative Review Exercises in the back of the book. Flip to page A10 to see the answers for this chapter.
- • Got the answer wrong? We can tell you where to go in the book for help! For example, scroll down page A10 to find the answer for the first exercise, which is 235 r17. You'll see that after this answer, there is an **objective reference** [1.5C]. This means that the question was taken from Chapter 1, Section 5, Objective C. Go here to restudy the objective.
- ■ **PREP TESTS:**   These tests are found at the beginning of every chapter and will help you see if you've mastered all of the skills needed for the new chapter.

Here's an example of how to use the Prep Test:

- • Turn to page 173 and look at the Prep Test for Chapter 4.
- • All of the answers to the Prep Tests are in the back of the book. You'll find them in the first set of answers in each answer section for a chapter. Turn to page A10 to see the answers for this Prep Test.
- • Restudy the objectives if you need some extra help.

Photodisc

- Before you start a new section, take a few minutes to read the **Objective Statement** for that section. Then, browse through the objective material. Especially note the words or phrases in bold type—these are important concepts that you'll need as you're moving along in the course.

- As you start moving through the chapter, pay special attention to the **rule boxes.** These rules give you the reasons certain types of problems are solved the way they are. When you see a rule, try to rewrite the rule in your own words.

The Order of Operations Agreement

Step 1. Do all the operations inside parentheses.

Step 2. Simplify any number expressions containing exponents.

Step 3. Do multiplications and divisions as they occur from left to right.

Step 4. Do additions and subtractions as they occur from left to right.

Page 110

Knowing what to pay attention to as you move through a chapter will help you study and prepare.

INTERACTION   We want you to be actively involved in learning mathematics and have given you many ways to get hands-on with this book.

- **HOW TO EXAMPLES** Take a look at page 150 shown here. See the HOW TO example? This contains an explanation by each step of the solution to a sample problem.

HOW TO · 1   Divide: $3.25)\overline{15.275}$

$3.\underset{\smile}{25}.)\overline{15.\underset{\smile}{27}.5}$ · Move the decimal point 2 places to the right in the divisor and then in the dividend. Place the decimal point in the quotient.

$$
\begin{array}{r}
4.7 \\
325.)\overline{1527.5} \\
-1300 \\
\hline
227\ 5 \\
-227\ 5 \\
\hline
0
\end{array}
$$
· Divide as with whole numbers.

Page 150

Grab a paper and pencil and work along as you're reading through each example. When you're done, get a clean sheet of paper. Write down the problem and try to complete the solution without looking at your notes or at the book. When you're done, check your answer. If you got it right, you're ready to move on.

- **EXAMPLE/YOU TRY IT PAIRS** You'll need hands-on practice to succeed in mathematics. When we show you an example, work it out beside our solution. Use the Example/You Try It pairs to get the practice you need.

Take a look at page 69, Example 5 and You Try It 5 shown here:

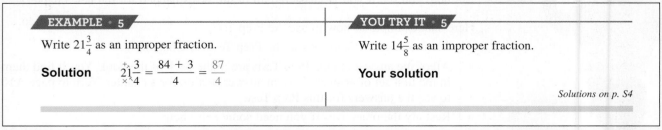

EXAMPLE · 5

Write $21\frac{3}{4}$ as an improper fraction.

Solution $\overset{+}{\underset{\times}{2}}1\frac{3}{4} = \frac{84 + 3}{4} = \frac{87}{4}$

YOU TRY IT · 5

Write $14\frac{5}{8}$ as an improper fraction.

Your solution

Solutions on p. S4

Page 69

You'll see that each Example is fully worked-out. Study this Example carefully by working through each step. Then, try your hand at it by completing the You Try It. If you get stuck, the solutions to the You Try Its are provided in the back of the book. There is a page number following the You Try It, which shows you where you can find the completely worked-out solution. Use the solution to get a hint for the step on which you are stuck. Then, try again!

When you've finished the solution, check your work against the solution in the back of the book. Turn to page S4 to see the solution for You Try It 5.

Remember that sometimes there can be more than one way to solve a problem. But, your answer should always match the answers we've given in the back of the book. If you have any questions about whether your method will always work, check with your instructor.

REVIEW We have provided many opportunities for you to practice and review the skills you have learned in each chapter.

- **SECTION EXERCISES** After you're done studying a section, flip to the end of the section and complete the exercises. If you immediately practice what you've learned, you'll find it easier to master the core skills. Want to know if you answered the questions correctly? The answers to the odd-numbered exercises are given in the back of the book.

- **CHAPTER SUMMARY** Once you've completed a chapter, look at the Chapter Summary. This is divided into two sections: *Key Words* and *Essential Rules and Procedures*. Flip to page 193 to see the Chapter Summary for Chapter 4. This summary shows all of the important topics covered in the chapter. See the reference following each topic? This shows you the objective reference and the page in the text where you can find more information on the concept.

- **CONCEPT REVIEW** Following the Chapter Summary for each chapter is the Concept Review. Flip to page 194 to see the Concept Review for Chapter 4. When you read each question, jot down a reminder note on the right about whatever you feel will be most helpful to remember if you need to apply that concept during an exam. You can also use the space on the right to mark what concepts your instructor expects you to know for the next test. If you are unsure of the answer to a concept review question, flip to the answers appendix at the back of the book.

- **CHAPTER REVIEW EXERCISES** You'll find the Chapter Review Exercises after the Concept Review. Flip to page 287 to see the Chapter Review Exercises for Chapter 6. When you do the review exercises, you're giving yourself an important opportunity to test your understanding of the chapter. The answer to each review exercise is given at the back of the book, along with the objective the question relates to. When you're done with the Chapter Review Exercises, check your answers. If you had trouble with any of the questions, you can restudy the objectives and retry some of the exercises in those objectives for extra help.

Photodisc

■ **CHAPTER TESTS** The Chapter Tests can be found after the Chapter Review Exercises and can be used to prepare for your exams. The answer to each test question is given at the back of the book, along with a reference to a How To, Example, or You Try It that the question relates to. Think of these tests as "practice runs" for your in-class tests. Take the test in a quiet place and try to work through it in the same amount of time you will be allowed for your exam.

Here are some strategies for success when you're taking your exams:

● Scan the entire test to get a feel for the questions (get the big picture).

● Read the directions carefully.

● Work the problems that are easiest for you first.

● Stay calm, and remember that you will have lots of opportunities for success in this class!

EXCEL Visit **www.cengage.com/math/aufmann** to learn about additional study tools!

■ *Enhanced WebAssign*® online practice exercises and homework problems match the textbook exercises.

■ **DVDs** Hosted by Dana Mosley, an experienced mathematics instructor, the DVDs will help you to get a better handle on topics that may be giving you trouble. A comprehensive set of DVDs for the entire course is available to order.

Get Involved

Have a question? Ask! Your professor and your classmates are there to help. Here are some tips to help you jump in to the action:

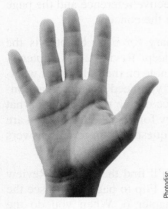

Photodisc

■ Raise your hand in class.

■ If your instructor prefers, email or call your instructor with your question. If your professor has a website where you can post your question, also look there for answers to previous questions from other students. Take advantage of these ways to get your questions answered.

■ Visit a **math center.** Ask your instructor for more information about the math center services available on your campus.

■ Your instructor will have **office hours** where he or she will be available to help you. Take note of where and when your instructor holds office hours. Use this time for one-on-one help, if you need it.

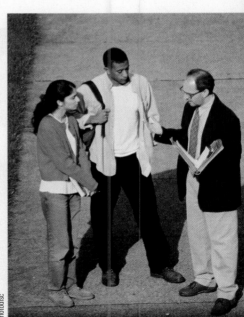

Photodisc

■ Form a **study group** with students from your class. This is a great way to prepare for tests, catch up on topics you may have missed, or get extra help on problems you're struggling with. Here are a few suggestions to make the most of your study group:

● **Test each other by asking questions.** Have each person bring a few sample questions when you get together.

- **Practice teaching each other.** We've found that you can learn a lot about what you know when you have to explain it to someone else.

- **Compare class notes.** Couldn't understand the last five minutes of class? Missed class because you were sick? Chances are someone in your group has the notes for the topics you missed.

- **Brainstorm test questions.**

- **Make a plan for your meeting.** Agree on what topics you'll talk about and how long you'll be meeting. When you make a plan, you'll be sure that you make the most of your meeting.

Ready, Set, Succeed! It takes hard work and commitment to succeed, but we know you can do it! Doing well in mathematics is just one step you'll take along the path to success.

I succeeded in Essential Mathematics!

We are confident that if you follow our suggestions, you will succeed. Good luck!

Whole Numbers

OBJECTIVES

SECTION 1.1
A To identify the order relation between two numbers
B To write whole numbers in words and in standard form
C To write whole numbers in expanded form
D To round a whole number to a given place value

SECTION 1.2
A To add whole numbers
B To solve application problems

SECTION 1.3
A To subtract whole numbers without borrowing
B To subtract whole numbers with borrowing
C To solve application problems

SECTION 1.4
A To multiply a number by a single digit
B To multiply larger whole numbers
C To solve application problems

SECTION 1.5
A To divide by a single digit with no remainder in the quotient
B To divide by a single digit with a remainder in the quotient
C To divide by larger whole numbers
D To solve application problems

SECTION 1.6
A To simplify expressions that contain exponents
B To use the Order of Operations Agreement to simplify expressions

SECTION 1.7
A To factor numbers
B To find the prime factorization of a number

ARE YOU READY?

Take the Chapter 1 Prep Test to find out if you are ready to learn to:

- Order whole numbers
- Round whole numbers
- Add, subtract, multiply, and divide whole numbers
- Simplify numerical expressions
- Factor numbers and find their prime factorization

PREP TEST

Do these exercises to prepare for Chapter 1.

1. Name the number of ◆s shown below.

    ◆ ◆ ◆ ◆ ◆ ◆ ◆ ◆   = 8 ◊

2. Write the numbers from 1 to 10.

    1 _2_ _3_ _4_ _5_ _6_ _7_ _8_ _9_ 10

3. Match the number with its word form.
    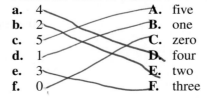
    a. 4   **A.** five
    b. 2   **B.** one
    c. 5   **C.** zero
    d. 1   **D.** four
    e. 3   **E.** two
    f. 0   **F.** three

SECTION

1.1   Introduction to Whole Numbers

OBJECTIVE A   **To identify the order relation between two numbers**

N = {1, 2, 3 . . . }

The **whole numbers** are 0, 1, 2, 3, 4, 5, 6, 7, 8, 9, 10, 11, 12, 13, 14, ...

W = {0, 1, 2, 3}

The three dots mean that the list continues on and on and that there is no largest whole number. *Continues the pattern*

Just as distances are associated with the markings on the edge of a ruler, the whole numbers can be associated with points on a line. This line is called the **number line.** The arrow on the number line below indicates that there is no largest whole number.

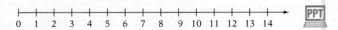

The **graph of a whole number** is shown by placing a heavy dot directly above that number on the number line. Here is the graph of 7 on the number line:

The number line can be used to show the order of whole numbers. A number that appears to the left of a given number **is less than (<)** the given number. A number that appears to the right of a given number **is greater than (>)** the given number.

Four is less than seven.
4 < 7

Twelve is greater than seven.
12 > 7

EXAMPLE • 1

Graph 11 on the number line.

Solution |—|—|—|—|—|—|—|—|—|—|—●|—|—|—→
        0 1 2 3 4 5 6 7 8 9 10 11 12 13 14

YOU TRY IT • 1

Graph 6 on the number line.

Your solution |—|—|—|—|—|—|—|—|—|—|—|—|—|—→
        0 1 2 3 4 5 6 7 8 9 10 11 12 13 14

EXAMPLE • 2

Place the correct symbol, < or >, between the two numbers.

a. 39 > 24

b. 0 < 51

Solution

a. 39 > 24

b. 0 > 51

YOU TRY IT • 2

Place the correct symbol, < or >, between the two numbers.

a. 45 > 29

b. 27 > 0

Your solution

a.

b.

Solutions on p. S1

OBJECTIVE B   **To write whole numbers in words and in standard form**

When a whole number is written using the digits 0, 1, 2, 3, 4, 5, 6, 7, 8, and 9, it is said to be in **standard form.** The position of each digit in the number determines the digit's **place value.** The diagram below shows a **place-value chart** naming the first 12 place values. The number 37,462 is in standard form and has been entered in the chart.

In the number 37,462, the position of the digit 3 determines that its place value is ten-thousands.

When a number is written in standard form, each group of digits separated from the other digits by a comma (or commas) is called a **period.** The number 3,786,451,294 has four periods. The period names are shown in red in the place-value chart above.

To write a number in words, start from the left. Name the number in each period. Then write the period name in place of the comma.

3,786,451,294 is read "three billion seven hundred eighty-six million four hundred fifty-one thousand two hundred ninety-four."

To write a whole number in standard form, write the number named in each period, and replace each period name with a comma.

Four million sixty-two thousand five hundred eighty-four is written 4,062,584. The zero is used as a place holder for the hundred-thousands place.

EXAMPLE · 3

Write 25,478,083 in words.

Solution
Twenty-five million four hundred seventy-eight thousand eighty-three

YOU TRY IT · 3

Write 36,462,075 in words.

Your solution
Thirty six thousand four hundred sixty two Seventy five

EXAMPLE · 4

Write three hundred three thousand three in standard form.

Solution
303,003

YOU TRY IT · 4

Write four hundred fifty-two thousand seven in standard form.

Your solution
452,007

Solutions on p. S1

OBJECTIVE C   **To write whole numbers in expanded form**

The whole number 26,429 can be written in **expanded form** as

20,000 + 6000 + 400 + 20 + 9.

The place-value chart can be used to find the expanded form of a number.

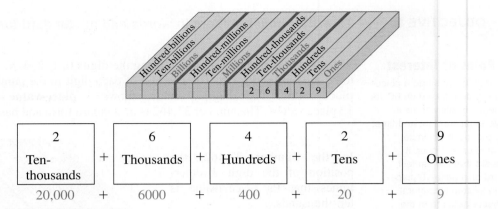

2 Ten-thousands	+	6 Thousands	+	4 Hundreds	+	2 Tens	+	9 Ones
20,000	+	6000	+	400	+	20	+	9

The number 420,806 is written in expanded form below. Note the effect of having zeros in the number.

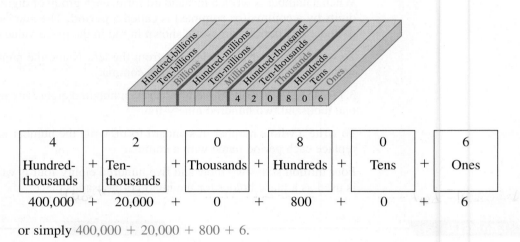

4 Hundred-thousands	+	2 Ten-thousands	+	0 Thousands	+	8 Hundreds	+	0 Tens	+	6 Ones
400,000	+	20,000	+	0	+	800	+	0	+	6

or simply $400,000 + 20,000 + 800 + 6$.

EXAMPLE • 5

Write 23,859 in expanded form.

Solution
$20,000 + 3000 + 800 + 50 + 9$

YOU TRY IT • 5

Write 68,281 in expanded form.

Your solution
Sixty thousand, + Eight thousand + two hundred + Eighty tens + One / 60,000 + 8,000 + 200 + 80 + 1

EXAMPLE • 6

Write 709,542 in expanded form.

Solution
$700,000 + 9000 + 500 + 40 + 2$

YOU TRY IT • 6

Write 109,207 in expanded form.

Your solution

Solutions on p. S1

OBJECTIVE D   **To round a whole number to a given place value**

When the distance to the moon is given as 240,000 miles, the number represents an approximation to the true distance. Taking an approximate value for an exact number is called **rounding.** A rounded number is always rounded to a given place value.

37 is closer to 40 than it is to 30. 37 rounded to the nearest ten is 40.

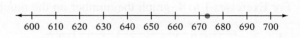

673 rounded to the nearest ten is 670. 673 rounded to the nearest hundred is 700.

A whole number is rounded to a given place value without using the number line by looking at the first digit to the right of the given place value.

HOW TO 1 Round 13,834 to the nearest hundred.

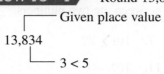

• If the digit to the right of the given place value is less than 5, that digit and all digits to the right are replaced by zeros.

13,834 rounded to the nearest hundred is 13,800.

HOW TO 2 Round 386,217 to the nearest ten-thousand.

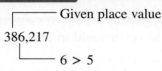

• If the digit to the right of the given place value is greater than or equal to 5, increase the digit in the given place value by 1, and replace all other digits to the right by zeros.

386,217 rounded to the nearest ten-thousand is 390,000.

EXAMPLE 7

Round 525,453 to the nearest ten-thousand.

Solution

┌─── Given place value
525,453
    └── 5 = 5

525,453 rounded to the nearest ten-thousand is 530,000.

YOU TRY IT 7

Round 368,492 to the nearest ten-thousand.

Your solution

370,000

EXAMPLE 8

Round 1972 to the nearest hundred.

Solution

┌─── Given place value
1972
   └── 7 > 5

1972 rounded to the nearest hundred is 2000.

YOU TRY IT 8

Round 3962 to the nearest hundred.

Your solution

4000

Solutions on p. S1

1.1 EXERCISES

Do only odd #'s

OBJECTIVE A   To identify the order relation between two numbers

For Exercises 1 to 4, graph the number on the number line.

1. 3   0 1 2 3 4 5 6 7 8 9 10 11 12

2. 5   0 1 2 3 4 5 6 7 8 9 10 11 12

3. 9   0 1 2 3 4 5 6 7 8 9 10 11 12

4. 0   0 1 2 3 4 5 6 7 8 9 10 11 12

For Exercises 5 to 12, place the correct symbol, $<$ or $>$, between the two numbers.

5. 37 $<$ 49

6. 58 $<$ 21

7. 101 $>$ 87

8. 245   158

9. 2701 $>$ 2071

10. 0 $<$ 45

11. 107   0

12. 815   928

13. Do the inequalities $21 < 30$ and $30 > 21$ express the same order relation?

OBJECTIVE B   To write whole numbers in words and in standard form

For Exercises 14 to 17, name the place value of the digit 3.

14. 83,479
     Thousand

15. 3,491,507
     Million

16. 2,634,958
     Tenth thousand

17. 76,319,204
     hundred Thousand

For Exercises 18 to 25, write the number in words.

18. 2675
     Two thousand Six hundred Seventy-five

19. 3790

20. 42,928

21. 58,473

22. 356,943

23. 498,512

24. 3,697,483

25. 6,842,715

For Exercises 26 to 31, write the number in standard form.

26. Eighty-five

27. Three hundred fifty-seven

28. Three thousand four hundred fifty-six

29. Sixty-three thousand seven hundred eighty

30. Six hundred nine thousand nine hundred forty-eight

31. Seven million twenty-four thousand seven hundred nine

32. What is the place value of the first number on the left in a seven-digit whole number?

OBJECTIVE C   To write whole numbers in expanded form

For Exercises 33 to 40, write the number in expanded form.

33. 5287

34. 6295

35. 58,943

36. 453,921

37. 200,583

38. 301,809

39. 403,705

40. 3,000,642

41. The expanded form of a number consists of four numbers added together. Must the number be a four-digit number?

OBJECTIVE D   To round a whole number to a given place value

For Exercises 42 to 53, round the number to the given place value.

42. 926   Tens

43. 845   Tens

44. 1439   Hundreds

45. 3973   Hundreds

46. 43,607   Thousands

47. 52,715   Thousands

48. 389,702   Thousands

49. 629,513   Thousands

50. 647,989   Ten-thousands

51. 253,678   Ten-thousands

52. 36,702,599   Millions

53. 71,834,250   Millions

54. True or false? If a number rounded to the nearest ten is less than the original number, then the ones digit of the original number is greater than 5.

Applying the Concepts

55. If 3846 is rounded to the nearest ten and then that number is rounded to the nearest hundred, is the result the same as what you get when you round 3846 to the nearest hundred? If not, which of the two methods is correct for rounding to the nearest hundred?

SECTION

1.2 Addition of Whole Numbers

OBJECTIVE A To add whole numbers

Addition is the process of finding the total of two or more numbers.

 Take Note

The numbers being added are called **addends**. The result is the **sum**.

By counting, we see that the total of $3 and $4 is $7.

$$\underset{\textbf{Addend}}{\$3} + \underset{\textbf{Addend}}{\$4} = \underset{\textbf{Sum}}{\$7}$$

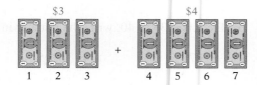

Addition can be illustrated on the number line by using arrows to represent the addends. The size, or magnitude, of a number can be represented on the number line by an arrow.

The number 3 can be represented anywhere on the number line by an arrow that is 3 units in length.

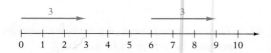

 Point of Interest

The first use of the plus sign appeared in 1489 in *Mercantile Arithmetic*. It was used to indicate a surplus, not as the symbol for addition. That use did not appear until about 1515.

To add on the number line, place the arrows representing the addends head to tail, with the first arrow starting at zero. The sum is represented by an arrow starting at zero and stopping at the tip of the last arrow.

$$3 + 4 = 7$$

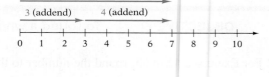

More than two numbers can be added on the number line.

$$3 + 2 + 4 = 9$$

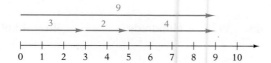

Some special properties of addition that are used frequently are given below.

Addition Property of Zero

Zero added to a number does not change the number.

$$4 + 0 = 4$$
$$0 + 7 = 7$$

Commutative Property of Addition

Two numbers can be added in either order; the sum will be the same.

$$4 + 8 = 8 + 4$$
$$12 = 12$$

 Take Note

This is the same addition problem shown on the number line above.

Associative Property of Addition

Grouping the addition in any order gives the same result. The parentheses are grouping symbols and have the meaning "Do the operations inside the parentheses first."

$$(3 + 2) + 4 = 3 + (2 + 4)$$
$$5 + 4 = 3 + 6$$
$$9 = 9$$

The number line is not useful for adding large numbers. The basic addition facts for adding one digit to one digit should be memorized. Addition of larger numbers requires the repeated use of the basic addition facts.

To add large numbers, begin by arranging the numbers vertically, keeping the digits of the same place value in the same column.

HOW TO 1 Add: 321 + 6472

$$
\begin{array}{r}
3\ 2\ 1 \\
+\ 6\ 4\ 7\ 2 \\
\hline
6\ 7\ 9\ 3
\end{array}
$$

(THOUSANDS, HUNDREDS, TENS, ONES)

• Add the digits in each column.

There are several words or phrases in English that indicate the operation of addition. Here are some examples:

added to	3 added to 5	5 + 3
more than	7 more than 5	5 + 7
the sum of	the sum of 3 and 9	3 + 9
increased by	4 increased by 6	4 + 6
the total of	the total of 8 and 3	8 + 3
plus	5 plus 10	5 + 10

Integrating Technology

Most scientific calculators use *algebraic logic:* the add (**+**), subtract (**−**), multiply (**x**), and divide (**÷**) keys perform the indicated operation on the number in the display and the next number keyed in. For instance, for the example at the right, enter 24 **+** 71 **=** . The display reads 95.

HOW TO 2 What is the sum of 24 and 71?

$$
\begin{array}{r}
24 \\
+\ 71 \\
\hline
95
\end{array}
$$

• The phrase *the sum of* means to add.

The sum of 24 and 71 is 95.

When the sum of the digits in a column exceeds 9, the addition will involve **carrying.**

HOW TO 3 Add: 487 + 369

$$
\begin{array}{r}
\overset{1}{}4\ 8\ 7 \\
+\ 3\ 6\ 9 \\
\hline
6
\end{array}
$$

(HUNDREDS, TENS, ONES)

• Add the ones column.
7 + 9 = 16 (1 ten + 6 ones).
Write the 6 in the ones column and carry the 1 ten to the tens column.

$$
\begin{array}{r}
\overset{1}{4}\ \overset{1}{8}\ 7 \\
+\ 3\ 6\ 9 \\
\hline
5\ 6
\end{array}
$$

• Add the tens column.
1 + 8 + 6 = 15 (1 hundred + 5 tens).
Write the 5 in the tens column and carry the 1 hundred to the hundreds column.

$$
\begin{array}{r}
\overset{1}{4}\ \overset{1}{8}\ 7 \\
+\ 3\ 6\ 9 \\
\hline
8\ 5\ 6
\end{array}
$$

• Add the hundreds column.
1 + 4 + 3 = 8 (8 hundreds).
Write the 8 in the hundreds column.

EXAMPLE · 1

Find the total of 17, 103, and 8.

Solution

$$\begin{array}{r} \overset{1}{17} \\ 103 \\ +\ \ 8 \\ \hline 128 \end{array}$$

• 7 + 3 + 8 = 18
Write the 8 in the ones
column. Carry the 1 to
the tens column.

YOU TRY IT · 1

What is 347 increased by 12,453?

Your solution

EXAMPLE · 2

Add: 89 + 36 + 98

Solution

$$\begin{array}{r} \overset{2}{89} \\ 36 \\ +\ 98 \\ \hline 223 \end{array}$$

• 9 + 6 + 8 = 23
Write the 3 in the ones
column. Carry the 2 to
the tens column.

YOU TRY IT · 2

Add: 95 + 88 + 67

Your solution

EXAMPLE · 3

Add: 41,395
         4,327
       497,625
     + 32,991

Solution

$$\begin{array}{r} \overset{1\,1\,2\ \ 2\,1}{41,395} \\ 4,327 \\ 497,625 \\ +\ \ 32,991 \\ \hline 576,338 \end{array}$$

YOU TRY IT · 3

Add: 392
         4,079
        89,035
      + 4,992

Your solution

Solutions on p. S1

**Integrating
Technology**

This example illustrates that
estimation is important when
one is using a calculator.

ESTIMATION

Estimation and Calculators

At some places in the text, you will be asked to use your calculator. Effective use
of a calculator requires that you estimate the answer to the problem. This helps
ensure that you have entered the numbers correctly and pressed the correct keys.

For example, if you use your calculator to find 22,347 + 5896 and the answer in
the calculator's display is 131,757,912, you should realize that you have entered
some part of the calculation incorrectly. In this case, you pressed ⟨x⟩ instead of
⟨+⟩. By estimating the answer to a problem, you can help ensure the accuracy of
your calculations. We have a special symbol for **approximately equal to** (≈).

For example, to estimate the answer to
22,347 + 5896, round each number to the same
place value. In this case, we will round to the
nearest thousand. Then add.

$$\begin{array}{r} 22,347 \approx \ \ \ \ 22,000 \\ +\ \ 5,896 \approx +\ \ 6,000 \\ \hline 28,000 \end{array}$$

The sum 22,347 + 5896 is approximately 28,000. Knowing this, you would know
that 131,757,912 is much too large and is therefore incorrect.

To estimate the sum of two numbers, first round each whole number to the same
place value and then add. Compare this answer with the calculator's answer.

OBJECTIVE B   To solve application problems

To solve an application problem, first read the problem carefully. The **strategy** involves identifying the quantity to be found and planning the steps that are necessary to find that quantity. The **solution of an application problem** involves performing each operation stated in the strategy and writing the answer.

HOW TO • 4

The table below displays the Wal-Mart store count and square footage in the United States as reported in the Wal-Mart 2008 Annual Report.

	Discount Stores	*Supercenters*	*Sam's Clubs*	*Neighborhood Markets*
Number of Units	941	2523	593	134
Square footage (in millions)	105	457	78	5

Find the total number of Wal-Mart discount stores and Supercenters in the United States.

Strategy   To find the total number of Wal-Mart discount stores and Supercenters in the United States, read the table to find the number of each type of store in the United States. Then add the numbers.

Solution
```
    941
+  2523
  ─────
   3464
```

Wal-Mart has a total of 3464 discount stores and Supercenters in the United States.

EXAMPLE • 4

Use the table above to find the total number of Sam's Clubs and neighborhood markets that Wal-Mart has in the United States.

Strategy
To determine the total number of Sam's Clubs and neighborhood markets, read the table to find the number of Sam's Clubs and the number of neighborhood markets. Then add the two numbers.

Solution
```
   593
+  134
  ────
   727
```

Wal-Mart has a total of 727 Sam's Clubs and neighborhood markets.

YOU TRY IT • 4

Use the table above to determine the total square footage of Wal-Mart stores in the United States.

Your strategy

Your solution

Solution on p. S1

1.2 EXERCISES

OBJECTIVE A　　**To add whole numbers**

For Exercises 1 to 32, add.

1.　17
　　+ 11

2.　25
　　+ 63

3.　83
　　+ 42

4.　63
　　+ 94

5.　77
　　+ 25

6.　63
　　+ 49

7.　56
　　+ 98

8.　86
　　+ 68

9.　658
　　+ 831

10.　842
　　+ 936

11.　735
　　+ 93

12.　189
　　+ 50

13.　859
　　+ 725

14.　637
　　+ 829

15.　470
　　+ 749

16.　427
　　+ 690

17.　36,925
　　+ 65,392

18.　56,772
　　+ 51,239

19.　50,873
　　+ 28,453

20.　34,872
　　+ 46,079

21.　878
　　737
　　+ 189

22.　768
　　461
　　+ 669

23.　319
　　348
　　+ 912

24.　292
　　579
　　+ 315

25.　9409
　　3253
　　+ 7078

26.　8188
　　8020
　　+ 7104

27.　2038
　　2243
　　+ 3139

28.　4252
　　6882
　　+ 5235

29.　67,428
　　32,171
　　+ 20,971

30.　52,801
　　11,664
　　+ 89,638

31.　76,290
　　43,761
　　+ 87,402

32.　43,901
　　98,301
　　+ 67,943

For Exercises 33 to 40, add.

33. 20,958 + 3218 + 42

34. 80,973 + 5168 + 29

35. 392 + 37 + 10,924 + 621

36. 694 + 62 + 70,129 + 217

37. 294 + 1029 + 7935 + 65

38. 692 + 2107 + 3196 + 92

39. 97 + 7234 + 69,532 + 276

40. 87 + 1698 + 27,317 + 727

41. What is 9874 plus 4509?

42. What is 7988 plus 5678?

43. What is 3487 increased by 5986?

44. What is 99,567 increased by 126,863?

45. What is 23,569 more than 9678?

46. What is 7894 more than 45,872?

47. What is 479 added to 4579?

48. What is 23,902 added to 23,885?

49. Find the total of 659, 55, and 1278.

50. Find the total of 4561, 56, and 2309.

51. Find the sum of 34, 329, 8, and 67,892.

52. Find the sum of 45, 1289, 7, and 32,876.

For Exercises 53 to 56, use a calculator to add. Then round the numbers to the nearest hundred, and use estimation to determine whether the sum is reasonable.

53. 1234 + 9780 + 6740

54. 919 + 3642 + 8796

55. 241 + 569 + 390 + 1672

56. 107 + 984 + 1035 + 2904

For Exercises 57 to 60, use a calculator to add. Then round the numbers to the nearest thousand, and use estimation to determine whether the sum is reasonable.

57.
```
  32,461
   9,844
+ 59,407
```

58.
```
  29,036
  22,904
+  7,903
```

59.
```
  25,432
  62,941
+ 70,390
```

60.
```
  66,541
  29,365
+ 98,742
```

For Exercises 61 to 64, use a calculator to add. Then round the numbers to the nearest ten-thousand, and use estimation to determine whether the sum is reasonable.

61.	67,421	**62.**	21,896	**63.**	281,421	**64.**	542,698
	82,984		4,235		9,874		97,327
	66,361		62,544		34,394		7,235
	10,792		21,892		526,398		73,667
	+ 34,037		+ 1,334		+ 94,631		+ 173,201

 65. Which property of addition (see page 8) allows you to use either arrangement shown at the right to find the sum of 691 and 452?

$$\begin{array}{r} 691 \\ + 452 \end{array} \qquad \begin{array}{r} 452 \\ + 691 \end{array}$$

OBJECTIVE B   **To solve application problems**

 66. Use the table of Wal-Mart data on page 11. What does the sum 105 + 457 represent?

67. Demographics In a recent year, according to the U.S. Department of Health and Human Services, there were 110,670 twin births in this country, 6919 triplet births, 627 quadruplet deliveries, and 79 quintuplet and other higher-order multiple births. Find the total number of multiple births during the year.

Laura Dwight/PhotoEdit, Inc.

68. Demographics The Census Bureau estimates that the U.S. population will grow by 296 million people from 2000 to 2100. Given that the U.S. population in 2000 was 281 million, find the Census Bureau's estimate of the U.S. population in 2100.

The Film Industry The graph at the right shows the domestic box-office income from the first four *Star Wars* movies. Use this information for Exercises 69 to 71.

69. Estimate the total income from the first four *Star Wars* movies.

70. Find the total income from the first four *Star Wars* movies.

71. a. Find the total income from the two movies with the lowest box-office incomes.
  b. Does the total income from the two movies with the lowest box-office incomes exceed the income from the 1977 *Star Wars* production?

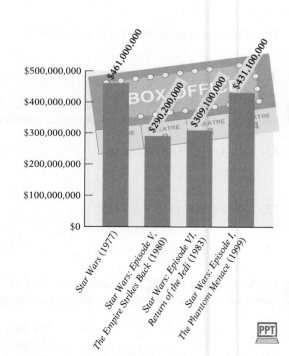

Source: www.worldwideboxoffice.com

72. **Geometry** The perimeter of a triangle is the sum of the lengths of the three sides of the triangle. Find the perimeter of a triangle that has sides that measure 12 inches, 14 inches, and 17 inches.

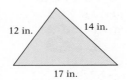

73. **Travel** The odometer on a moving van reads 68,692. The driver plans to drive 515 miles the first day, 492 miles the second day, and 278 miles the third day.
    a. How many miles will be driven during the three days? *1,285*
    b. What will the odometer reading be at the end of the trip? *69,977*

74. **Internet** Thirty-one million U.S. households do not have Internet access. Eighty-three million U.S. households do have Internet access. How many households are there in the United States? (*Source:* U.S. Bureau of the Census)

75. **Trail** Although 685 miles of the Northern Forest Canoe Trail can be paddled, there are another 55 miles of land over which a canoe must be carried. Find the total length of the Northern Forest Canoe Trail. (*Source: Yankee,* May/June 2007)

Northern Forest Canoe Trail

76. **Energy** In a recent year, the United States produced 5,102,000 barrels of crude oil per day and imported 10,118,000 barrels of crude oil per day. Find the total number of barrels of crude oil produced and imported per day in the United States. (*Source:* Energy Information Administration)

Applying the Concepts

77. If you roll two ordinary six-sided dice and add the two numbers that appear on top, how many different sums are possible?

78. If you add two *different* whole numbers, is the sum always greater than either one of the numbers? If not, give an example.

79. If you add two whole numbers, is the sum always greater than either one of the numbers? If not, give an example. (Compare this with the previous exercise.)

80. Make up a word problem for which the answer is the sum of 34 and 28.

81. Call a number "lucky" if it ends in a 7. How many lucky numbers are less than 100?

SECTION

1.3 Subtraction of Whole Numbers

OBJECTIVE A To subtract whole numbers without borrowing

Subtraction is the process of finding the difference between two numbers.

✓ **Take Note**

The **minuend** is the number from which another number is subtracted. The **subtrahend** is the number that is subtracted from another number. The result is the **difference**.

By counting, we see that the difference between $8 and $5 is $3.

$$\$8 \quad - \quad \$5 \quad = \quad \$3$$

Minuend Subtrahead Difference

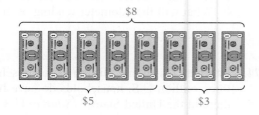

The difference $8 - 5$ can be shown on the number line.

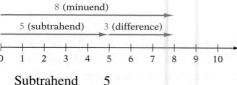

Note from the number line that addition and subtraction are related.

Subtrahend	5
+ Difference	+ 3
= Minuend	8

 Point of Interest

The use of the minus sign dates from the same period as the plus sign, around 1515.

The fact that the sum of the subtrahend and the difference equals the minuend can be used to check subtraction.

To subtract large numbers, begin by arranging the numbers vertically, keeping the digits that have the same place value in the same column. Then subtract the digits in each column.

HOW TO 1 Subtract $8955 - 2432$ and check.

```
  THOUSANDS
   HUNDREDS
      TENS
      ONES
    8 9 5 5
  - 2 4 3 2
    6 5 2 3
```

Check:
Subtrahend	2432
+ Difference	+ 6523
= Minuend	8955

EXAMPLE · 1

Subtract $6594 - 3271$ and check.

Solution
```
    6594      Check:    3271
  - 3271              + 3323
    3323                6594
```

YOU TRY IT · 1

Subtract $8925 - 6413$ and check.

Your solution

EXAMPLE · 2

Subtract $15,762 - 7541$ and check.

Solution
```
    15,762    Check:    7,541
   - 7,541            + 8,221
     8,221             15,762
```

YOU TRY IT · 2

Subtract $17,504 - 9302$ and check.

Your solution

Solutions on p. S1

OBJECTIVE B   **To subtract whole numbers with borrowing**

In all the subtraction problems in the previous objective, for each place value the lower digit was not larger than the upper digit. When the lower digit is larger than the upper digit, subtraction will involve **borrowing.**

HOW TO 2   Subtract: $692 - 378$

$$
\begin{array}{c|c|c}
\text{HUNDREDS} & \text{TENS} & \text{ONES}
\end{array}
$$

	8+1				8+1 10				8 12				8 12
6	9̸	2		6	9̸	2		6	9̸	2̸		6	9̸	2̸
− 3	7	8		− 3	7	8		− 3	7	8		− 3	7	8
												3 1 4

Because $8 > 2$, borrowing is necessary. 9 tens = 8 tens + 1 ten.

Borrow 1 ten from the tens column and write 10 in the ones column.

Add the borrowed 10 to 2.

Subtract the digits in each column.

The phrases below are used to indicate the operation of subtraction. An example is shown at the right of each phrase.

minus	8 minus 5	$8 - 5$
less	9 less 3	$9 - 3$
less than	2 less than 7	$7 - 2$
the difference between	the difference between 8 and 2	$8 - 2$
decreased by	5 decreased by 1	$5 - 1$

HOW TO 3   Find the difference between 1234 and 485, and check.
"The difference between 1234 and 485" means $1234 - 485$.

$$
\begin{array}{r}
{}^{2}\ {}^{14} \\
1\ 2\ \cancel{3}\ \cancel{4} \\
-\quad 4\ 8\ 5 \\
\hline
9
\end{array}
\qquad
\begin{array}{r}
{}^{1}\ {}^{12}\ {}^{14} \\
\cancel{1}\ \cancel{2}\ \cancel{3}\ \cancel{4} \\
-\quad 4\ 8\ 5 \\
\hline
4\ 9
\end{array}
\qquad
\begin{array}{r}
{}^{0}\ {}^{11}\ {}^{12}\ {}^{14} \\
1\ \cancel{2}\ \cancel{3}\ \cancel{4} \\
-\quad 4\ 8\ 5 \\
\hline
7\ 4\ 9
\end{array}
\qquad
\textit{Check:}\
\begin{array}{r}
{}^{1\ 1} \\
485 \\
+\ 749 \\
\hline
1234
\end{array}
$$

Subtraction with a zero in the minuend involves repeated borrowing.

HOW TO 4   Subtract: $3904 - 1775$

$$
\begin{array}{r}
{}^{8}\ {}^{10} \\
3\ \cancel{9}\ \cancel{0}\ 4 \\
-\ 1\ 7\ 7\ 5
\end{array}
\qquad
\begin{array}{r}
\ \ \ {}^{9} \\
{}^{8}\ {}^{10}\ {}^{14} \\
3\ \cancel{9}\ \cancel{0}\ \cancel{4} \\
-\ 1\ 7\ 7\ 5
\end{array}
\qquad
\begin{array}{r}
\ \ \ {}^{9} \\
{}^{8}\ {}^{10}\ {}^{14} \\
3\ \cancel{9}\ \cancel{0}\ \cancel{4} \\
-\ 1\ 7\ 7\ 5 \\
\hline
2\ 1\ 2\ 9
\end{array}
$$

$5 > 4$
There is a 0 in the tens column. Borrow 1 hundred (= 10 tens) from the hundreds column and write 10 in the tens column.

Borrow 1 ten from the tens column and add 10 to the 4 in the ones column.

Subtract the digits in each column.

Tips for Success
The HOW TO feature indicates an example with explanatory remarks. Using paper and pencil, you should work through the example. See *AIM for Success* at the front of the book.

EXAMPLE · 3

Subtract 4392 − 678 and check.

Solution

$$
\begin{array}{r}
\overset{3}{\cancel{4}}\ \overset{13}{\cancel{3}}\ \overset{8}{\cancel{9}}\ \overset{12}{\cancel{2}} \\
-\ \ 6\ 7\ 8 \\
\hline
3\ 7\ 1\ 4
\end{array}
\qquad
\begin{array}{r}
Check:\quad 678 \\
+\ 3714 \\
\hline
4392
\end{array}
$$

YOU TRY IT · 3

Subtract 3481 − 865 and check.

Your solution

EXAMPLE · 4

Find 23,954 less than 63,221 and check.

Solution

$$
\begin{array}{r}
\overset{5}{\cancel{6}}\ \overset{12}{\cancel{3}},\ \overset{11}{\cancel{2}}\ \overset{11}{\cancel{2}}\ \overset{11}{\cancel{1}} \\
-\ 2\ 3,9\ 5\ 4 \\
\hline
3\ 9,2\ 6\ 7
\end{array}
\qquad
\begin{array}{r}
Check:\quad 23,954 \\
+\ 39,267 \\
\hline
63,221
\end{array}
$$

YOU TRY IT · 4

Find 54,562 decreased by 14,485 and check.

Your solution

EXAMPLE · 5

Subtract 46,005 − 32,167 and check.

Solution

$$
\begin{array}{r}
\overset{5}{\cancel{4}}\ 6,\ \overset{10}{\cancel{0}}\ 0\ 5 \\
-\ 3\ 2,\ 1\ 6\ 7
\end{array}
$$

• There are two zeros in the minuend. Borrow 1 thousand from the thousands column and write 10 in the hundreds column.

$$
\begin{array}{r}
\overset{5}{\cancel{4}}\ 6,\ \overset{\overset{9}{\cancel{10}}}{\cancel{0}}\ \overset{10}{\cancel{0}}\ 5 \\
-\ 3\ 2,\ 1\ 6\ 7
\end{array}
$$

• Borrow 1 hundred from the hundreds column and write 10 in the tens column.

$$
\begin{array}{r}
\overset{5}{\cancel{4}}\ 6,\ \overset{\overset{9}{\cancel{10}}}{\cancel{0}}\ \overset{\overset{9}{\cancel{10}}}{\cancel{0}}\ \overset{15}{\cancel{5}} \\
-\ 3\ 2,\ 1\ 6\ 7 \\
\hline
1\ 3,\ 8\ 3\ 8
\end{array}
$$

• Borrow 1 ten from the tens column and add 10 to the 5 in the ones column.

$$
\begin{array}{r}
Check:\quad 32,167 \\
+\ 13,838 \\
\hline
46,005
\end{array}
$$

YOU TRY IT · 5

Subtract 64,003 − 54,936 and check.

Your solution

Solutions on pp. S1–S2

ESTIMATION

Estimating the Difference Between Two Whole Numbers

Calculate 323,502 − 28,912. Then use estimation to determine whether the difference is reasonable.

Subtract to find the exact difference. To estimate the difference, round each number to the same place value. Here we have rounded to the nearest ten-thousand. Then subtract. The estimated answer is 290,000, which is very close to the exact difference 294,590.

$$
\begin{array}{r}
323,502 \approx \quad 320,000 \\
-\ 28,912 \approx -\ 30,000 \\
\hline
294,590 \qquad 290,000
\end{array}
$$

OBJECTIVE C   To solve application problems

The table at the right shows the number of personnel on active duty in the branches of the U.S. military in 1940 and 1945. Use this table for Example 6 and You Try It 6.

Branch	1940	1945
U.S. Army	267,767	8,266,373
U.S. Navy	160,997	3,380,817
U.S. Air Force	51,165	2,282,259
U.S. Marine Corps	28,345	474,680

Source: Dept. of the Army, Dept. of the Navy, Air Force Dept., Dept. of the Marines, U.S. Dept. of Defense

---

EXAMPLE • 6

Find the difference between the number of U.S. Army personnel on active duty in 1945 and the number in 1940.

Strategy
To find the difference, subtract the number of U.S. Army personnel on active duty in 1940 (267,767) from the number on active duty in 1945 (8,266,373).

Solution

$$\begin{array}{r} 8,266,373 \\ -\ \ \ 267,767 \\ \hline 7,998,606 \end{array}$$

There were 7,998,606 more personnel on active duty in the U.S. Army in 1945 than in 1940.

YOU TRY IT • 6

Find the difference between the number of personnel on active duty in the Navy and the number in the Air Force in 1945.

Your strategy

Your solution

EXAMPLE • 7

You had a balance of $415 on your student debit card. You then used the card, deducting $197 for books, $48 for art supplies, and $24 for theater tickets. What is your new student debit card balance?

Strategy
To find your new debit card balance:
• Add to find the total of the three deductions (197 + 48 + 24).
• Subtract the total of the three deductions from the old balance (415).

Solution

$$\begin{array}{r} 197 \\ 48 \\ +\ 24 \\ \hline 269 \end{array}\ \text{total deductions} \qquad \begin{array}{r} 415 \\ -\ 269 \\ \hline 146 \end{array}$$

Your new debit card balance is $146.

YOU TRY IT • 7

Your total weekly salary is $638. Deductions of $127 for taxes, $18 for insurance, and $35 for savings are taken from your pay. Find your weekly take-home pay.

Your strategy

Your solution

Solutions on p. S2

1.3 EXERCISES

OBJECTIVE A   To subtract whole numbers without borrowing

For Exercises 1 to 35, subtract.

1.
$$\begin{array}{r} 9 \\ -\ 5 \\ \hline \end{array}$$

2.
$$\begin{array}{r} 8 \\ -\ 7 \\ \hline \end{array}$$

3.
$$\begin{array}{r} 8 \\ -\ 4 \\ \hline \end{array}$$

4.
$$\begin{array}{r} 7 \\ -\ 3 \\ \hline \end{array}$$

5.
$$\begin{array}{r} 10 \\ -\ 0 \\ \hline \end{array}$$

6.
$$\begin{array}{r} 11 \\ -\ 4 \\ \hline \end{array}$$

7.
$$\begin{array}{r} 12 \\ -\ 8 \\ \hline \end{array}$$

8.
$$\begin{array}{r} 19 \\ -\ 8 \\ \hline \end{array}$$

9.
$$\begin{array}{r} 15 \\ -\ 6 \\ \hline \end{array}$$

10.
$$\begin{array}{r} 16 \\ -\ 7 \\ \hline \end{array}$$

11.
$$\begin{array}{r} 25 \\ -\ 3 \\ \hline \end{array}$$

12.
$$\begin{array}{r} 55 \\ -\ 4 \\ \hline \end{array}$$

13.
$$\begin{array}{r} 68 \\ -\ 8 \\ \hline \end{array}$$

14.
$$\begin{array}{r} 77 \\ -\ 3 \\ \hline \end{array}$$

15.
$$\begin{array}{r} 89 \\ -\ 23 \\ \hline \end{array}$$

16.
$$\begin{array}{r} 54 \\ -\ 21 \\ \hline \end{array}$$

17.
$$\begin{array}{r} 88 \\ -\ 57 \\ \hline \end{array}$$

18.
$$\begin{array}{r} 1202 \\ -\ 701 \\ \hline \end{array}$$

19.
$$\begin{array}{r} 1305 \\ -\ 404 \\ \hline \end{array}$$

20.
$$\begin{array}{r} 1763 \\ -\ 801 \\ \hline \end{array}$$

21.
$$\begin{array}{r} 1497 \\ -\ 706 \\ \hline \end{array}$$

22.
$$\begin{array}{r} 8974 \\ -\ 3972 \\ \hline \end{array}$$

23.
$$\begin{array}{r} 2836 \\ -\ 1711 \\ \hline \end{array}$$

24.
$$\begin{array}{r} 8976 \\ -\ 7463 \\ \hline \end{array}$$

25.
$$\begin{array}{r} 9273 \\ -\ 6142 \\ \hline \end{array}$$

26. $77 - 36$

27. $129 - 82$

28. $132 - 61$

29. $969 - 44$

30. $1347 - 103$

31. $4865 - 304$

32. $1525 - 702$

33. $9999 - 6794$

34. $7806 - 3405$

35. $8843 - 7621$

36. Suppose three whole numbers, called *minuend*, *subtrahend*, and *difference*, are related by the subtraction statement *minuend* − *subtrahend* = *difference*. State whether the given relationship *must be true*, *might be true*, or *cannot be true*.
a. minuend > difference   **b.** subtrahend < difference

OBJECTIVE B   To subtract whole numbers with borrowing

For Exercises 37 to 80, subtract.

37.
$$\begin{array}{r} 71 \\ -\ 18 \\ \hline \end{array}$$

38.
$$\begin{array}{r} 93 \\ -\ 28 \\ \hline \end{array}$$

39.
$$\begin{array}{r} 47 \\ -\ 18 \\ \hline \end{array}$$

40.
$$\begin{array}{r} 44 \\ -\ 27 \\ \hline \end{array}$$

41.
$$\begin{array}{r} 37 \\ -\ 29 \\ \hline \end{array}$$

42.
$$\begin{array}{r} 50 \\ -\ 27 \\ \hline \end{array}$$

43.
$$\begin{array}{r} 70 \\ -\ 33 \\ \hline \end{array}$$

44.
$$\begin{array}{r} 993 \\ -\ 537 \\ \hline \end{array}$$

45. 250
   − 192

46. 840
   − 783

47. 768
   − 194

48. 770
   − 395

49. 674 − 337

50. 3526 − 387

51. 1712 − 289

52. 4350 − 729

53. 1702 − 948

54. 1607 − 869

55. 5933 − 3754

56. 7293 − 3748

57. 9407 − 2918

58. 3706 − 2957

59. 8605 − 7716

60. 8052 − 2709

61. 80,305 − 9176

62. 70,702 − 4239

63. 10,004 − 9306

64. 80,009 − 63,419

65. 70,618 − 41,213

66. 80,053 − 27,649

67. 70,700 − 21,076

68. 80,800 − 42,023

69. 2600
   − 1972

70. 8400
   − 3762

71. 9003
   − 2471

72. 6004
   − 2392

73. 8202
   − 3916

74. 7050
   − 4137

75. 7015
   − 2973

76. 4207
   − 1624

77. 7005
   − 1796

78. 8003
   − 2735

79. 20,005
   − 9,627

80. 80,004
   − 8,237

 81. Which of the following phrases represent the subtraction 673 − 571?
   (i) 571 less 673   **(ii)** 571 less than 673   **(iii)** 673 decreased by 571

82. Find 10,051 less 9027.

83. Find 17,031 less 5792.

84. Find the difference between 1003 and 447.

85. What is 29,874 minus 21,392?

86. What is 29,797 less than 68,005?

87. What is 69,379 less than 70,004?

88. What is 25,432 decreased by 7994?

89. What is 86,701 decreased by 9976?

For Exercises 90 to 93, use the relationship between addition and subtraction to complete the statement.

90. ___ + 39 = 104   **91.** 67 + ___ = 90   **92.** ___ + 497 = 862   **93.** 253 + ___ = 4901

For Exercises 94 to 99, use a calculator to subtract. Then round the numbers to the nearest ten-thousand and use estimation to determine whether the difference is reasonable.

94. 80,032
      − 19,605

95. 90,765
      − 60,928

96. 32,574
      − 10,961

97. 96,430
      − 59,762

98. 567,423
      − 208,444

99. 300,712
      − 198,714

OBJECTIVE C   To solve application problems

100. Banking You have $304 in your checking account. If you write a check for $139, how much is left in your checking account?

101. Insects The table at the right shows the number of taste genes and the number of smell genes in the mosquito, fruit fly, and honey bee.
  a. How many more smell genes does the honey bee have than the mosquito?
  b. How many more taste genes does the mosquito have than the fruit fly?
  c. Which of these insects has the best sense of smell?
  d. Which of these insects has the worst sense of taste?

	Mosquito	Fruit Fly	Honey Bee
Taste genes	76	68	10
Smell genes	79	62	170

Source: www.sciencedaily.com

102. Car Sales The graph at the right shows the number of cars sold in India for each year from 2003 to 2007.
  a. Has the number of cars sold increased each year from 2003 to 2007?
  b. How many more cars were sold in India in 2007 than in 2003?
  c. Between which two years shown did car sales increase the most?

Tata Motors' One Lakh Car

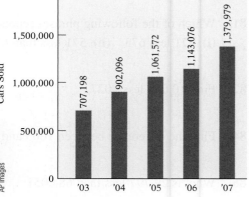
Cars Sold in India
Source: Society of Indian Automobile Manufacturers

103. **Earth Science** Use the graph at the right to find the difference between the maximum height to which Great Fountain geyser erupts and the maximum height to which Valentine erupts.

104. **Earth Science** According to the graph at the right, how much higher is the eruption of the Giant than that of Old Faithful?

105. **Education** In a recent year, 775,424 women and 573,079 men earned a bachelor's degree. How many more women than men earned a bachelor's degree in that year? (*Source: The National Center for Education Statistics*)

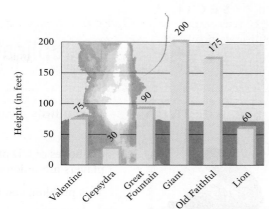

The Maximum Heights of the Eruptions of Six Geysers at Yellowstone National Park

Demographics The graph at the right shows the expected U.S. population aged 100 and over for every 2 years from 2010 to 2020. Use this information for Exercises 106 to 108.

106. What is the expected growth in the population aged 100 and over during the 10-year period?

107. **a.** Which 2-year period has the smallest expected increase in the number of people aged 100 and over?
b. Which 2-year period has the greatest expected increase?

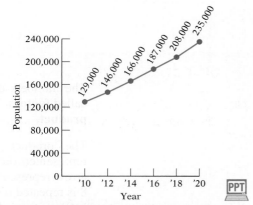

Expected U.S. Population Aged 100 and Over
Source: Census Bureau

108. What does the difference 208,000 − 166,000 represent?

109. **Finances** You had a credit card balance of $409 before you used the card to purchase books for $168, CDs for $36, and a pair of shoes for $97. You then made a payment to the credit card company of $350. Find your new credit card balance.

Applying the Concepts

110. Answer true or false.
a. The phrases "the difference between 9 and 5" and "5 less than 9" mean the same thing.
b. $9 - (5 - 3) = (9 - 5) - 3$
c. Subtraction is an associative operation. *Hint:* See part (b) of this exercise.

111. Make up a word problem for which the difference between 15 and 8 is the answer.

SECTION

1.4 Multiplication of Whole Numbers

OBJECTIVE A To multiply a number by a single digit

Six boxes of CD players are ordered. Each box contains eight CD players. How many CD players are ordered?

This problem can be worked by adding 6 eights.

$$8 + 8 + 8 + 8 + 8 + 8 = 48$$

This problem involves repeated addition of the same number and can be worked by a shorter process called **multiplication.** Multiplication is the repeated addition of the same number.

$$8 + 8 + 8 + 8 + 8 + 8 = 48$$

The numbers that are multiplied are called **factors.** The result is called the **product.**

| | | | or | | |

$$6 \quad \times \quad 8 \quad = \quad 48$$

Factor   **Factor**   **Product**

The product of 6×8 can be represented on the number line. The arrow representing the whole number 8 is repeated 6 times. The result is the arrow representing 48.

The times sign "$\times$" is only one symbol that is used to indicate multiplication. Each of the expressions that follow represents multiplication.

$$7 \times 8 \qquad 7 \cdot 8 \qquad 7(8) \qquad (7)(8) \qquad (7)8$$

As with addition, there are some useful properties of multiplication.

Multiplication Property of Zero

The product of a number and zero is zero.

$0 \times 4 = 0$
$7 \times 0 = 0$

Multiplication Property of One

The product of a number and one is the number.

$1 \times 6 = 6$
$8 \times 1 = 8$

Commutative Property of Multiplication

Two numbers can be multiplied in either order. The product will be the same.

$4 \times 3 = 3 \times 4$
$12 = 12$

Associative Property of Multiplication

Grouping the numbers to be multiplied in any order gives the same result. Do the multiplication inside the parentheses first.

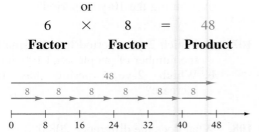

The basic facts for multiplying one-digit numbers should be memorized. Multiplication of larger numbers requires the repeated use of the basic multiplication facts.

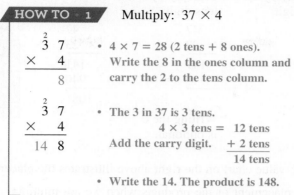

HOW TO · 1   Multiply: 37×4

$$\overset{2}{3}\,7 \\ \times \quad 4 \\ \overline{\qquad 8}$$

- $4 \times 7 = 28$ (2 tens + 8 ones). Write the 8 in the ones column and carry the 2 to the tens column.

$$\overset{2}{3}\,7 \\ \times \quad 4 \\ \overline{14\ 8}$$

- The 3 in 37 is 3 tens.

  $4 \times 3 \text{ tens} = \quad 12 \text{ tens}$
  Add the carry digit. $\underline{+ \ 2 \text{ tens}}$
                              14 tens

- Write the 14. The product is 148.

The phrases below are used to indicate the operation of multiplication. An example is shown at the right of each phrase.

times	7 times 3	$7 \cdot 3$
the product of	the product of 6 and 9	$6 \cdot 9$
multiplied by	8 multiplied by 2	$2 \cdot 8$

EXAMPLE · 1

Multiply: 735×9

Solution

$$\overset{3\ 4}{735} \\ \times \quad 9 \\ \overline{6615}$$

- $9 \times 5 = 45$
  Write the 5 in the ones column. Carry the 4 to the tens column.
  $9 \times 3 = 27,\ 27 + 4 = 31$
  $9 \times 7 = 63,\ 63 + 3 = 66$

YOU TRY IT · 1

Multiply: 648×7

Your solution

Solution on p. S2

OBJECTIVE B   To multiply larger whole numbers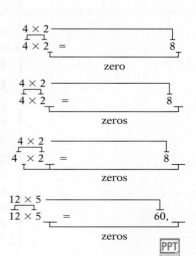

Note the pattern when the following numbers are multiplied.

Multiply the nonzero part of the factors.

Now attach the same number of zeros to the product as the total number of zeros in the factors.

4×2
$4 \times 2 =$ 8
zero

4×2
$4 \times 2 =$ 8
zeros

4×2
$4 \times 2 =$ 8
zeros

12×5
$12 \times 5 =$ 60,
zeros

PPT

HOW TO · 2 Find the product of 47 and 23.

Multiply by the ones digit.	Multiply by the tens digit.	Add.

$$\begin{array}{r} 47 \\ \times\ 23 \\ \hline 141 \end{array} \ (= 47 \times 3)$$

$$\begin{array}{r} 47 \\ \times\ 23 \\ \hline 141 \\ 940 \end{array} \ (= 47 \times 20)$$

$$\begin{array}{r} 47 \\ \times\ 23 \\ \hline 141 \\ 940 \\ \hline 1081 \end{array}$$

Writing the 0 is optional.

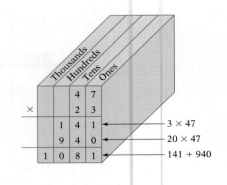

3×47
20×47
$141 + 940$

The place-value chart on the right above illustrates the placement of the products.

Note the placement of the products when we are multiplying by a factor that contains a zero.

HOW TO · 3 Multiply: 439×206

$$\begin{array}{r} 439 \\ \times\ 206 \\ \hline 2634 \\ 000 \\ 878 \\ \hline 90{,}434 \end{array}$$

0×439

When working the problem, we usually write only one zero. Writing this zero ensures the proper placement of the products.

$$\begin{array}{r} 439 \\ \times\ 206 \\ \hline 2634 \\ 8780 \\ \hline 90{,}434 \end{array}$$

EXAMPLE · 2

Find 829 multiplied by 603.

Solution

$$\begin{array}{r} 829 \\ \times\ 603 \\ \hline 2487 \\ 49740 \\ \hline 499{,}887 \end{array}$$

- $3 \times 829 = 2487$
- Write a zero in the tens column for 0×829.
- $6 \times 829 = 4974$

YOU TRY IT · 2

Multiply: 756×305

Your solution

Solution on p. S2

ESTIMATION

Estimating the Product of Two Whole Numbers

Calculate 3267×389. Then use estimation to determine whether the product is reasonable.

Multiply to find the exact product.

3267 **x** 389 **=** $1{,}270{,}863$

To estimate the product, round each number so that it has only one nonzero digit. Then multiply. The estimated answer is 1,200,000, which is very close to the exact product 1,270,863.

$$\begin{array}{r} 3267 \approx\ \ \ 3000 \\ \times\ 389 \approx\ \times\ 400 \\ \hline 1{,}200{,}000 \end{array}$$

OBJECTIVE C   To solve application problems

EXAMPLE • 3

An auto mechanic receives a salary of $1050 each week. How much does the auto mechanic earn in 4 weeks?

Strategy

To find the mechanic's earnings for 4 weeks, multiply the weekly salary (1050) by the number of weeks (4).

Solution

```
  1050
×    4
  4200
```

The mechanic earns $4200 in 4 weeks.

YOU TRY IT • 3

A new-car dealer receives a shipment of 37 cars each month. Find the number of cars the dealer will receive in 12 months.

Your strategy

Your solution

EXAMPLE • 4

A press operator earns $640 for working a 40-hour week. This week the press operator also worked 7 hours of overtime at $26 an hour. Find the press operator's total pay for the week.

Strategy

To find the press operator's total pay for the week:

• Find the overtime pay by multiplying the hours of overtime (7) by the overtime rate of pay (26).
• Add the weekly salary (640) to the overtime pay.

Solution

```
   26                  640
×   7               + 182
  182  overtime pay    822
```

The press operator earned $822 this week.

YOU TRY IT • 4

The buyer for Ross Department Store can buy 80 men's suits for $4800. Each sports jacket will cost the store $23. The manager orders 80 men's suits and 25 sports jackets. What is the total cost of the order?

Your strategy

Your solution

Solutions on p. S2

1.4 EXERCISES

OBJECTIVE A   To multiply a number by a single digit

For Exercises 1 to 4, write the expression as a product.

1. $2 + 2 + 2 + 2 + 2 + 2$ **2.** $4 + 4 + 4 + 4 + 4$ **3.** $7 + 7 + 7 + 7$ **4.** $18 + 18 + 18$

For Exercises 5 to 39, multiply.

5. $\begin{array}{r} 3 \\ \times\,4 \\ \hline \end{array}$ **6.** $\begin{array}{r} 2 \\ \times\,8 \\ \hline \end{array}$ **7.** $\begin{array}{r} 5 \\ \times\,7 \\ \hline \end{array}$ **8.** $\begin{array}{r} 6 \\ \times\,4 \\ \hline \end{array}$ **9.** $\begin{array}{r} 5 \\ \times\,5 \\ \hline \end{array}$

10. $\begin{array}{r} 7 \\ \times\,7 \\ \hline \end{array}$ **11.** $\begin{array}{r} 0 \\ \times\,7 \\ \hline \end{array}$ **12.** $\begin{array}{r} 8 \\ \times\,0 \\ \hline \end{array}$ **13.** $\begin{array}{r} 8 \\ \times\,9 \\ \hline \end{array}$ **14.** $\begin{array}{r} 7 \\ \times\,6 \\ \hline \end{array}$

15. $\begin{array}{r} 66 \\ \times\,3 \\ \hline \end{array}$ **16.** $\begin{array}{r} 70 \\ \times\,4 \\ \hline \end{array}$ **17.** $\begin{array}{r} 67 \\ \times\,5 \\ \hline \end{array}$ **18.** $\begin{array}{r} 127 \\ \times\,9 \\ \hline \end{array}$ **19.** $\begin{array}{r} 623 \\ \times\,4 \\ \hline \end{array}$

20. $\begin{array}{r} 802 \\ \times\,5 \\ \hline \end{array}$ **21.** $\begin{array}{r} 607 \\ \times\,9 \\ \hline \end{array}$ **22.** $\begin{array}{r} 300 \\ \times\,5 \\ \hline \end{array}$ **23.** $\begin{array}{r} 600 \\ \times\,7 \\ \hline \end{array}$ **24.** $\begin{array}{r} 906 \\ \times\,8 \\ \hline \end{array}$

25. $\begin{array}{r} 703 \\ \times\,9 \\ \hline \end{array}$ **26.** $\begin{array}{r} 127 \\ \times\,5 \\ \hline \end{array}$ **27.** $\begin{array}{r} 632 \\ \times\,3 \\ \hline \end{array}$ **28.** $\begin{array}{r} 559 \\ \times\,4 \\ \hline \end{array}$ **29.** $\begin{array}{r} 632 \\ \times\,8 \\ \hline \end{array}$

30. $\begin{array}{r} 524 \\ \times\,4 \\ \hline \end{array}$ **31.** $\begin{array}{r} 337 \\ \times\,5 \\ \hline \end{array}$ **32.** $\begin{array}{r} 841 \\ \times\,6 \\ \hline \end{array}$ **33.** $\begin{array}{r} 6709 \\ \times\,7 \\ \hline \end{array}$ **34.** $\begin{array}{r} 3608 \\ \times\,5 \\ \hline \end{array}$

35. $\begin{array}{r} 8568 \\ \times\,7 \\ \hline \end{array}$ **36.** $\begin{array}{r} 5495 \\ \times\,4 \\ \hline \end{array}$ **37.** $\begin{array}{r} 4780 \\ \times\,4 \\ \hline \end{array}$ **38.** $\begin{array}{r} 3690 \\ \times\,5 \\ \hline \end{array}$ **39.** $\begin{array}{r} 9895 \\ \times\,2 \\ \hline \end{array}$

 40. True or false? The product of two one-digit whole numbers must be a two-digit whole number.

41. Find the product of 5, 7, and 4.

42. Find the product of 6, 2, and 9.

43. What is 3208 multiplied by 7?

44. What is 5009 multiplied by 4?

45. What is 3105 times 6?

46. What is 8957 times 8?

OBJECTIVE B   To multiply larger whole numbers

For Exercises 47 to 78, multiply. — Evaluate( )

47.
$$\begin{array}{r} 16 \\ \times\ 21 \\ \hline \end{array}$$

48.
$$\begin{array}{r} 18 \\ \times\ 24 \\ \hline \end{array}$$

49.
$$\begin{array}{r} 35 \\ \times\ 26 \\ \hline \end{array}$$

50.
$$\begin{array}{r} 27 \\ \times\ 72 \\ \hline \end{array}$$

51.
$$\begin{array}{r} 693 \\ \times\ 91 \\ \hline \end{array}$$

52.
$$\begin{array}{r} 581 \\ \times\ 72 \\ \hline \end{array}$$

53.
$$\begin{array}{r} 419 \\ \times\ 80 \\ \hline \end{array}$$

54.
$$\begin{array}{r} 727 \\ \times\ 60 \\ \hline \end{array}$$

55.
$$\begin{array}{r} 8279 \\ \times\ 46 \\ \hline \end{array}$$

56.
$$\begin{array}{r} 9577 \\ \times\ 35 \\ \hline \end{array}$$

57.
$$\begin{array}{r} 6938 \\ \times\ 78 \\ \hline \end{array}$$

58.
$$\begin{array}{r} 8875 \\ \times\ 67 \\ \hline \end{array}$$

59.
$$\begin{array}{r} 7035 \\ \times\ 57 \\ \hline \end{array}$$

60.
$$\begin{array}{r} 6702 \\ \times\ 48 \\ \hline \end{array}$$

61.
$$\begin{array}{r} 3009 \\ \times\ 35 \\ \hline \end{array}$$

62.
$$\begin{array}{r} 6003 \\ \times\ 57 \\ \hline \end{array}$$

63.
$$\begin{array}{r} 809 \\ \times\ 530 \\ \hline \end{array}$$

64.
$$\begin{array}{r} 607 \\ \times\ 460 \\ \hline \end{array}$$

65.
$$\begin{array}{r} 800 \\ \times\ 325 \\ \hline \end{array}$$

66.
$$\begin{array}{r} 700 \\ \times\ 274 \\ \hline \end{array}$$

67.
$$\begin{array}{r} 987 \\ \times\ 349 \\ \hline \end{array}$$

68.
$$\begin{array}{r} 688 \\ \times\ 674 \\ \hline \end{array}$$

69.
$$\begin{array}{r} 312 \\ \times\ 134 \\ \hline \end{array}$$

70.
$$\begin{array}{r} 423 \\ \times\ 427 \\ \hline \end{array}$$

71.
$$\begin{array}{r} 379 \\ \times\ 500 \\ \hline \end{array}$$

72.
$$\begin{array}{r} 684 \\ \times\ 700 \\ \hline \end{array}$$

73.
$$\begin{array}{r} 985 \\ \times\ 408 \\ \hline \end{array}$$

74.
$$\begin{array}{r} 758 \\ \times\ 209 \\ \hline \end{array}$$

75.
$$\begin{array}{r} 3407 \\ \times\ 309 \\ \hline \end{array}$$

76.
$$\begin{array}{r} 5207 \\ \times\ 902 \\ \hline \end{array}$$

77.
$$\begin{array}{r} 4258 \\ \times\ 986 \\ \hline \end{array}$$

78.
$$\begin{array}{r} 6327 \\ \times\ 876 \\ \hline \end{array}$$

79. Find a one-digit number and a two-digit number whose product is a number that ends in two zeros.

80. What is 5763 times 45?

81. What is 7349 times 27?

82. Find the product of 2, 19, and 34.

83. Find the product of 6, 73, and 43.

84. What is 376 multiplied by 402?

85. What is 842 multiplied by 309?

 For Exercises 86 to 93, use a calculator to multiply. Then use estimation to determine whether the product is reasonable.

86.   8745
     $\times\ \ \ 63$

87.   4732
     $\times\ \ \ 93$

88.   2937
     $\times\ \ 206$

89.   8941
     $\times\ \ 726$

90.   3097
     $\times\ 1025$

91.   6379
     $\times\ 2936$

92.   32,508
     $\times\ \ \ \ 591$

93.   62,504
     $\times\ \ \ \ 923$

OBJECTIVE C   To solve application problems

 94. The price of Braeburn apples is $1.29 per pound, and the price of Cameo apples is $1.79 per pound. Which of the following represents the price of 3 pounds of Braeburn apples and 2 pounds of Cameo apples?
(i) $(3 \times 1.29) + (3 \times 1.79)$ **(ii)** $(2 \times 1.29) + (3 \times 1.79)$
(iii) $5 \times (1.29 + 1.79)$ **(iv)** $(3 \times 1.29) + (2 \times 1.79)$

95. Fuel Efficiency Rob Hill owns a compact car that averages 43 miles on 1 gallon of gas. How many miles could the car travel on 12 gallons of gas?

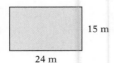
16 mi

96. Fuel Efficiency A plane flying from Los Angeles to Boston uses 865 gallons of jet fuel each hour. How many gallons of jet fuel were used on a 6-hour flight?

15 m

24 m

97. Geometry The perimeter of a square is equal to four times the length of a side of the square. Find the perimeter of a square whose side measures 16 miles.

98. Geometry The area of a rectangle is equal to the product of the length of the rectangle times its width. Find the area of a rectangle that has a length of 24 meters and a width of 15 meters. The area will be in square meters.

99. Matchmaking Services See the news clipping at the right. **a.** How many marriages occur between eHarmony members each week? **b.** How many marriages occur each year? Use a 365-day year.

In the News

Find Your Match Online

eHarmony, the online matchmaking service, boasts marriages among its members at the rate of 90 a day.

Source: Time, January 17, 2008

100. **College Education** See the news clipping at the right. **a.** Find the average cost of tuition, room, and board for 4 years at a public college. **b.** Find the average cost of tuition, room, and board for 4 years at a private college. **c.** Find the difference in cost for tuition, room, and board between 4 years at a private college and 4 years at a public college.

> **In the News**
>
> **Comparing Tuition Costs**
>
> The average annual cost of tuition, room, and board at a four-year public college is $12,796. At a four-year private college, the average cost is $30,367.
>
> *Source:* Kiplinger.com, January 24, 2007

Construction The table at the right shows the hourly wages of four different job classifications at a small construction company. Use this table for Exercises 101 to 103.

101. The owner of this company wants to provide the electrical installation for a new house. On the basis of the architectural plans for the house, it is estimated that it will require 3 electricians, each working 50 hours, to complete the job. What is the estimated cost for the electricians' labor?

Type of Work	Wage per Hour
Electrician	$34
Plumber	$30
Clerk	$16
Bookkeeper	$20

102. Carlos Vasquez, a plumbing contractor, hires 4 plumbers from this company at the hourly wage given in the table. If each plumber works 23 hours, what are the total wages paid by Carlos?

103. The owner of this company estimates that remodeling a kitchen will require 1 electrician working 30 hours and 1 plumber working 33 hours. This project also requires 3 hours of clerical work and 4 hours of bookkeeping. What is the total cost for these four components of this remodeling?

Applying the Concepts

104. Determine whether each of the following statements is always true, sometimes true, or never true.
a. A whole number times zero is zero.
b. A whole number times one is the whole number.
c. The product of two whole numbers is greater than either one of the whole numbers.

105. **Safety** According to the National Safety Council, in a recent year a death resulting from an accident occurred at the rate of 1 every 5 minutes. At this rate, how many accidental deaths occurred each hour? Each day? Throughout the year? Explain how you arrived at your answers.

106. **Demographics** According to the Population Reference Bureau, in the world today, 261 people are born every minute and 101 people die every minute. Using this statistic, what is the increase in the world's population every hour? Every day? Every week? Every year? Use a 365-day year. Explain how you arrived at your answers.

© Blaine Harrington III/Corbis

1.5 Division of Whole Numbers

OBJECTIVE A | **To divide by a single digit with no remainder in the quotient**

Division is used to separate objects into equal groups.

A store manager wants to display 24 new objects equally on 4 shelves. From the diagram, we see that the manager would place 6 objects on each shelf.

The manager's division problem can be written as follows:

> ✓ **Take Note**
> The **divisor** is the number that is divided into another number. The **dividend** is the number into which the divisor is divided. The result is the **quotient**.

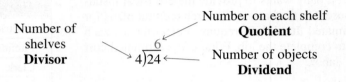

Note that the quotient multiplied by the divisor equals the dividend.

$$\overset{6}{4\overline{)24}} \quad \text{because} \quad \boxed{\text{Quotient}} \times \boxed{\text{Divisor}} = \boxed{\text{Dividend}}$$

$$\overset{6}{4\overline{)24}} \qquad\qquad\qquad 6 \qquad\qquad 4 \qquad\qquad 24$$

$$\overset{6}{9\overline{)54}} \quad \text{because} \quad 6 \quad \times \quad 9 \quad = \quad 54$$

$$\overset{5}{8\overline{)40}} \quad \text{because} \quad 5 \quad \times \quad 8 \quad = \quad 40$$

Here are some important quotients and the properties of zero in division:

Properties of One in Division

Any whole number, except zero, divided by itself is 1.

$$\overset{1}{8\overline{)8}} \qquad \overset{1}{14\overline{)14}} \qquad \overset{1}{10\overline{)10}}$$

> 🖩 **Integrating Technology**
> Enter 8 ÷ 0 = on your calculator. An error message is displayed because division by zero is not allowed.

Any whole number divided by 1 is the whole number.

$$\overset{9}{1\overline{)9}} \qquad \overset{27}{1\overline{)27}} \qquad \overset{10}{1\overline{)10}}$$

Properties of Zero in Division

Zero divided by any other whole number is zero.

$$\overset{0}{7\overline{)0}} \qquad \overset{0}{13\overline{)0}} \qquad \overset{0}{10\overline{)0}}$$

Division by zero is not allowed.

$$\overset{?}{0\overline{)8}}$$

There is no number whose product with 0 is 8.

When the dividend is a larger whole number, the digits in the quotient are found in steps.

HOW TO · 1   Divide $4\overline{)3192}$ and check.

$$
\begin{array}{r}
7 \\
4\overline{)3192} \\
-28 \\
\hline
39
\end{array}
$$

• Think $4\overline{)31}$.
• Subtract 7×4.
• Bring down the 9.

$$
\begin{array}{r}
79 \\
4\overline{)3192} \\
-28 \\
\hline
39 \\
-36 \\
\hline
32
\end{array}
$$

• Think $4\overline{)39}$.
• Subtract 9×4.
• Bring down the 2.

$$
\begin{array}{r}
798 \\
4\overline{)3192} \\
-28 \\
\hline
39 \\
-36 \\
\hline
32 \\
-32 \\
\hline
0
\end{array}
$$

• Think $\overline{}$.
• Subtract 8×4.

Check:
$$
\begin{array}{r}
798 \\
\times\quad 4 \\
\hline
3192
\end{array}
$$

The place-value chart can be used to show why this method works.

$$
\begin{array}{r}
\text{HUNDREDS} \;\; \text{TENS} \;\; \text{ONES} \\
7 \quad 9 \quad 8 \\
4\overline{)3 \quad 1 \quad 9 \quad 2} \\
-2 \quad 8 \quad 0 \quad 0 \\
\hline
3 \quad 9 \quad 2 \\
-3 \quad 6 \quad 0 \\
\hline
3 \quad 2 \\
-3 \quad 2 \\
\hline
0
\end{array}
$$

7 hundreds $\times$ 4

9 tens $\times$ 4

8 ones $\times$ 4

There are other ways of expressing division.

54 divided by 9 equals 6.

$54 \div 9$ equals 6.

$\dfrac{54}{9}$ equals 6.

EXAMPLE • 1

Divide 7)56 and check.

Solution

$$\begin{array}{r} 8 \\ 7\overline{)56} \end{array}$$

Check: $8 \times 7 = 56$

YOU TRY IT • 1

Divide 9)63 and check.

Your solution

EXAMPLE • 2

Divide $2808 \div 8$ and check.

Solution

$$\begin{array}{r} 351 \\ 8\overline{)2808} \\ -24 \\ \hline 40 \\ -40 \\ \hline 08 \\ -8 \\ \hline 0 \end{array}$$

Check: $351 \times 8 = 2808$

YOU TRY IT • 2

Divide $4077 \div 9$ and check.

Your solution

EXAMPLE • 3

Divide 7)2856 and check.

Solution

$$\begin{array}{r} 408 \\ 7\overline{)2856} \\ -28 \\ \hline 05 \\ -0 \\ \hline 56 \\ -56 \\ \hline 0 \end{array}$$

- Think 7)5. Place 0 in quotient.
- Subtract 0×7.
- Bring down the 6.

Check: $408 \times 7 = 2856$

YOU TRY IT • 3

Divide 9)6345 and check.

Your solution

Solutions on pp. S2–S3

OBJECTIVE B   To divide by a single digit with a remainder in the quotient

Sometimes it is not possible to separate objects into a whole number of equal groups.

A baker has 14 muffins to pack into 3 boxes. Each box holds 4 muffins. From the diagram, we see that after the baker places 4 muffins in each box, there are 2 left over. The 2 is called the **remainder.**

The baker's division problem could be written

$$
\begin{array}{r}
\text{Quotient} \\
4 \leftarrow \text{(Number in each box)}
\end{array}
$$

Divisor $\longrightarrow$ 3) 14 $\leftarrow$ **Dividend**
(Number of boxes) $\underline{-12}$ (Total number of objects)
                            2 $\leftarrow$ **Remainder**
                            (Number left over)

The answer to a division problem with a remainder is frequently written

$$3\overline{)14}\,^{4\text{ r}2}$$

Note that $\boxed{\underset{\text{Quotient} \times \text{Divisor}}{4 \qquad 3}} + \boxed{\underset{\text{Remainder}}{2}} = \boxed{\underset{\text{Dividend}}{14}}$.

EXAMPLE · 4

Divide 4)2522 and check.

Solution

$$
\begin{array}{r}
630 \text{ r}2 \\
4)\overline{2522} \\
-24 \\
\hline
12 \\
-12 \\
\hline
02 \\
-0 \\
\hline
2
\end{array}
$$

• **Think 4)2. Place 0 in quotient.**
• **Subtract 0 × 4.**

Check: $(630 \times 4) + 2 =$
         $2520 \quad + 2 = 2522$

YOU TRY IT · 4

Divide 6)5225 and check.

Your solution

EXAMPLE · 5

Divide 9)27,438 and check.

Solution

$$
\begin{array}{r}
3,048 \text{ r}6 \\
9)\overline{27,438} \\
-27 \\
\hline
0\,4 \\
-0 \\
\hline
43 \\
-36 \\
\hline
78 \\
-72 \\
\hline
6
\end{array}
$$

• **Think 9)4.**
• **Subtract 0 × 9.**

Check: $(3048 \times 9) + 6 =$
         $27,432 \quad + 6 = 27,438$

YOU TRY IT · 5

Divide 7)21,409 and check.

Your solution

Solutions on p. S3

OBJECTIVE C　　To divide by larger whole numbers

When the divisor has more than one digit, estimate at each step by using the first digit of the divisor. If that product is too large, lower the guess by 1 and try again.

> **HOW TO　2**　Divide 34)1598 and check.
>
> ```
>        5
> 34) 1598      • Think ___.
>    −170        • Subtract 5 × 34.
> ```
>
> ```
>        4
> 34) 1598
>    −136        • Subtract 4 × 34.
>     238
> ```
>
> 170 is too large. Lower the guess by 1 and try again.
>
> ```
>        47
> 34) 1598
>    −136
>     238        • Think 3)23.
>    −238        • Subtract 7 × 34.
>       0
> ```
>
> *Check:*
> ```
>        47
>       ×34
>       188
>       141
>      1598
> ```

Tips for Success

One of the key instructional features of this text is the Example/You Try It pairs. Each Example is completely worked. You are to solve the You Try It problems. When you are ready, check your solution against the one in the Solutions section. The solution for You Try It 6 below is on page S3 (see the reference at the bottom right of the You Try It). See *AIM for Success* at the front of the book.

The phrases below are used to indicate the operation of division. An example is shown at the right of each phrase.

the quotient of	the quotient of 9 and 3	$9 \div 3$
divided by	6 divided by 2	$6 \div 2$

EXAMPLE · 6

Find 7077 divided by 34 and check.

Solution

```
       208 r5
34) 7077
   −68
     27        • Think 34)27.
    −0         • Place 0 in quotient.
    277        • Subtract 0 × 34.
   −272
      5
```

Check: (208 × 34) + 5 =
　　　　　7072　+ 5 = 7077

YOU TRY IT · 6

Divide 4578 ÷ 42 and check.

Your solution

Solution on p. S3

EXAMPLE · 7

Find the quotient of 21,312 and 56 and check.

Solution

$$
\begin{array}{r}
380 \ r32 \\
56\overline{)21{,}312} \\
-16\,8 \\
\hline
4\,51 \\
-4\,48 \\
\hline
32 \\
-0 \\
\hline
32
\end{array}
$$

• Think $5\overline{)21}$.
  4×56 is too large.
  Try 3.

Check: $(380 \times 56) + 32 =$
$\qquad 21{,}280 \quad + 32 = 21{,}312$

YOU TRY IT · 7

Divide $18{,}359 \div 39$ and check.

Your solution

EXAMPLE · 8

Divide $427\overline{)24{,}782}$ and check.

Solution

$$
\begin{array}{r}
58 \ r16 \\
427\overline{)24{,}782} \\
-21\,35 \\
\hline
3\,432 \\
-3\,416 \\
\hline
16
\end{array}
$$

Check: $(58 \times 427) + 16 =$
$\qquad 24{,}766 \quad + 16 = 24{,}782$

YOU TRY IT · 8

Divide $534\overline{)33{,}219}$ and check.

Your solution

EXAMPLE · 9

Divide $386\overline{)206{,}149}$ and check.

Solution

$$
\begin{array}{r}
534 \ r25 \\
386\overline{)206{,}149} \\
-193\,0 \\
\hline
13\,14 \\
-11\,58 \\
\hline
1\,569 \\
-1\,544 \\
\hline
25
\end{array}
$$

Check: $(534 \times 386) + 25 =$
$\qquad 206{,}124 \quad + 25 = 206{,}149$

YOU TRY IT · 9

Divide $515\overline{)216{,}848}$ and check.

Your solution

Solutions on p. S3

ESTIMATION

Estimating the Quotient of Two Whole Numbers

Calculate 36,936 ÷ 54. Then use estimation to determine whether the quotient is reasonable.

Divide to find the exact quotient.

To estimate the quotient, round each number so that it contains one nonzero digit. Then divide. The estimated answer is 800, which is close to the exact quotient 684.

$$36{,}936 \div 54 = 684$$

$$36{,}936 \div 54 \approx$$
$$40{,}000 \div 50 = 800$$

OBJECTIVE D To solve application problems

The **average** of several numbers is the sum of all the numbers divided by the number of those numbers.

$$\text{Average test score} = \frac{81 + 87 + 80 + 85 + 79 + 86}{6} = \frac{498}{6} = 83$$

HOW TO 3

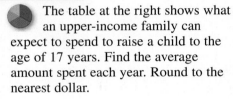 The table at the right shows what an upper-income family can expect to spend to raise a child to the age of 17 years. Find the average amount spent each year. Round to the nearest dollar.

Expenses to Raise a Child	
Housing	$89,580
Food	$35,670
Transportation	$32,760
Child care/education	$26,520
Clothing	$13,770
Health care	$13,380
Other	$30,090

Source: Department of Agriculture, *Expenditures on Children by Families*

Strategy

To find the average amount spent each year:

- Add all the numbers in the table to find the total amount spent during the 17 years.
- Divide the sum by 17.

Solution

```
  89,580
  35,670
  32,760
  26,520
  13,770
  13,380
+ 30,090
 ─────────
 241,770   Sum of all
           the costs
```

```
         14,221
    17) 241,770
        −17
        ───
         71
        −68
        ───
          37
         −34
         ───
          37
         −34
         ───
           30
          −17
          ───
           13
```

- **When rounding to the nearest whole number, compare twice the remainder to the divisor. If twice the remainder is less than the divisor, drop the remainder. If twice the remainder is greater than or equal to the divisor, add 1 to the units digit of the quotient.**

- **Twice the remainder is 2 × 13 = 26. Because 26 > 17, add 1 to the units digit of the quotient.**

The average amount spent each year to raise a child to the age of 17 is $14,222.

EXAMPLE · 10

Ngan Hui, a freight supervisor, shipped
192,600 bushels of wheat in 9 railroad cars. Find
the amount of wheat shipped in each car.

Strategy
To find the amount of wheat shipped in each car,
divide the number of bushels (192,600) by the
number of cars (9).

Solution

$$
\begin{array}{r}
21{,}400 \\
9{\overline{\smash{\big)}\,192{,}600}} \\
-18 \\
\hline
12 \\
-9 \\
\hline
36 \\
-36 \\
\hline
0
\end{array}
$$

Each car carried 21,400 bushels of wheat.

YOU TRY IT · 10

Suppose a Michelin retail outlet can store 270 tires
on 15 shelves. How many tires can be stored on
each shelf?

Your strategy

Your solution

EXAMPLE · 11

The used car you are buying costs $11,216. A down
payment of $2000 is required. The remaining balance
is paid in 48 equal monthly payments. What is the
monthly payment?

Strategy
To find the monthly payment:

• Find the remaining balance by subtracting the
down payment (2000) from the total cost of the car
(11,216).
• Divide the remaining balance by the number of
equal monthly payments (48).

Solution

$$
\begin{array}{r}
11{,}216 \\
-\ 2{,}000 \\
\hline
9{,}216
\end{array}
$$

Remaining balance

$$
\begin{array}{r}
192 \\
48{\overline{\smash{\big)}\,9216}} \\
-48 \\
\hline
441 \\
-432 \\
\hline
96 \\
-96 \\
\hline
0
\end{array}
$$

The monthly payment is $192.

YOU TRY IT · 11

A soft-drink manufacturer produces 12,600 cans
of soft drink each hour. Cans are packed 24 to a
case. How many cases of soft drink are produced
in 8 hours?

Your strategy

Your solution

Solutions on p. S3

1.5 EXERCISES

OBJECTIVE A   To divide by a single digit with no remainder in the quotient

For Exercises 1 to 20, divide.

1. $4\overline{)8}$

2. $3\overline{)9}$

3. $6\overline{)36}$

4. $9\overline{)81}$

5. $7\overline{)49}$

6. $5\overline{)80}$

7. $6\overline{)96}$

8. $6\overline{)480}$

9. $4\overline{)840}$

10. $3\overline{)690}$

11. $7\overline{)308}$

12. $7\overline{)203}$

13. $9\overline{)6327}$

14. $4\overline{)2120}$

15. $8\overline{)7280}$

16. $9\overline{)8118}$

17. $3\overline{)64,680}$

18. $4\overline{)50,760}$

19. $6\overline{)21,480}$

20. $5\overline{)18,050}$

21. What is 7525 divided by 7?

22. What is 32,364 divided by 4?

 23. If the dividend and the divisor in a division problem are the same number, what is the quotient?

For Exercises 24 to 27, use the relationship between multiplication and division to complete the multiplication problem.

24. ___ × 7 = 364

25. 8 × ___ = 376

26. 5 × ___ = 170

27. ___ × 4 = 92

OBJECTIVE B   To divide by a single digit with a remainder in the quotient

For Exercises 28 to 50, divide.

28. $4\overline{)9}$

29. $2\overline{)7}$

30. $5\overline{)27}$

31. $9\overline{)88}$

32. $3\overline{)40}$

33. $6\overline{)97}$

34. $8\overline{)83}$

35. $5\overline{)54}$

36. $7\overline{)632}$

37. $4\overline{)363}$

38. 4)921   **39.** 7)845   **40.** 8)1635   **41.** 5)1548   **42.** 7)9432

43. 7)8124   **44.** 3)5162   **45.** 5)3542   **46.** 8)3274

47. 4)15,301   **48.** 7)43,500   **49.** 8)72,354   **50.** 5)43,542

51. What is 45,738 divided by 4? Round to the nearest ten.

52. What is 37,896 divided by 9? Round to the nearest hundred.

53. What is 3572 divided by 7? Round to the nearest ten.

54. What is 78,345 divided by 4? Round to the nearest hundred.

 55. True or false? When a three-digit number is divided by a one-digit number, the quotient can be a one-digit number.

OBJECTIVE C   To divide by larger whole numbers

For Exercises 56 to 83, divide.

56. 27)96   **57.** 44)82   **58.** 42)87   **59.** 67)93

60. 41)897   **61.** 32)693   **62.** 23)784   **63.** 25)772

64. 74)600   **65.** 92)500   **66.** 70)329   **67.** 50)467

68. 36)7225   **69.** 44)8821   **70.** 19)3859   **71.** 32)9697

72. 88)3127   **73.** 92)6177   **74.** 33)8943   **75.** 27)4765

76. 22)98,654   **77.** 77)83,629   **78.** 64)38,912   **79.** 78)31,434

80. $206\overline{)3097}$ **81.** $504\overline{)6504}$ **82.** $654\overline{)1217}$ **83.** $546\overline{)2344}$

84. Find the quotient of 5432 and 21.   **85.** Find the quotient of 8507 and 53.

86. What is 37,294 divided by 72?   **87.** What is 76,788 divided by 46?

88. Find 23,457 divided by 43. Round to the nearest hundred.   **89.** Find 341,781 divided by 43. Round to the nearest ten.

90. True or false? If the remainder of a division problem is 210, then the divisor was less than 210.

For Exercises 91 to 102, use a calculator to divide. Then use estimation to determine whether the quotient is reasonable.

91. $76\overline{)389,804}$ **92.** $53\overline{)117,925}$ **93.** $29\overline{)637,072}$ **94.** $67\overline{)738,072}$

95. $38\overline{)934,648}$ **96.** $34\overline{)906,304}$ **97.** $309\overline{)876,324}$ **98.** $642\overline{)323,568}$

99. $209\overline{)632,016}$ **100.** $614\overline{)332,174}$ **101.** $179\overline{)5,734,444}$ **102.** $374\overline{)7,712,254}$

OBJECTIVE D   **To solve application problems**

Insurance The table at the right shows the sources of insurance claims for losses of laptop computers in a recent year. Claims have been rounded to the nearest ten thousand dollars. Use this information for Exercises 103 and 104.

103. What was the average monthly claim for theft?

104. For all sources combined, find the average claims per month.

Source	Claims
Accidents	$560,000
Theft	$300,000
Power surge	$80,000
Lightning	$50,000
Transit	$20,000
Water/flood	$20,000
Other	$110,000

Source: Safeware, The Insurance Company

Work Hours The table at the right shows, for different countries, the average number of hours per year that employees work. Use this information for Exercises 105 and 106. Use a 50-week year. Round answers to the nearest whole number.

Country	Annual Number of Hours Worked
Britian	1731
France	1656
Japan	1889
Norway	1399
United States	1966

Source: International Labor Organization

105. What is the average number of hours worked per week by employees in Britain?

106. On average, how many more hours per week do employees in the United States work than employees in France?

107. **Coins** The U.S. Mint estimates that about 114,000,000,000 of the 312,000,000,000 pennies it has minted over the last 30 years are in active circulation. That works out to how many pennies in circulation for each of the 300,000,000 people living in the United States?

© blickwinkel/Alamy

108. **Toy Sales** Every hour, 25,200 sets of Legos® are sold by retailers worldwide. (*Source: Time,* February 11, 2008) How many sets of Legos are sold each second by retailers worldwide?

109. **U.S. Postal Service** There are 114 households in the United States. Use the information in the news clipping at the right to determine, on average, how many pieces of mail each household will receive between Thanksgiving and Christmas this year. Round to the nearest whole number.

110. **Arlington National Cemetery** There are approximately 10,200 funerals each year at Arlington National Cemetery. (*Source:* www.arlingtoncemetery.org) Calculate the average number of funerals each day at Arlington National Cemetery. Round to the nearest whole number.

© 2009 Jupiterimages

Arlington National Cemetery

111. Which problems below require division to solve?
  (i) Four friends want to share a restaurant bill of $45.65 equally. Find the amount that each friend should pay.
  (ii) On average, Sam spends $30 a week on gas. Find Sam's average yearly expenditure for gas.
  (iii) Emma's 12 phone bills for last year totaled $660. Find Emma's average monthly phone bill.

Applying the Concepts

112. **Wages** A sales associate earns $374 for working a 40-hour week. Last week the associate worked an additional 9 hours at $13 an hour. Find the sales associate's total pay for last week's work.

113. Payroll Deductions Your paycheck shows deductions of $225 for savings, $98 for taxes, and $27 for insurance. Find the total of the three deductions.

 Dairy Products The topic of the graph at the right is the eggs produced in the United States in a recent year. It shows where the eggs that were produced went or how they were used. Use this table for Exercises 114 and 115.

114. Use the graph to determine the total number of cases of eggs produced during the year.

115. How many more cases of eggs were sold by retail stores than were used for non-shell products?

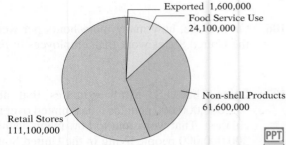

Exported 1,600,000
Food Service Use 24,100,000
Non-shell Products 61,600,000
Retail Stores 111,100,000

Eggs Produced in the United States (in cases)
Source: American Egg Board

Finance The graph at the right shows the annual expenditures, in a recent year, of the average household in the United States. Use this information for Exercises 116 to 118. Round answers to the nearest whole number.

116. What is the total amount spent annually by the average household in the United States?

117. What is the average monthly expense for housing?

118. What is the difference between the average monthly expense for food and the average monthly expense for health care?

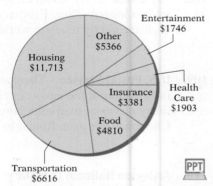

Entertainment $1746
Other $5366
Housing $11,713
Insurance $3381
Health Care $1903
Food $4810
Transportation $6616

Average Annual Household Expenses
Source: Bureau of Labor Statistics Consumer Expenditure Survey

The Military The graph at the right shows the basic monthly pay for Army officers with over 20 years of service. Use this graph for Exercises 119 and 120.

119. What is a major's annual pay?

120. What is the difference between a colonel's annual pay and a lieutenant colonel's annual pay?

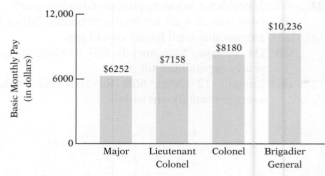

Basic Montly Pay for Army Officers
Source: Department of Defense

121. Finances You purchase a used car with a down payment of $2500 and monthly payments of $195 for 48 months. Find the total amount paid for the car.

SECTION

1.6

Exponential Notation and the Order of Operations Agreement

OBJECTIVE A   **To simplify expressions that contain exponents**

Repeated multiplication of the same factor can be written in two ways:

$$3 \cdot 3 \cdot 3 \cdot 3 \cdot 3 \quad \text{or} \quad 3^5 \leftarrow \textbf{Exponent}$$

The **exponent** indicates how many times the factor occurs in the multiplication. The expression 3^5 is in **exponential notation.**

It is important to be able to read numbers written in exponential notation.

$$6 = 6^1 \quad \text{is read "six to the first \textbf{power}" or just "six." Usually the exponent 1 is not written.}$$
$$6 \cdot 6 = 6^2 \quad \text{is read "six squared" or "six to the second power."}$$
$$6 \cdot 6 \cdot 6 = 6^3 \quad \text{is read "six cubed" or "six to the third power."}$$
$$6 \cdot 6 \cdot 6 \cdot 6 = 6^4 \quad \text{is read "six to the fourth power."}$$
$$6 \cdot 6 \cdot 6 \cdot 6 \cdot 6 = 6^5 \quad \text{is read "six to the fifth power."}$$

Each place value in the place-value chart can be expressed as a power of 10.

$$
\begin{aligned}
\text{Ten} &= 10 &=& 10 &=& 10^1 \\
\text{Hundred} &= 100 &=& 10 \cdot 10 &=& 10^2 \\
\text{Thousand} &= 1000 &=& 10 \cdot 10 \cdot 10 &=& 10^3 \\
\text{Ten-thousand} &= 10{,}000 &=& 10 \cdot 10 \cdot 10 \cdot 10 &=& 10^4 \\
\text{Hundred-thousand} &= 100{,}000 &=& 10 \cdot 10 \cdot 10 \cdot 10 \cdot 10 &=& 10^5 \\
\text{Million} &= 1{,}000{,}000 &=& 10 \cdot 10 \cdot 10 \cdot 10 \cdot 10 \cdot 10 &=& 10^6
\end{aligned}
$$

Integrating Technology

A calculator can be used to evaluate an exponential expression. The y^x key (or, on some calculators, an x^y or $\wedge$ key) is used to enter the exponent. For instance, for the example at the right, enter 4 y^x 3 $=$. The display reads 64.

To simplify a numerical expression containing exponents, write each factor as many times as indicated by the exponent and carry out the indicated multiplication.

$$4^3 = 4 \cdot 4 \cdot 4 = 64$$

$$2^2 \cdot 3^4 = (2 \cdot 2) \cdot (3 \cdot 3 \cdot 3 \cdot 3) = 4 \cdot 81 = 324$$

EXAMPLE · 1

Write $3 \cdot 3 \cdot 3 \cdot 5 \cdot 5$ in exponential notation.

Solution $3 \cdot 3 \cdot 3 \cdot 5 \cdot 5 = 3^3 \cdot 5^2$

YOU TRY IT · 1

Write $2 \cdot 2 \cdot 2 \cdot 2 \cdot 3 \cdot 3 \cdot 3$ in exponential notation.

Your solution

EXAMPLE · 2

Write as a power of 10: $10 \cdot 10 \cdot 10 \cdot 10$

Solution $10 \cdot 10 \cdot 10 \cdot 10 = 10^4$

YOU TRY IT · 2

Write as a power of 10: $10 \cdot 10 \cdot 10 \cdot 10 \cdot 10 \cdot 10 \cdot 10$

Your solution

EXAMPLE · 3

Simplify $3^2 \cdot 5^3$.

Solution
$$3^2 \cdot 5^3 = (3 \cdot 3) \cdot (5 \cdot 5 \cdot 5)$$
$$= 9 \cdot 125 = 1125$$

YOU TRY IT · 3

Simplify $2^3 \cdot 5^2$.

Your solution

Solutions on p. S4

OBJECTIVE B

To use the Order of Operations Agreement to simplify expressions

More than one operation may occur in a numerical expression. The answer may be different, depending on the order in which the operations are performed. For example, consider $3 + 4 \times 5$.

Multiply first, then add.

$$3 + \underbrace{4 \times 5}$$
$$\underbrace{3 + 20}$$
$$23$$

Add first, then multiply.

$$\underbrace{3 + 4} \times 5$$
$$\underbrace{7 \times 5}$$
$$35$$

An Order of Operations Agreement is used so that only one answer is possible.

The Order of Operations Agreement

Step 1. Do all the operations inside parentheses.

Step 2. Simplify any number expressions containing exponents.

Step 3. Do multiplication and division as they occur from left to right.

Step 4. Do addition and subtraction as they occur from left to right.

Integrating Technology

Many scientific calculators have an x^2 key. This key is used to square the displayed number. For example, after the user presses 2 x^2 = , the display reads 4.

HOW TO • 1 $3 \times (2 + 1) - 2^2 + 4 \div 2$ by using the Order of Operations Agreement.

$3 \times \underbrace{(2 + 1)} - 2^2 + 4 \div 2$ **1.** Perform operations in parentheses.

$3 \times 3 - \underbrace{2^2} + 4 \div 2$ **2.** Simplify expressions with exponents.

$\underbrace{3 \times 3} - 4 + \underbrace{4 \div 2}$ **3.** Do multiplication and division as they occur from left to right.

$\underbrace{9 - 4} + 2$

$\underbrace{5 + 2}$ **4.** Do addition and subtraction as they occur from left to right.

7

One or more of these steps may not be needed to simplify an expression. In that case, proceed to the next step in the Order of Operations Agreement.

HOW TO • 2 Simplify $5 + 8 \div 2$.   There are no parentheses or exponents. Proceed to Step 3 of the agreement.

$5 + \underbrace{8 \div 2}$ **3.** Do multiplication or division.

$\underbrace{5 + 4}$ **4.** Do addition or subtraction.

9

EXAMPLE • 4

Simplify: $64 \div (8 - 4)^2 \cdot 9 - 5^2$

Solution

$64 \div (8 - 4)^2 \cdot 9 - 5^2$
$= 64 \div 4^2 \cdot 9 - 5^2$ • Parentheses
$= 64 \div 16 \cdot 9 - 25$ • Exponents
$= 4 \cdot 9 - 25$ • Division and
$= 36 - 25$ multiplication
$= 11$ • Subtraction

YOU TRY IT • 4

Simplify: $5 \cdot (8 - 4)^2 \div 4 - 2$

Your solution

Solution on p. S4

1.6 EXERCISES

OBJECTIVE A   To simplify expressions that contain exponents

For Exercises 1 to 12, write the number in exponential notation.

1. $2 \cdot 2 \cdot 2$

2. $7 \cdot 7 \cdot 7 \cdot 7 \cdot 7$

3. $6 \cdot 6 \cdot 6 \cdot 7 \cdot 7 \cdot 7 \cdot 7$

4. $6 \cdot 6 \cdot 9 \cdot 9 \cdot 9 \cdot 9$

5. $2 \cdot 2 \cdot 2 \cdot 3 \cdot 3 \cdot 3$

6. $3 \cdot 3 \cdot 10 \cdot 10$

7. $5 \cdot 7 \cdot 7 \cdot 7 \cdot 7 \cdot 7$

8. $4 \cdot 4 \cdot 4 \cdot 5 \cdot 5 \cdot 5$

9. $3 \cdot 3 \cdot 3 \cdot 6 \cdot 6 \cdot 6 \cdot 6$

10. $2 \cdot 2 \cdot 5 \cdot 5 \cdot 5 \cdot 8$

11. $3 \cdot 3 \cdot 3 \cdot 5 \cdot 9 \cdot 9 \cdot 9$

12. $2 \cdot 2 \cdot 2 \cdot 4 \cdot 7 \cdot 7 \cdot 7$

For Exercises 13 to 37, simplify.

13. 2^3

14. 2^6

15. $2^4 \cdot 5^2$

16. $2^6 \cdot 3^2$

17. $3^2 \cdot 10^2$

18. $2^3 \cdot 10^4$

19. $6^2 \cdot 3^3$

20. $4^3 \cdot 5^2$

21. $5 \cdot 2^3 \cdot 3$

22. $6 \cdot 3^2 \cdot 4$

23. $2^2 \cdot 3^2 \cdot 10$

24. $3^2 \cdot 5^2 \cdot 10$

25. $0^2 \cdot 4^3$

26. $6^2 \cdot 0^3$

27. $3^2 \cdot 10^4$

28. $5^3 \cdot 10^3$

29. $2^2 \cdot 3^3 \cdot 5$

30. $5^2 \cdot 7^3 \cdot 2$

31. $2 \cdot 3^4 \cdot 5^2$

32. $6 \cdot 2^6 \cdot 7^2$

33. $5^2 \cdot 3^2 \cdot 7^2$

34. $4^2 \cdot 9^2 \cdot 6^2$

35. $3^4 \cdot 2^6 \cdot 5$

36. $4^3 \cdot 6^3 \cdot 7$

37. $4^2 \cdot 3^3 \cdot 10^4$

38. Rewrite the expression using the numbers 3 and 5 exactly once. Then simplify the expression.
   a. $3 + 3 + 3 + 3 + 3$
   b. $3 \cdot 3 \cdot 3 \cdot 3 \cdot 3$

OBJECTIVE B   To use the Order of Operations Agreement to simplify expressions

For Exercises 39 to 77, simplify by using the Order of Operations Agreement.

39. $4 - 2 + 3$

40. $6 - 3 + 2$

41. $6 \div 3 + 2$

42. $8 \div 4 + 8$

43. $6 \cdot 3 + 5$ **44.** $5 \cdot 9 + 2$ **45.** $3^2 - 4$ **46.** $5^2 - 17$

47. $4 \cdot (5 - 3) + 2$ **48.** $3 + (4 + 2) \div 3$ **49.** $5 + (8 + 4) \div 6$ **50.** $8 - 2^2 + 4$

51. $16 \cdot (3 + 2) \div 10$ **52.** $12 \cdot (1 + 5) \div 12$ **53.** $10 - 2^3 + 4$ **54.** $5 \cdot 3^2 + 8$

55. $16 + 4 \cdot 3^2$ **56.** $12 + 4 \cdot 2^3$ **57.** $16 + (8 - 3) \cdot 2$ **58.** $7 + (9 - 5) \cdot 3$

59. $2^2 + 3 \cdot (6 - 2)^2$ **60.** $3^3 + 5 \cdot (8 - 6)^3$ **61.** $2^2 \cdot 3^2 + 2 \cdot 3$ **62.** $4 \cdot 6 + 3^2 \cdot 4^2$

63. $16 - 2 \cdot 4$ **64.** $12 + 3 \cdot 5$ **65.** $3 \cdot (6 - 2) + 4$

66. $5 \cdot (8 - 4) - 6$ **67.** $8 - (8 - 2) \div 3$ **68.** $12 - (12 - 4) \div 4$

69. $8 + 2 - 3 \cdot 2 \div 3$ **70.** $10 + 1 - 5 \cdot 2 \div 5$ **71.** $3 \cdot (4 + 2) \div 6$

72. $(7 - 3)^2 \div 2 - 4 + 8$ **73.** $20 - 4 \div 2 \cdot (3 - 1)^3$ **74.** $12 \div 3 \cdot 2^2 + (7 - 3)^2$

75. $(4 - 2) \cdot 6 \div 3 + (5 - 2)^2$ **76.** $18 - 2 \cdot 3 + (4 - 1)^3$ **77.** $100 \div (2 + 3)^2 - 8 \div 2$

 For Exercises 78 to 80, insert parentheses as needed in the expression $8 - 2 \cdot 3 + 1$ in order to make the statement true.

78. $8 - 2 \cdot 3 + 1 = 3$ **79.** $8 - 2 \cdot 3 + 1 = 0$ **80.** $8 - 2 \cdot 3 + 1 = 24$

Applying the Concepts

 81. Explain the difference that the order of operations makes between **a.** $(14 - 2) \div 2 \cdot 3$ and **b.** $(14 - 2) \div (2 \cdot 3)$. Work the two problems. What is the difference between the larger answer and the smaller answer?

SECTION 1.7 Prime Numbers and Factoring

OBJECTIVE A   To factor numbers

Whole-number **factors of a number** divide that number evenly (there is no remainder).

1, 2, 3, and 6 are whole-number factors of 6 because they divide 6 evenly.

$$\frac{6}{1)6} \quad \frac{3}{2)6} \quad \frac{2}{3)6} \quad \frac{1}{6)6}$$

Note that both the divisor and the quotient are factors of the dividend.

To find the factors of a number, try dividing the number by 1, 2, 3, 4, 5, ... Those numbers that divide the number evenly are its factors. Continue this process until the factors start to repeat.

> **HOW TO • 1**   Find all the factors of 42.
>
> | | |
> |---|---|
> | $42 \div 1 = 42$ | 1 and 42 are factors. |
> | $42 \div 2 = 21$ | 2 and 21 are factors. |
> | $42 \div 3 = 14$ | 3 and 14 are factors. |
> | $42 \div 4$ | Will not divide evenly |
> | $42 \div 5$ | Will not divide evenly |
> | $42 \div 6 = 7$ | 6 and 7 are factors. ⎫ Factors are repeating; all the |
> | $42 \div 7 = 6$ | 7 and 6 are factors. ⎭ factors of 42 have been found. |
>
> 1, 2, 3, 6, 7, 14, 21, and 42 are factors of 42.

The following rules are helpful in finding the factors of a number.

2 is a factor of a number if the last digit of the number is 0, 2, 4, 6, or 8.

436 ends in 6; therefore, 2 is a factor of 436. (436 ÷ 2 = 218)

3 is a factor of a number if the sum of the digits of the number is divisible by 3.

The sum of the digits of 489 is 4 + 8 + 9 = 21. 21 is divisible by 3. Therefore, 3 is a factor of 489. (489 ÷ 3 = 163)

5 is a factor of a number if the last digit of the number is 0 or 5.

520 ends in 0; therefore, 5 is a factor of 520. (520 ÷ 5 = 104)

EXAMPLE • 1

Find all the factors of 30.

Solution

$30 \div 1 = 30$
$30 \div 2 = 15$
$30 \div 3 = 10$
$30 \div 4$ Will not divide evenly
$30 \div 5 = 6$
$30 \div 6 = 5$ Factors repeating

1 2, 3, 5, 6, 10, 15, and 30 are factors of 30.

YOU TRY IT • 1

Find all the factors of 40.

Your solution

Solution on p. S4

| OBJECTIVE B | To find the prime factorization of a number |

Point of Interest

Prime numbers are an important part of cryptology, the study of secret codes. To make it less likely that codes can be broken, cryptologists use prime numbers that have hundreds of digits.

A number is a **prime number** if its only whole-number factors are 1 and itself. 7 is prime because its only factors are 1 and 7. If a number is not prime, it is called a **composite number.** Because 6 has factors of 2 and 3, 6 is a composite number. The number 1 is not considered a prime number; therefore, it is not included in the following list of prime numbers less than 50.

$$2, 3, 5, 7, 11, 13, 17, 19, 23, 29, 31, 37, 41, 43, 47$$

The **prime factorization** of a number is the expression of the number as a product of its prime factors. We use a "T-diagram" to find the prime factors of 60. Begin with the smallest prime number as a trial divisor, and continue with prime numbers as trial divisors until the final quotient is 1.

$$
\begin{array}{c}
60 \\
\hline
\begin{array}{c|c}
2 & 30 \\
2 & 15 \\
3 & 5 \\
5 & 1
\end{array}
\end{array}
\qquad
\begin{array}{l}
60 \div 2 = 30 \\
30 \div 2 = 15 \\
15 \div 3 = 5 \\
5 \div 5 = 1
\end{array}
$$

The prime factorization of 60 is $2 \cdot 2 \cdot 3 \cdot 5$.

Finding the prime factorization of larger numbers can be more difficult. Try each prime number as a trial divisor. Stop when the square of the trial divisor is greater than the number being factored.

HOW TO • 2 Find the prime factorization of 106.

$$
\begin{array}{c}
106 \\
\hline
\begin{array}{c|c}
2 & 53 \\
53 & 1
\end{array}
\end{array}
$$

• 53 cannot be divided evenly by 2, 3, 5, 7, or 11. Prime numbers greater than 11 need not be tested because 11^2 is greater than 53.

The prime factorization of 106 is $2 \cdot 53$.

EXAMPLE • 2

Find the prime factorization of 315.

Solution

$$
\begin{array}{c}
315 \\
\hline
\begin{array}{c|c}
3 & 105 \\
3 & 35 \\
5 & 7 \\
7 & 1
\end{array}
\end{array}
$$

• $315 \div 3 = 105$
• $105 \div 3 = 35$
• $35 \div 5 = 7$
• $7 \div 7 = 1$

$315 = 3 \cdot 3 \cdot 5 \cdot 7$

YOU TRY IT • 2

Find the prime factorization of 44.

Your solution

EXAMPLE • 3

Find the prime factorization of 201.

Solution

$$
\begin{array}{c}
201 \\
\hline
\begin{array}{c|c}
3 & 67 \\
67 & 1
\end{array}
\end{array}
$$

• Try only 2, 3, 5, 7, and 11 because $11^2 > 67$.

$201 = 3 \cdot 67$

YOU TRY IT • 3

Find the prime factorization of 177.

Your solution

Solutions on p. S4

1.7 EXERCISES

OBJECTIVE A   To factor numbers

For Exercises 1 to 40, find all the factors of the number.

1. 4   **2.** 6   **3.** 10   **4.** 20

5. 7   **6.** 12   **7.** 9   **8.** 8

9. 13   **10.** 17   **11.** 18   **12.** 24

13. 56   **14.** 36   **15.** 45   **16.** 28

17. 29   **18.** 33   **19.** 22   **20.** 26

21. 52   **22.** 49   **23.** 82   **24.** 37

25. 57   **26.** 69   **27.** 48   **28.** 64

29. 95   **30.** 46   **31.** 54   **32.** 50

33. 66   **34.** 77   **35.** 80   **36.** 100

37. 96   **38.** 85   **39.** 90   **40.** 101

 41. True or false? A number can have an odd number of factors.

 42. True or false? If a number has exactly four factors, then the product of those four factors must be the number.

OBJECTIVE B   To find the prime factorization of a number

For Exercises 43 to 86, find the prime factorization.

43. 6   **44.** 14   **45.** 17   **46.** 83

47. 24　　　　**48.** 12　　　　**49.** 27　　　　**50.** 9

51. 36　　　　**52.** 40　　　　**53.** 19　　　　**54.** 37

55. 90　　　　**56.** 65　　　　**57.** 115　　　**58.** 80

59. 18　　　　**60.** 26　　　　**61.** 28　　　　**62.** 49

63. 31　　　　**64.** 42　　　　**65.** 62　　　　**66.** 81

67. 22　　　　**68.** 39　　　　**69.** 101　　　**70.** 89

71. 66　　　　**72.** 86　　　　**73.** 74　　　　**74.** 95

75. 67　　　　**76.** 78　　　　**77.** 55　　　　**78.** 46

79. 120　　　**80.** 144　　　**81.** 160　　　**82.** 175

83. 216　　　**84.** 400　　　**85.** 625　　　**86.** 225

87. True or false? The prime factorization of 102 is $2 \cdot 51$.

Applying the Concepts

88. In 1742, Christian Goldbach conjectured that every even number greater than 2 could be expressed as the sum of two prime numbers. Show that this conjecture is true for 8, 24, and 72. (*Note:* Mathematicians have not yet been able to determine whether Goldbach's conjecture is true or false.)

89. Explain why 2 is the only even prime number.

FOCUS ON PROBLEM SOLVING

Questions to Ask

© Brownie Harris/Corbis

You encounter problem-solving situations every day. Some problems are easy to solve, and you may mentally solve these problems without considering the steps you are taking in order to draw a conclusion. Others may be more challenging and may require more thought and consideration.

Suppose a friend suggests that you both take a trip over spring break. You'd like to go. What questions go through your mind? You might ask yourself some of the following questions:

How much will the trip cost? What will be the cost for travel, hotel rooms, meals, and so on?

Are some costs going to be shared by both me and my friend?

Can I afford it?

How much money do I have in the bank?

How much more money than I have now do I need?

How much time is there to earn that much money?

How much can I earn in that amount of time?

How much money must I keep in the bank in order to pay the next tuition bill (or some other expense)?

These questions require different mathematical skills. Determining the cost of the trip requires **estimation;** for example, you must use your knowledge of air fares or the cost of gasoline to arrive at an estimate of these costs. If some of the costs are going to be shared, you need to **divide** those costs by 2 in order to determine your share of the expense. The question regarding how much more money you need requires **subtraction:** the amount needed minus the amount currently in the bank. To determine how much money you can earn in the given amount of time requires **multiplication**—for example, the amount you earn per week times the number of weeks to be worked. To determine if the amount you can earn in the given amount of time is sufficient, you need to use your knowledge of **order relations** to compare the amount you can earn with the amount needed.

Facing the problem-solving situation described above may not seem difficult to you. The reason may be that you have faced similar situations before and, therefore, know how to work through this one. You may feel better prepared to deal with a circumstance such as this one because you know what questions to ask. An important aspect of learning to solve problems is learning what questions to ask. As you work through application problems in this text, try to become more conscious of the mental process you are going through. You might begin the process by asking yourself the following questions whenever you are solving an application problem.

1. Have I read the problem enough times to be able to understand the situation being described?

2. Will restating the problem in different words help me to understand the problem situation better?

3. What facts are given? (You might make a list of the information contained in the problem.)

4. What information is being asked for?

5. What relationship exists among the given facts? What relationship exists between the given facts and the solution?

6. What mathematical operations are needed in order to solve the problem?

Try to focus on the problem-solving situation, not on the computation or on getting the answer quickly. And remember, the more problems you solve, the better able you will be to solve other problems in the future, partly because you are learning what questions to ask.

PROJECTS AND GROUP ACTIVITIES

Order of Operations

Does your calculator use the Order of Operations Agreement? To find out, try this problem:

$$2 + 4 \cdot 7$$

If your answer is 30, then the calculator uses the Order of Operations Agreement. If your answer is 42, it does not use that agreement.

Even if your calculator does not use the Order of Operations Agreement, you can still correctly evaluate numerical expressions. The parentheses keys, $($ and $)$, are used for this purpose.

Remember that $2 + 4 \cdot 7$ means $2 + (4 \cdot 7)$ because the multiplication must be completed before the addition. To evaluate this expression, enter the following:

Enter: 2   **+**   **(**   4   **x**   7   **)**   **=**

Display: 2   2   **(**   4   4   7   28   30

When using your calculator to evaluate numerical expressions, insert parentheses around multiplications and around divisions. This has the effect of forcing the calculator to do the operations in the order you want.

For Exercises 1 to 10, evaluate.

1. $3 \cdot 8 - 5$

2. $6 + 8 \div 2$

3. $3 \cdot (8 - 2)^2$

4. $24 - (4 - 2)^2 \div 4$

5. $3 + (6 \div 2 + 4)^2 - 2$

6. $16 \div 2 + 4 \cdot (8 - 12 \div 4)^2 - 50$

7. $3 \cdot (15 - 2 \cdot 3) - 36 \div 3$

8. $4 \cdot 2^2 - (12 + 24 \div 6) + 5$

9. $16 \div 4 \cdot 3 + (3 \cdot 4 - 5) + 2$

10. $15 \cdot 3 \div 9 + (2 \cdot 6 - 3) + 4$

Patterns in Mathematics

For the circle at the left, use a straight line to connect each dot on the circle with every other dot on the circle. How many different straight lines are there?

Follow the same procedure for each of the circles shown below. How many different straight lines are there in each?

Find a pattern to describe the number of dots on a circle and the corresponding number of different lines drawn. Use the pattern to determine the number of different lines that would be drawn in a circle with 7 dots and in a circle with 8 dots.

Now use the pattern to answer the following question. You are arranging a tennis tournament with 9 players. How many singles matches will be played among the 9 players if each player plays each of the other players only once?

Search the World Wide Web

Go to www.census.gov on the Internet.

Jonathan Nourak/PhotoEdit, Inc.

1. Find a projection for the total U.S. population 10 years from now and a projection for the total population 20 years from now. Record the two numbers.

2. Use the data from Exercise 1 to determine the expected growth in the population over the next 10 years.

3. Use the answer from Exercise 2 to find the average increase in the U.S. population per year over the next 10 years. Round to the nearest million.

4. Use data in the population table you found to write two word problems. Then state whether addition, subtraction, multiplication, or division is required to solve each of the problems.

CHAPTER 1

SUMMARY

KEY WORDS	EXAMPLES
The *whole numbers* are 0, 1, 2, 3, 4, 5, 6, 7, 8, 9, 10, ... [1.1A, p. 2]	
The *graph of a whole number* is shown by placing a heavy dot directly above that number on the number line. [1.1A, p. 2]	This is the graph of 4 on the number line. 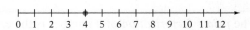
The symbol for *is less than* is <. The symbol for *is greater than* is >. These symbols are used to show the order relation between two numbers. [1.1A, p. 2]	3 < 7 9 > 2

When a whole number is written using the digits 0, 1, 2, 3, 4, 5, 6, 7, 8, and 9, it is said to be in *standard form*. The position of each digit in the number determines the digit's *place value*. The place values are used to write the expanded form of a number. [1.1B, p. 3]

The number 598,317 is in standard form. The digit 8 is in the thousands place. The number 598,317 is written in expanded form as 500,000 + 90,000 + 8000 + 300 + 10 + 7.

Addition is the process of finding the total of two or more numbers. The numbers being added are called *addends*. The result is the *sum*. [1.2A, p. 8]

$$\begin{array}{r} \overset{1\ \ 1\,1}{8{,}762} \\ +\ 1{,}359 \\ \hline 10{,}121 \end{array}$$

Subtraction is the process of finding the difference between two numbers. The *minuend* minus the *subtrahend* equals the *difference*. [1.3A, p. 16]

$$\begin{array}{r} \overset{4}{\cancel{8}}\ \overset{11}{\cancel{2}}{,}\overset{11}{\cancel{1}}\ \overset{6}{\cancel{7}}\ \overset{13}{\cancel{8}} \\ -\ 3\ 4{,}9\ 6\ 8 \\ \hline 1\ 7{,}2\ 0\ 5 \end{array}$$

Multiplication is the repeated addition of the same number. The numbers that are multiplied are called *factors*. The result is the *product*. [1.4A, p. 24]

$$\begin{array}{r} \overset{4\,5}{358} \\ \times\ \ \ 7 \\ \hline 2506 \end{array}$$

Division is used to separate objects into equal groups. The *dividend* divided by the *divisor* equals the *quotient*. [1.5A, p. 32]
For any division problem,
(*quotient* · *divisor*) + *remainder* = *dividend*. [1.5B, p. 35]

$$\begin{array}{r} 93\ \text{r}3 \\ 7\overline{)\ 654} \\ -63 \\ \hline 24 \\ -21 \\ \hline 3 \end{array}$$

Check: (7 · 93) + 3 = 651 + 3 = 654

The expression 4^3 is in *exponential notation*. The *exponent*, 3, indicates how many times 4 occurs as a factor in the multiplication. [1.6A, p. 45]

$5^4 = 5 \cdot 5 \cdot 5 \cdot 5 = 625$

Whole-number *factors of a number* divide that number evenly (there is no remainder). [1.7A, p. 49]

$18 \div 1 = 18$
$18 \div 2 = 9$
$18 \div 3 = 6$
$18 \div 4$ ⠀⠀4 does not divide 18 evenly.
$18 \div 5$ ⠀⠀5 does not divide 18 evenly.
$18 \div 6 = 3$ ⠀The factors are repeating.
The factors of 18 are 1, 2, 3, 6, 9, and 18.

A number greater than 1 is a *prime number* if its only whole-number factors are 1 and itself. If a number is not prime, it is a *composite number*. [1.7B, p. 50]

The prime numbers less than 20 are 2, 3, 5, 7, 11, 13, 17, and 19.
The composite numbers less than 20 are 4, 6, 8, 9, 10, 12, 14, 15, 16, and 18.

The *prime factorization* of a number is the expression of the number as a product of its prime factors. [1.7B, p. 50]

$$\begin{array}{r} 42 \\ \hline 2\,\vert\,21 \\ 3\,\vert\ \ 7 \\ 7\,\vert\ \ 1 \end{array}$$

The prime factorization of 42 is $2 \cdot 3 \cdot 7$.

ESSENTIAL RULES AND PROCEDURES	EXAMPLES

To round a number to a given place value: If the digit to the right of the given place value is less than 5, replace that digit and all digits to the right by zeros. If the digit to the right of the given place value is greater than or equal to 5, increase the digit in the given place value by 1, and replace all other digits to the right by zeros. [1.1D, p. 5]

36,178 rounded to the nearest thousand is 36,000.

4592 rounded to the nearest thousand is 5000.

Properties of Addition [1.2A, p. 8]

Addition Property of Zero
Zero added to a number does not change the number.

$7 + 0 = 7$

Commutative Property of Addition
Two numbers can be added in either order; the sum will be the same.

$8 + 3 = 3 + 8$

Associative Property of Addition
Numbers to be added can be grouped in any order; the sum will be the same.

$(2 + 4) + 6 = 2 + (4 + 6)$

To estimate the answer to an addition calculation: Round each number to the same place value. Perform the calculation using the rounded numbers. [1.2A, p. 10]

$$\begin{array}{rr} 39{,}471 & 40{,}000 \\ 12{,}586 & +\ 10{,}000 \\ \hline & 50{,}000 \end{array}$$

50,000 is an estimate of the sum of 39,471 and 12,586.

Properties of Multiplication [1.4A, p. 24]

Multiplication Property of Zero
The product of a number and zero is zero.

$3 \cdot 0 = 0$

Multiplication Property of One
The product of a number and one is the number.

$6 \cdot 1 = 6$

Commutative Property of Multiplication
Two numbers can be multiplied in either order; the product will be the same.

$2 \cdot 8 = 8 \cdot 2$

Associative Property of Multiplication
Grouping numbers to be multiplied in any order gives the same result.

$(2 \cdot 4) \cdot 6 = 2 \cdot (4 \cdot 6)$

Division Properties of Zero and One [1.5A, p. 32]
Any whole number, except zero, divided by itself is 1.
Any whole number divided by 1 is the whole number.
Zero divided by any other whole number is zero.
Division by zero is not allowed.

$3 \div 3 = 1$
$3 \div 1 = 3$
$0 \div 3 = 0$
$3 \div 0$ is not allowed.

Order of Operations Agreement [1.6B, p. 46]

Step 1 Do all the operations inside parentheses.

Step 2 Simplify any number expressions containing exponents.

Step 3 Do multiplications and divisions as they occur from left to right.

Step 4 Do addition and subtraction as they occur from left to right.

$$5^2 - 3(2 + 4) = 5^2 - 3(6)$$
$$= 25 - 3(6)$$
$$= 25 - 18$$
$$= 7$$

CHAPTER 1

CONCEPT REVIEW

Test your knowledge of the concepts presented in this chapter. Answer each question.
Then check your answers against the ones provided in the Answer Section.

1.   What is the difference between the symbols $<$ and $>$?

2.   How do you round a four-digit whole number to the nearest hundred?

3.   What is the difference between the Commutative Property of Addition and the Associative Property of Addition?

4.   How do you estimate the sum of two numbers?

5.   When is it necessary to borrow when performing subtraction?

6.   What is the difference between the Multiplication Property of Zero and the Multiplication Property of One?

7.   How do you multiply a whole number by 100?

8.   How do you estimate the product of two numbers?

9.   What is the difference between $0 \div 9$ and $9 \div 0$?

10.   How do you check the answer to a division problem that has a remainder?

11.   What are the steps in the Order of Operations Agreement?

12.   How do you know if a number is a factor of another number?

13.   What is a quick way to determine if 3 is a factor of a number?

REVIEW EXERCISES

1. Simplify: $3 \cdot 2^3 \cdot 5^2$

2. Write 10,327 in expanded form.

3. Find all the factors of 18.

4. Find the sum of 5894, 6301, and 298.

5. Subtract: $\begin{array}{r} 4926 \\ -\ 3177 \end{array}$

6. Divide: $7\overline{)14{,}945}$

7. Place the correct symbol, $<$ or $>$, between the two numbers: 101   87

8. Write $5 \cdot 5 \cdot 7 \cdot 7 \cdot 7 \cdot 7 \cdot 7$ in exponential notation.

9. What is 2019 multiplied by 307?

10. What is 10,134 decreased by 4725?

11. Add: $\begin{array}{r} 298 \\ 461 \\ +\ 322 \end{array}$

12. Simplify: $2^3 - 3 \cdot 2$

13. Round 45,672 to the nearest hundred.

14. Write 276,057 in words.

15. Find the quotient of 109,763 and 84.

16. Write two million eleven thousand forty-four in standard form.

17. What is 3906 divided by 8?

18. Simplify: $3^2 + 2^2 \cdot (5 - 3)$

19. Simplify: $8 \cdot (6 - 2)^2 \div 4$

20. Find the prime factorization of 72.

21. What is 3895 minus 1762?

22. Multiply: 843
                × 27

23. Wages Vincent Meyers, a sales assistant, earns $480 for working a 40-hour week. Last week Vincent worked an additional 12 hours at $24 an hour. Find Vincent's total pay for last week's work.

24. Fuel Efficiency Louis Reyes, a sales executive, drove a car 351 miles on 13 gallons of gas. Find the number of miles driven per gallon of gasoline.

25. Consumerism A car is purchased for $29,880, with a down payment of $3000. The balance is paid in 48 equal monthly payments. Find the monthly car payment.

26. Compensation An insurance account executive received commissions of $723, $544, $812, and $488 during a 4-week period. Find the total income from commissions for the 4 weeks.

27. Banking You had a balance of $516 in your checking account before making deposits of $88 and $213. Find the total amount deposited, and determine your new account balance.

28. Compensation You have a car payment of $246 per month. What is the total of the car payments over a 12-month period?

Athletics The table at the right shows the athletic participation by males and females at U.S. colleges in 1972 and 2005. Use this information for Exercises 29 to 32.

Year	Male Athletes	Female Athletes
1972	170,384	29,977
2005	291,797	205,492

Source: U.S. Department of Education commission report

29. In which year, 1972 or 2005, were there more males involved in sports at U.S. colleges?

30. What is the difference between the number of males involved in sports and the number of females involved in sports at U.S. colleges in 1972?

31. Find the increase in the number of females involved in sports in U.S. colleges from 1972 to 2005.

© Pete Saloutos/Corbis

32. How many more U.S. college students were involved in athletics in 2005 than in 1972?

CHAPTER 1

TEST

1. Simplify: $3^3 \cdot 4^2$

2. Write 207,068 in words.

3. Subtract:
$$\begin{array}{r} 17{,}495 \\ -\ 8{,}162 \end{array}$$

4. Find all the factors of 20.

5. Multiply:
$$\begin{array}{r} 9736 \\ \times\ 704 \end{array}$$

6. Simplify: $4^2 \cdot (4 - 2) \div 8 + 5$

7. Write 906,378 in expanded form.

8. Round 74,965 to the nearest hundred.

9. Divide: $97\overline{)108{,}764}$

10. Write $3 \cdot 3 \cdot 3 \cdot 7 \cdot 7$ in exponential form.

11. Find the sum of 8756, 9094, and 37,065.

12. Find the prime factorization of 84.

13. Simplify: $16 \div 4 \cdot 2 - (7 - 5)^2$

14. Find the product of 8 and 90,763.

15. Write one million two hundred four thousand six in standard form.

16. Divide: $7\overline{)60{,}972}$

17. Place the correct symbol, $<$ or $>$, between the two numbers: 21   19

18. Find the quotient of 5624 and 8.

19. Add: 25,492
            +71,306

20. Find the difference between 29,736 and 9814.

Education The table at the right shows the projected enrollment in public and private elementary and secondary schools in the fall of 2013 and the fall of 2016. Use this information for Exercises 21 and 22.

Year	Pre-Kindergarten through Grade 8	Grades 9 through 12
2013	41,873,000	16,000,000
2016	43,097,000	16,684,000

Source: The National Center for Education Statistics

21. Find the difference between the total enrollment in 2016 and that in 2013.

22. Find the average enrollment in each of grades 9 through 12 in 2016.

23. **Farming** A farmer harvested 48,290 pounds of lemons from one grove and 23,710 pounds of lemons from another grove. The lemons were packed in boxes with 24 pounds of lemons in each box. How many boxes were needed to pack the lemons?

24. **Investments** An investor receives $237 each month from a corporate bond fund. How much will the investor receive over a 12-month period?

25. **Travel** A family drives 425 miles the first day, 187 miles the second day, and 243 miles the third day of their vacation. The odometer read 47,626 miles at the start of the vacation.
   a. How many miles were driven during the 3 days?
   b. What is the odometer reading at the end of the 3 days?

Fractions

ARE YOU READY?

Take the Chapter 2 Prep Test to find out if you are ready to learn to:

- Write equivalent fractions
- Write fractions in simplest form
- Add, subtract, multiply, and divide fractions
- Compare fractions

PREP TEST

Do these exercises to prepare for Chapter 2.

For Exercises 1 to 6, add, subtract, multiply, or divide.

1. 4×5

2. $2 \cdot 2 \cdot 2 \cdot 3 \cdot 5$

3. 9×1

4. $6 + 4$

5. $10 - 3$

6. $63 \div 30$

7. Which of the following numbers divide evenly into 12?
1 2 3 4 5 6 7 8 9 10 11 12

8. Simplify: $8 \times 7 + 3$

9. Complete: $8 = ? + 1$

10. Place the correct symbol, $<$ or $>$, between the two numbers.
44   48

SECTION

2.1

The Least Common Multiple and Greatest Common Factor

OBJECTIVE A   **To find the least common multiple (LCM)**

 Tips for Success

Before you begin a new chapter, you should take some time to review previously learned skills. One way to do this is to complete the Prep Test. See page 63. This test focuses on the particular skills that will be required for the new chapter.

The **multiples of a number** are the products of that number and the numbers 1, 2, 3, 4, 5, ...

$3 \times 1 = 3$
$3 \times 2 = 6$
$3 \times 3 = 9$
$3 \times 4 = 12$ The multiples of 3 are 3, 6, 9, 12, 15, ...
$3 \times 5 = 15$
⋮
⋮

A number that is a multiple of two or more numbers is a **common multiple** of those numbers.

The multiples of 4 are 4, 8, 12, 16, 20, 24, 28, 32, 36, ...
The multiples of 6 are 6, 12, 18, 24, 30, 36, 42, ...
Some common multiples of 4 and 6 are 12, 24, and 36.

The **least common multiple (LCM)** is the smallest common multiple of two or more numbers.

The least common multiple of 4 and 6 is 12.

Listing the multiples of each number is one way to find the LCM. Another way to find the LCM uses the prime factorization of each number.

To find the LCM of 450 and 600, find the prime factorization of each number and write the factorization of each number in a table. Circle the greatest product in each column. The LCM is the product of the circled numbers.

	2	3	5
450 =	2	③·③	⑤·⑤
600 =	②·②·②	3	5·5

• In the column headed by 5, the products are equal. Circle just one product.

The LCM is the product of the circled numbers.
The LCM = $2 \cdot 2 \cdot 2 \cdot 3 \cdot 3 \cdot 5 \cdot 5 = 1800$.

EXAMPLE • 1

Find the LCM of 24, 36, and 50.

Solution

	2	3	5
24 =	②·②·②	3	
36 =	2·2	③·③	
50 =	2		⑤·⑤

The LCM = $2 \cdot 2 \cdot 2 \cdot 3 \cdot 3 \cdot 5 \cdot 5 = 1800$.

YOU TRY IT • 1

Find the LCM of 12, 27, and 50.

Your solution

Solution on p. S4

OBJECTIVE B **To find the greatest common factor (GCF)**

Recall that a number that divides another number evenly is a factor of that number. The number 64 can be evenly divided by 1, 2, 4, 8, 16, 32, and 64, so the numbers 1, 2, 4, 8, 16, 32, and 64 are factors of 64.

A number that is a factor of two or more numbers is a **common factor** of those numbers.

The factors of 30 are 1, 2, 3, 5, 6, 10, 15, and 30.
The factors of 105 are 1, 3, 5, 7, 15, 21, 35, and 105.
The common factors of 30 and 105 are 1, 3, 5, and 15.

The **greatest common factor (GCF)** is the largest *common factor* of two or more numbers.

The greatest common factor of 30 and 105 is 15.

Listing the factors of each number is one way of finding the GCF. Another way to find the GCF is to use the prime factorization of each number.

To find the GCF of 126 and 180, find the prime factorization of each number and write the factorization of each number in a table. Circle the least product in each column that does not have a blank. The GCF is the product of the circled numbers.

	2	3	5	7
126 =	②	③ · 3		7
180 =	2 · 2	3 · 3	5	

• In the column headed by 3, the products are equal. Circle just one product. Columns 5 and 7 have a blank, so 5 and 7 are not common factors of 126 and 180. Do not circle any number in these columns.

The GCF is the product of the circled numbers.
The GCF = 2 · 3 · 3 = 18.

EXAMPLE · 2

Find the GCF of 90, 168, and 420.

Solution

	2	3	5	7
90 =	②	3 · 3	5	
168 =	2 · 2 · 2	③		7
420 =	2 · 2	3	5	7

The GCF = 2 · 3 = 6.

YOU TRY IT · 2

Find the GCF of 36, 60, and 72.

Your solution

EXAMPLE · 3

Find the GCF of 7, 12, and 20.

Solution

	2	3	5	7
7 =				7
12 =	2 · 2	3		
20 =	2 · 2		5	

Because no numbers are circled, the GCF = 1.

YOU TRY IT · 3

Find the GCF of 11, 24, and 30.

Your solution

Solutions on p. S4

2.1 EXERCISES

OBJECTIVE A — To find the least common multiple (LCM)

For Exercises 1 to 34, find the LCM.

1. 5, 8   **2.** 3, 6   **3.** 3, 8   **4.** 2, 5   **5.** 5, 6

6. 5, 7   **7.** 4, 6   **8.** 6, 8   **9.** 8, 12   **10.** 12, 16

11. 5, 12   **12.** 3, 16   **13.** 8, 14   **14.** 6, 18   **15.** 3, 9

16. 4, 10   **17.** 8, 32   **18.** 7, 21   **19.** 9, 36   **20.** 14, 42

21. 44, 60   **22.** 120, 160   **23.** 102, 184   **24.** 123, 234   **25.** 4, 8, 12

26. 5, 10, 15   **27.** 3, 5, 10   **28.** 2, 5, 8   **29.** 3, 8, 12   **30.** 5, 12, 18

31. 9, 36, 64   **32.** 18, 54, 63   **33.** 16, 30, 84   **34.** 9, 12, 15

35. True or false? If two numbers have no common factors, then the LCM of the two numbers is their product.

36. True or false? If one number is a multiple of a second number, then the LCM of the two numbers is the second number.

OBJECTIVE B — To find the greatest common factor (GCF)

For Exercises 37 to 70, find the GCF.

37. 3, 5   **38.** 5, 7   **39.** 6, 9   **40.** 18, 24   **41.** 15, 25

42. 14, 49   **43.** 25, 100   **44.** 16, 80   **45.** 32, 51   **46.** 21, 44

47. 12, 80   **48.** 8, 36   **49.** 16, 140   **50.** 12, 76

51. 24, 30   **52.** 48, 144   **53.** 44, 96   **54.** 18, 32

55. 3, 5, 11   **56.** 6, 8, 10   **57.** 7, 14, 49   **58.** 6, 15, 36

59. 10, 15, 20   **60.** 12, 18, 20   **61.** 24, 40, 72   **62.** 3, 17, 51

63. 17, 31, 81   **64.** 14, 42, 84   **65.** 25, 125, 625   **66.** 12, 68, 92

67. 28, 35, 70   **68.** 1, 49, 153   **69.** 32, 56, 72   **70.** 24, 36, 48

 71. True or false? If two numbers have a GCF of 1, then the LCM of the two numbers is their product.

 72. True or false? If the LCM of two numbers is one of the two numbers, then the GCF of the numbers is the other of the two numbers.

Applying the Concepts

73. **Work Schedules** Joe Salvo, a lifeguard, works 3 days and then has a day off. Joe's friend works 5 days and then has a day off. How many days after Joe and his friend have a day off together will they have another day off together?

 74. Find the LCM of each of the following pairs of numbers: 2 and 3, 5 and 7, and 11 and 19. Can you draw a conclusion about the LCM of two prime numbers? Suggest a way of finding the LCM of three distinct prime numbers.

75. Find the GCF of each of the following pairs of numbers: 3 and 5, 7 and 11, and 29 and 43. Can you draw a conclusion about the GCF of two prime numbers? What is the GCF of three distinct prime numbers?

76. Using the pattern for the first two triangles at the right, determine the center number of the last triangle.

The Least Common Multiple and

SECTION

2.2 Introduction to Fractions

OBJECTIVE A **To write a fraction that represents part of a whole**

 ✓ Take Note

The **fraction bar** separates the numerator from the denominator. The **numerator** is the part of the fraction that appears above the fraction bar. The **denominator** is the part of the fraction that appears below the fraction bar.

◎ Point of Interest

The fraction bar was first used in 1050 by al-Hassar. It is also called a vinculum.

A **fraction** can represent the number of equal parts of a whole.

The shaded portion of the circle is represented by the fraction $\frac{4}{7}$. Four of the seven equal parts of the circle (that is, four-sevenths of it) are shaded.

Each part of a fraction has a name.

Fraction bar → $\dfrac{4}{7}$ ← **Numerator**
← **Denominator**

A **proper fraction** is a fraction less than 1. The numerator of a proper fraction is smaller than the denominator. The shaded portion of the circle can be represented by the proper fraction $\frac{3}{4}$.

A **mixed number** is a number greater than 1 with a whole-number part and a fractional part. The shaded portion of the circles can be represented by the mixed number $2\frac{1}{4}$.

An **improper fraction** is a fraction greater than or equal to 1. The numerator of an improper fraction is greater than or equal to the denominator. The shaded portion of the circles can be represented by the improper fraction $\frac{9}{4}$. The shaded portion of the square can be represented by $\frac{4}{4}$.

 $\frac{4}{7}$

 $\frac{3}{4}$

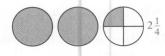 $2\frac{1}{4}$

 $\frac{9}{4}$

 $\frac{4}{4}$

EXAMPLE • 1

Express the shaded portion of the circles as a mixed number.

Solution $3\frac{2}{5}$

EXAMPLE • 2

Express the shaded portion of the circles as an improper fraction.

Solution $\frac{17}{5}$

YOU TRY IT • 1

Express the shaded portion of the circles as a mixed number.

Your solution

YOU TRY IT • 2

Express the shaded portion of the circles as an improper fraction.

Your solution

Solutions on p. S4

OBJECTIVE B   To write an improper fraction as a mixed number or a whole number, and a mixed number as an improper fraction

Note from the diagram that the mixed number $2\frac{3}{5}$ and the improper fraction $\frac{13}{5}$ both represent the shaded portion of the circles.

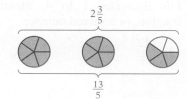

$$2\frac{3}{5} = \frac{13}{5}$$

An improper fraction can be written as a mixed number or a whole number.

HOW TO 1   Write $\frac{13}{5}$ as a mixed number.

Divide the numerator by the denominator.

$$\begin{array}{r} 2 \\ 5{\overline{\smash{\big)}\,13}} \\ \underline{-10} \\ 3 \end{array}$$

To write the fractional part of the mixed number, write the remainder over the divisor.

$$\begin{array}{r} 2\frac{3}{5} \\ 5{\overline{\smash{\big)}\,13}} \\ \underline{-10} \\ 3 \end{array}$$

Write the answer.

$$\frac{13}{5} = 2\frac{3}{5}$$

To write a mixed number as an improper fraction, multiply the denominator of the fractional part by the whole-number part. The sum of this product and the numerator of the fractional part is the numerator of the improper fraction. The denominator remains the same.

HOW TO 2   Write $7\frac{3}{8}$ as an improper fraction.

$$7\frac{3}{8} = \frac{(8 \times 7) + 3}{8} = \frac{56 + 3}{8} = \frac{59}{8} \qquad 7\frac{3}{8} = \frac{59}{8}$$

EXAMPLE 3

Write $\frac{21}{4}$ as a mixed number.

Solution

$$\begin{array}{r} 5 \\ 4{\overline{\smash{\big)}\,21}} \\ \underline{-20} \\ 1 \end{array} \qquad \frac{21}{4} = 5\frac{1}{4}$$

YOU TRY IT 3

Write $\frac{22}{5}$ as a mixed number.

Your solution

EXAMPLE 4

Write $\frac{18}{6}$ as a whole number.

Solution $\frac{18}{6} = 18 \div 6 = 3$

YOU TRY IT 4

Write $\frac{28}{7}$ as a whole number.

Your solution

EXAMPLE 5

Write $21\frac{3}{4}$ as an improper fraction.

Solution $21\frac{3}{4} = \frac{84 + 3}{4} = \frac{87}{4}$

YOU TRY IT 5

Write $14\frac{5}{8}$ as an improper fraction.

Your solution

Solutions on p. S4

2.2 EXERCISES

OBJECTIVE A   To write a fraction that represents part of a whole

For Exercises 1 to 4, identify the fraction as a proper fraction, an improper fraction, or a mixed number.

1. $\dfrac{12}{7}$ **2.** $5\dfrac{2}{11}$ **3.** $\dfrac{29}{40}$ **4.** $\dfrac{19}{13}$

For Exercises 5 to 8, express the shaded portion of the circle as a fraction.

5.   **6.**   **7.**   **8.**

For Exercises 9 to 14, express the shaded portion of the circles as a mixed number.

9.   **10.**

11.    **12.**

13.    **14.**

For Exercises 15 to 20, express the shaded portion of the circles as an improper fraction.

15. 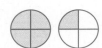   **16.**

17.    **18.**

19. 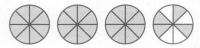   **20.**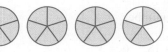

21. Shade $1\dfrac{2}{5}$ of 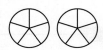   **22.** Shade $1\dfrac{3}{4}$ of

23. Shade $\dfrac{6}{5}$ of 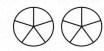   **24.** Shade $\dfrac{7}{3}$ of

25. True or false? The fractional part of a mixed number is an improper fraction.

OBJECTIVE B   **To write an improper fraction as a mixed number or a whole number, and a mixed number as an improper fraction**

For Exercises 26 to 49, write the improper fraction as a mixed number or a whole number.

26. $\dfrac{11}{4}$ **27.** $\dfrac{16}{3}$ **28.** $\dfrac{20}{4}$ **29.** $\dfrac{18}{9}$ **30.** $\dfrac{9}{8}$ **31.** $\dfrac{13}{4}$

32. $\dfrac{23}{10}$ **33.** $\dfrac{29}{2}$ **34.** $\dfrac{48}{16}$ **35.** $\dfrac{51}{3}$ **36.** $\dfrac{8}{7}$ **37.** $\dfrac{16}{9}$

38. $\dfrac{7}{3}$ **39.** $\dfrac{9}{5}$ **40.** $\dfrac{16}{1}$ **41.** $\dfrac{23}{1}$ **42.** $\dfrac{17}{8}$ **43.** $\dfrac{31}{16}$

44. $\dfrac{12}{5}$ **45.** $\dfrac{19}{3}$ **46.** $\dfrac{9}{9}$ **47.** $\dfrac{40}{8}$ **48.** $\dfrac{72}{8}$ **49.** $\dfrac{3}{3}$

For Exercises 50 to 73, write the mixed number as an improper fraction.

50. $2\dfrac{1}{3}$ **51.** $4\dfrac{2}{3}$ **52.** $6\dfrac{1}{2}$ **53.** $8\dfrac{2}{3}$ **54.** $6\dfrac{5}{6}$ **55.** $7\dfrac{3}{8}$

56. $9\dfrac{1}{4}$ **57.** $6\dfrac{1}{4}$ **58.** $10\dfrac{1}{2}$ **59.** $15\dfrac{1}{8}$ **60.** $8\dfrac{1}{9}$ **61.** $3\dfrac{5}{12}$

62. $5\dfrac{3}{11}$ **63.** $3\dfrac{7}{9}$ **64.** $2\dfrac{5}{8}$ **65.** $12\dfrac{2}{3}$ **66.** $1\dfrac{5}{8}$ **67.** $5\dfrac{3}{7}$

68. $11\dfrac{1}{9}$ **69.** $12\dfrac{3}{5}$ **70.** $3\dfrac{3}{8}$ **71.** $4\dfrac{5}{9}$ **72.** $6\dfrac{7}{13}$ **73.** $8\dfrac{5}{14}$

 74. True or false? If an improper fraction is equivalent to 1, then the numerator and the denominator are the same number.

Applying the Concepts

 75. Name three situations in which fractions are used. Provide an example of a fraction that is used in each situation.

SECTION

2.3 Writing Equivalent Fractions

OBJECTIVE A **To find equivalent fractions by raising to higher terms**

Equal fractions with different denominators are called **equivalent fractions.**

$\frac{4}{6}$ is equivalent to $\frac{2}{3}$.

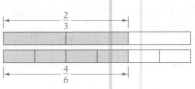

Remember that the Multiplication Property of One states that the product of a number and one is the number. This is true for fractions as well as whole numbers. This property can be used to write equivalent fractions.

$$\frac{2}{3} \times 1 = \frac{2}{3} \times \frac{1}{1} = \frac{2 \cdot 1}{3 \cdot 1} = \frac{2}{3}$$

$$\frac{2}{3} \times 1 = \frac{2}{3} \times \boxed{\frac{2}{2}} = \frac{2 \cdot 2}{3 \cdot 2} = \frac{4}{6} \qquad \frac{4}{6} \text{ is equivalent to } \frac{2}{3}.$$

$$\frac{2}{3} \times 1 = \frac{2}{3} \times \boxed{\frac{4}{4}} = \frac{2 \cdot 4}{3 \cdot 4} = \frac{8}{12} \qquad \frac{8}{12} \text{ is equivalent to } \frac{2}{3}.$$

$\frac{2}{3}$ was rewritten as the equivalent fractions $\frac{4}{6}$ and $\frac{8}{12}$.

HOW TO • 1 Write a fraction that is equivalent to $\frac{5}{8}$ and has a denominator of 32.

$32 \div 8 = 4$ • Divide the larger denominator by the smaller.

$\frac{5}{8} = \frac{5 \cdot 4}{8 \cdot 4} = \frac{20}{32}$ • Multiply the numerator and denominator of the given fraction by the quotient (4).

$\frac{20}{32}$ is equivalent to $\frac{5}{8}$.

EXAMPLE • 1

Write $\frac{2}{3}$ as an equivalent fraction that has a denominator of 42.

Solution $42 \div 3 = 14$ $\quad \frac{2}{3} = \frac{2 \cdot 14}{3 \cdot 14} = \frac{28}{42}$

$\frac{28}{42}$ is equivalent to $\frac{2}{3}$.

YOU TRY IT • 1

Write $\frac{3}{5}$ as an equivalent fraction that has a denominator of 45.

Your solution

EXAMPLE • 2

Write 4 as a fraction that has a denominator of 12.

Solution Write 4 as $\frac{4}{1}$.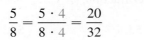

$12 \div 1 = 12$ $\quad 4 = \frac{4 \cdot 12}{1 \cdot 12} = \frac{48}{12}$

$\frac{48}{12}$ is equivalent to 4.

YOU TRY IT • 2

Write 6 as a fraction that has a denominator of 18.

Your solution

Solutions on p. S4

OBJECTIVE B **To write a fraction in simplest form**

Writing the **simplest form of a fraction** means writing it so that the numerator and denominator have no common factors other than 1.

The fractions $\frac{4}{6}$ and $\frac{2}{3}$ are equivalent fractions.

$\frac{4}{6}$ has been written in simplest form as $\frac{2}{3}$.

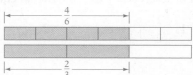

The Multiplication Property of One can be used to write fractions in simplest form. Write the numerator and denominator of the given fraction as a product of factors. Write factors common to both the numerator and denominator as an improper fraction equivalent to 1.

$$\frac{4}{6} = \frac{2 \cdot 2}{2 \cdot 3} = \frac{2}{2} \cdot \frac{2}{3} = 1 \cdot \frac{2}{3} = \frac{2}{3}$$

The process of eliminating common factors is displayed with slashes through the common factors as shown at the right.

To write a fraction in simplest form, eliminate the common factors.

An improper fraction can be changed to a mixed number.

$$\frac{4}{6} = \frac{\cancel{2} \cdot 2}{\cancel{2} \cdot 3} = \frac{2}{3}$$

$$\frac{18}{30} = \frac{\cancel{2} \cdot \cancel{3} \cdot 3}{\cancel{2} \cdot \cancel{3} \cdot 5} = \frac{3}{5}$$

$$\frac{22}{6} = \frac{\cancel{2} \cdot 11}{\cancel{2} \cdot 3} = \frac{11}{3} = 3\frac{2}{3}$$

EXAMPLE • 3

Write $\frac{15}{40}$ in simplest form.

Solution

$$\frac{15}{40} = \frac{3 \cdot \cancel{5}}{2 \cdot 2 \cdot 2 \cdot \cancel{5}} = \frac{3}{8}$$

YOU TRY IT •

Write $\frac{16}{24}$ in simplest form.

Your solution

EXAMPLE • 4

Write $\frac{6}{42}$ in simplest form.

Solution

$$\frac{6}{42} = \frac{\cancel{2} \cdot \cancel{3}}{\cancel{2} \cdot \cancel{3} \cdot 7} = \frac{1}{7}$$

YOU TRY IT • 4

Write $\frac{8}{56}$ in simplest form.

Your solution

EXAMPLE • 5

Write $\frac{8}{9}$ in simplest form.

Solution

$$\frac{8}{9} = \frac{2 \cdot 2 \cdot 2}{3 \cdot 3} = \frac{8}{9}$$

$\frac{8}{9}$ is already in simplest form because there are no common factors in the numerator and denominator.

YOU TRY IT • 5

Write $\frac{15}{32}$ in simplest form.

Your solution

EXAMPLE • 6

Write $\frac{30}{12}$ in simplest form.

Solution

$$\frac{30}{12} = \frac{\cancel{2} \cdot \cancel{3} \cdot 5}{\cancel{2} \cdot 2 \cdot \cancel{3}} = \frac{5}{2} = 2\frac{1}{2}$$

YOU TRY IT • 6

Write $\frac{48}{36}$ in simplest form.

Your solution

Solutions on p. S4

2.3 EXERCISES

OBJECTIVE A   To find equivalent fractions by raising to higher terms

For Exercises 1 to 35, write an equivalent fraction with the given denominator.

1. $\dfrac{1}{2} = \dfrac{}{10}$ 2. $\dfrac{1}{4} = \dfrac{}{16}$ 3. $\dfrac{3}{16} = \dfrac{}{48}$ 4. $\dfrac{5}{9} = \dfrac{}{81}$ 5. $\dfrac{3}{8} = \dfrac{}{32}$

6. $\dfrac{7}{11} = \dfrac{}{33}$ 7. $\dfrac{3}{17} = \dfrac{}{51}$ 8. $\dfrac{7}{10} = \dfrac{}{90}$ 9. $\dfrac{3}{4} = \dfrac{}{16}$ 10. $\dfrac{5}{8} = \dfrac{}{32}$

11. $3 = \dfrac{}{9}$ 12. $5 = \dfrac{}{25}$ 13. $\dfrac{1}{3} = \dfrac{}{60}$ 14. $\dfrac{1}{16} = \dfrac{}{48}$ 15. $\dfrac{11}{15} = \dfrac{}{60}$

16. $\dfrac{3}{50} = \dfrac{}{300}$ 17. $\dfrac{2}{3} = \dfrac{}{18}$ 18. $\dfrac{5}{9} = \dfrac{}{36}$ 19. $\dfrac{5}{7} = \dfrac{}{49}$ 20. $\dfrac{7}{8} = \dfrac{}{32}$

21. $\dfrac{5}{9} = \dfrac{}{18}$ 22. $\dfrac{11}{12} = \dfrac{}{36}$ 23. $7 = \dfrac{21}{3}$ 24. $9 = \dfrac{}{4}$ 25. $\dfrac{7}{9} = \dfrac{}{45}$

26. $\dfrac{5}{6} = \dfrac{}{42}$ 27. $\dfrac{15}{16} = \dfrac{}{64}$ 28. $\dfrac{11}{18} = \dfrac{}{54}$ 29. $\dfrac{3}{14} = \dfrac{}{98}$ 30. $\dfrac{5}{6} = \dfrac{}{144}$

31. $\dfrac{5}{8} = \dfrac{}{48}$ 32. $\dfrac{7}{12} = \dfrac{}{96}$ 33. $\dfrac{5}{14} = \dfrac{}{42}$ 34. $\dfrac{2}{3} = \dfrac{}{42}$ 35. $\dfrac{17}{24} = \dfrac{}{144}$

36. When you multiply the numerator and denominator of a fraction by the same number, you are actually multiplying the fraction by the number _____.

OBJECTIVE B   To write a fraction in simplest form

For Exercises 37 to 71, write the fraction in simplest form.

37. $\dfrac{4}{12}$ 38. $\dfrac{8}{22}$ 39. $\dfrac{22}{44}$ 40. $\dfrac{2}{14}$ 41. $\dfrac{2}{12}$

42. $\dfrac{50}{75}$ 43. $\dfrac{40}{36}$ 44. $\dfrac{12}{8}$ 45. $\dfrac{0}{30}$ 46. $\dfrac{10}{10}$

47. $\dfrac{9}{22}$ 48. $\dfrac{14}{35}$ 49. $\dfrac{75}{25}$ 50. $\dfrac{8}{60}$ 51. $\dfrac{16}{84}$

52. $\dfrac{20}{44}$ 53. $\dfrac{12}{35}$ 54. $\dfrac{8}{36}$ 55. $\dfrac{28}{44}$ 56. $\dfrac{12}{16}$

57. $\dfrac{16}{12}$ 58. $\dfrac{24}{18}$ 59. $\dfrac{24}{40}$ 60. $\dfrac{44}{60}$ 61. $\dfrac{8}{88}$

62. $\dfrac{9}{90}$ 63. $\dfrac{144}{36}$ 64. $\dfrac{140}{297}$ 65. $\dfrac{48}{144}$ 66. $\dfrac{32}{120}$

67. $\dfrac{60}{100}$ 68. $\dfrac{33}{110}$ 69. $\dfrac{36}{16}$ 70. $\dfrac{80}{45}$ 71. $\dfrac{32}{160}$

 72. Suppose the denominator of a fraction is a multiple of the numerator. When the fraction is written in simplest form, what number is its numerator?

Applying the Concepts

73. Make a list of five different fractions that are equivalent to $\dfrac{2}{3}$.

74. Show that $\dfrac{15}{24} = \dfrac{5}{8}$ by using a diagram.

75. **a. Geography** What fraction of the states in the United States of America have names that begin with the letter M?
    b. What fraction of the states have names that begin and end with a vowel?

SECTION

2.4 Addition of Fractions and Mixed Numbers

OBJECTIVE A To add fractions with the same denominator

Fractions with the same denominator are added by adding the numerators and placing the sum over the common denominator. After adding, write the sum in simplest form.

HOW TO • 1 Add: $\frac{2}{7} + \frac{4}{7}$

$$\frac{2}{7}$$
$$+\frac{4}{7}$$
$$\frac{6}{7}$$

• Add the numerators and place the sum over the common denominator.

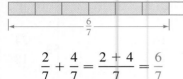

$$\frac{2}{7} + \frac{4}{7} = \frac{2+4}{7} = \frac{6}{7}$$

EXAMPLE • 1

Add: $\frac{5}{12} + \frac{11}{12}$

Solution

$$\frac{5}{12}$$
$$+\frac{11}{12}$$
$$\frac{16}{12} = \frac{4}{3} = 1\frac{1}{3}$$

• The denominators are the same. Add the numerators. Place the sum over the common denominator.

YOU TRY IT • 1

Add: $\frac{3}{8} + \frac{7}{8}$

Your solution

Solution on p. S5

OBJECTIVE B To add fractions with different denominators

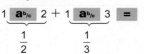

Integrating Technology

Some scientific calculators have a fraction key, $\boxed{a^{b/c}}$. It is used to perform operations on fractions. To use this key to simplify the expression at the right, enter

$\underbrace{1 \boxed{a^{b/c}} 2}_{\frac{1}{2}} + \underbrace{1 \boxed{a^{b/c}} 3}_{\frac{1}{3}} \boxed{=}$

To add fractions with different denominators, first rewrite the fractions as equivalent fractions with a common denominator. The common denominator is the LCM of the denominators of the fractions.

HOW TO • 2 Find the total of $\frac{1}{2}$ and $\frac{1}{3}$.

The common denominator is the LCM of 2 and 3. The LCM = 6. The LCM of denominators is sometimes called the **least common denominator (LCD).**

Write equivalent fractions using the LCM.

$$\frac{1}{2} = \frac{3}{6}$$
$$+\frac{1}{3} = \frac{2}{6}$$

Add the fractions.

$$\frac{1}{2} = \frac{3}{6}$$
$$+\frac{1}{3} = \frac{2}{6}$$
$$\frac{5}{6}$$

EXAMPLE · 2

Find $\frac{7}{12}$ more than $\frac{3}{8}$.

Solution

$$\frac{3}{8} = \frac{9}{24}$$

$$+ \frac{7}{12} = \frac{14}{24}$$

$$\frac{23}{24}$$

• The LCM of 8 and 12 is 24.

YOU TRY IT · 2

Find the sum of $\frac{5}{12}$ and $\frac{9}{16}$.

Your solution

EXAMPLE · 3

Add: $\frac{5}{8} + \frac{7}{9}$

Solution

$$\frac{5}{8} = \frac{45}{72}$$

$$+ \frac{7}{9} = \frac{56}{72}$$

$$\frac{101}{72} = 1\frac{29}{72}$$

YOU TRY IT · 3

Add: $\frac{7}{8} + \frac{11}{15}$

Your solution

EXAMPLE · 4

Add: $\frac{2}{3} + \frac{3}{5} + \frac{5}{6}$

Solution

$$\frac{2}{3} = \frac{20}{30}$$

$$\frac{3}{5} = \frac{18}{30}$$

$$+ \frac{5}{6} = \frac{25}{30}$$

$$\frac{63}{30} = 2\frac{3}{30} = 2\frac{1}{10}$$

• The LCM of 3, 5, and 6 is 30.

YOU TRY IT · 4

Add: $\frac{3}{4} + \frac{4}{5} + \frac{5}{8}$

Your solution

Solutions on p. S5

OBJECTIVE C **To add whole numbers, mixed numbers, and fractions**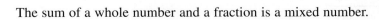

✓ **Take Note**

The procedure at the right illustrates why $2 + \frac{2}{3} = 2\frac{2}{3}$. You do not need to show these steps when adding a whole number and a fraction. Here are two more examples:

$$7 + \frac{1}{5} = 7\frac{1}{5}$$

$$6 + \frac{3}{4} = 6\frac{3}{4}$$

The sum of a whole number and a fraction is a mixed number.

HOW TO · 3 Add: $2 + \frac{2}{3}$

$$\boxed{2} + \frac{2}{3} = \boxed{\frac{6}{3}} + \frac{2}{3} = \frac{8}{3} = 2\frac{2}{3}$$

To add a whole number and a mixed number, write the fraction and then add the whole numbers.

HOW TO · 4 Add: $7\frac{2}{5} + 4$

Write the fraction.

$$7\frac{2}{5}$$
$$+ 4$$
$$\frac{2}{5}$$

Add the whole numbers.

$$7\frac{2}{5}$$
$$+ 4$$
$$11\frac{2}{5}$$

Integrating Technology

Use the fraction key on a calculator to enter mixed numbers. For the example at the right, enter

5 $\boxed{a^{b/c}}$ 4 $\boxed{a^{b/c}}$ 9 $\boxed{+}$

$5\frac{4}{9}$

6 $\boxed{a^{b/c}}$ 14 $\boxed{a^{b/c}}$ 15 $\boxed{=}$

$6\frac{14}{15}$

To add two mixed numbers, add the fractional parts and then add the whole numbers. Remember to reduce the sum to simplest form.

HOW TO • 5 What is $6\frac{14}{15}$ added to $5\frac{4}{9}$?

The LCM of 9 and 15 is 45.

Add the fractional parts.

$$5\frac{4}{9} = 5\frac{20}{45}$$
$$+ 6\frac{14}{15} = 6\frac{42}{45}$$
$$\overline{\phantom{5\frac{4}{9} = }}\ \frac{62}{45}$$

Add the whole numbers.

$$5\frac{4}{9} = 5\frac{20}{45}$$
$$+ 6\frac{14}{15} = 6\frac{42}{45}$$
$$\overline{\phantom{5\frac{4}{9} = }}\ 11\frac{62}{45} = 11 + 1\frac{17}{45} = 12\frac{17}{45}$$

EXAMPLE • 5

Add: $5 + \frac{3}{8}$

Solution $5 + \frac{3}{8} = 5\frac{3}{8}$

YOU TRY IT • 5

What is 7 added to $\frac{6}{11}$?

Your solution

EXAMPLE • 6

Find 17 increased by $3\frac{3}{8}$.

Solution $17 + 3\frac{3}{8} = 20\frac{3}{8}$

YOU TRY IT • 6

Find the sum of 29 and $17\frac{5}{12}$.

Your solution

EXAMPLE • 7

Add: $5\frac{2}{3} + 11\frac{5}{6} + 12\frac{7}{9}$

Solution

$$5\frac{2}{3} = 5\frac{12}{18} \qquad \bullet \text{ LCM} = 18$$
$$11\frac{5}{6} = 11\frac{15}{18}$$
$$+ 12\frac{7}{9} = 12\frac{14}{18}$$
$$\overline{\phantom{+ 12\frac{7}{9} = }}\ 28\frac{41}{18} = 30\frac{5}{18}$$

YOU TRY IT • 7

Add: $7\frac{4}{5} + 6\frac{7}{10} + 13\frac{11}{15}$

Your solution

EXAMPLE • 8

Add: $11\frac{5}{8} + 7\frac{5}{9} + 8\frac{7}{15}$

Solution

$$11\frac{5}{8} = 11\frac{225}{360} \qquad \bullet \text{ LCM} = 360$$
$$7\frac{5}{9} = 7\frac{200}{360}$$
$$+ 8\frac{7}{15} = 8\frac{168}{360}$$
$$\overline{\phantom{+ 8\frac{7}{15} = }}\ 26\frac{593}{360} = 27\frac{233}{360}$$

YOU TRY IT • 8

Add: $9\frac{3}{8} + 17\frac{7}{12} + 10\frac{14}{15}$

Your solution

Solutions on p. S5

OBJECTIVE D   **To solve application problems**

EXAMPLE · 9

A rain gauge collected $2\frac{1}{3}$ inches of rain in October, $5\frac{1}{2}$ inches in November, and $3\frac{3}{8}$ inches in December. Find the total rainfall for the 3 months.

Strategy

To find the total rainfall for the 3 months, add the three amounts of rainfall $\left(2\frac{1}{3}, 5\frac{1}{2}, \text{ and } 3\frac{3}{8}\right)$.

Solution

$$2\frac{1}{3} = 2\frac{8}{24}$$

$$5\frac{1}{2} = 5\frac{12}{24}$$

$$+\ 3\frac{3}{8} = 3\frac{9}{24}$$

$$10\frac{29}{24} = 11\frac{5}{24}$$

The total rainfall for the 3 months was $11\frac{5}{24}$ inches.

YOU TRY IT · 9

On Monday, you spent $4\frac{1}{2}$ hours in class, $3\frac{3}{4}$ hours studying, and $1\frac{1}{3}$ hours driving. Find the total number of hours spent on these three activities.

Your strategy

Your solution

EXAMPLE · 10

Barbara Walsh worked 4 hours, $2\frac{1}{3}$ hours, and $5\frac{2}{3}$ hours this week at a part-time job. Barbara is paid $9 an hour. How much did she earn this week?

Strategy

To find how much Barbara earned:
• Find the total number of hours worked.
• Multiply the total number of hours worked by the hourly wage (9).

Solution

$$
\begin{array}{ll}
4 & 12 \\
2\frac{1}{3} & \underline{\times\ \ 9} \\
+\ 5\frac{2}{3} & 108 \\
\hline
11\frac{3}{3} = 12 \text{ hours worked}
\end{array}
$$

Barbara earned $108 this week.

YOU TRY IT · 10

Jeff Sapone, a carpenter, worked $1\frac{2}{3}$ hours of overtime on Monday, $3\frac{1}{3}$ hours of overtime on Tuesday, and 2 hours of overtime on Wednesday. At an overtime hourly rate of $36, find Jeff's overtime pay for these 3 days.

Your strategy

Your solution

Solutions on p. S5

2.4 EXERCISES

OBJECTIVE A   To add fractions with the same denominator

For Exercises 1 to 16, add.

1. $\dfrac{2}{7} + \dfrac{1}{7}$

2. $\dfrac{3}{11} + \dfrac{5}{11}$

3. $\dfrac{1}{2} + \dfrac{1}{2}$

4. $\dfrac{1}{3} + \dfrac{2}{3}$

5. $\dfrac{8}{11} + \dfrac{7}{11}$

6. $\dfrac{9}{13} + \dfrac{7}{13}$

7. $\dfrac{8}{5} + \dfrac{9}{5}$

8. $\dfrac{5}{3} + \dfrac{7}{3}$

9. $\dfrac{3}{5} + \dfrac{8}{5} + \dfrac{3}{5}$

10. $\dfrac{3}{8} + \dfrac{5}{8} + \dfrac{7}{8}$

11. $\dfrac{3}{4} + \dfrac{1}{4} + \dfrac{5}{4}$

12. $\dfrac{2}{7} + \dfrac{4}{7} + \dfrac{5}{7}$

13. $\dfrac{3}{8} + \dfrac{7}{8} + \dfrac{1}{8}$

14. $\dfrac{5}{12} + \dfrac{7}{12} + \dfrac{1}{12}$

15. $\dfrac{4}{15} + \dfrac{7}{15} + \dfrac{11}{15}$

16. $\dfrac{5}{7} + \dfrac{4}{7} + \dfrac{5}{7}$

17. Find the sum of $\dfrac{5}{12}$, $\dfrac{1}{12}$, and $\dfrac{11}{12}$.

18. Find the total of $\dfrac{5}{8}$, $\dfrac{3}{8}$, and $\dfrac{7}{8}$.

 For Exercises 19 to 22, each statement concerns a pair of fractions that have the same denominator. State whether the sum of the fractions is a proper fraction, the number 1, a mixed number, or a whole number other than 1.

19. The sum of the numerators is a multiple of the denominator.

20. The sum of the numerators is one more than the denominator.

21. The sum of the numerators is the denominator.

22. The sum of the numerators is smaller than the denominator.

OBJECTIVE B   To add fractions with different denominators

For Exercises 23 to 42, add.

23. $\dfrac{1}{2} + \dfrac{2}{3}$

24. $\dfrac{2}{3} + \dfrac{1}{4}$

25. $\dfrac{3}{14} + \dfrac{5}{7}$

26. $\dfrac{3}{5} + \dfrac{7}{10}$

27. $\dfrac{8}{15} + \dfrac{7}{20}$

28. $\dfrac{1}{6} + \dfrac{7}{9}$

29. $\dfrac{3}{8} + \dfrac{9}{14}$

30. $\dfrac{5}{12} + \dfrac{5}{16}$

31. $\dfrac{3}{20} + \dfrac{7}{30}$ 　　　**32.** $\dfrac{5}{12} + \dfrac{7}{30}$ 　　　**33.** $\dfrac{1}{3} + \dfrac{5}{6} + \dfrac{7}{9}$ 　　　**34.** $\dfrac{2}{3} + \dfrac{5}{6} + \dfrac{7}{12}$

35. $\dfrac{5}{6} + \dfrac{1}{12} + \dfrac{5}{16}$ 　　　**36.** $\dfrac{2}{9} + \dfrac{7}{15} + \dfrac{4}{21}$ 　　　**37.** $\dfrac{2}{3} + \dfrac{1}{5} + \dfrac{7}{12}$ 　　　**38.** $\dfrac{3}{4} + \dfrac{4}{5} + \dfrac{7}{12}$

39. $\dfrac{2}{3} + \dfrac{3}{5} + \dfrac{7}{8}$ 　　　**40.** $\dfrac{3}{10} + \dfrac{14}{15} + \dfrac{9}{25}$ 　　　**41.** $\dfrac{2}{3} + \dfrac{5}{8} + \dfrac{7}{9}$ 　　　**42.** $\dfrac{1}{3} + \dfrac{2}{9} + \dfrac{7}{8}$

43. What is $\dfrac{3}{8}$ added to $\dfrac{3}{5}$?

44. What is $\dfrac{5}{9}$ added to $\dfrac{7}{12}$?

45. Find the sum of $\dfrac{3}{8}$, $\dfrac{5}{6}$, and $\dfrac{7}{12}$.

46. Find the total of $\dfrac{1}{2}$, $\dfrac{5}{8}$, and $\dfrac{7}{9}$.

 47. Which statement describes a pair of fractions for which the least common denominator is the product of the denominators?
  (i) The denominator of one fraction is a multiple of the denominator of the second fraction.
  (ii) The denominators of the two fractions have no common factors.

OBJECTIVE C　　To add whole numbers, mixed numbers, and fractions

For Exercises 48 to 69, add.

48. $2\dfrac{2}{5}$
$+\ 3\dfrac{3}{10}$

49. $4\dfrac{1}{2}$
$+\ 5\dfrac{7}{12}$

50. $3\dfrac{3}{8}$
$+\ 2\dfrac{5}{16}$

51. 4
$+\ 5\dfrac{2}{7}$

52. $6\dfrac{8}{9}$
$+\ 12$

53. $7\dfrac{5}{12} + 2\dfrac{9}{16}$ 　　　**54.** $9\dfrac{1}{2} + 3\dfrac{3}{11}$ 　　　**55.** $6 + 2\dfrac{3}{13}$ 　　　**56.** $8\dfrac{21}{40} + 6$

57. $8\dfrac{29}{30} + 7\dfrac{11}{40}$ 　　　**58.** $17\dfrac{5}{16} + 3\dfrac{11}{24}$ 　　　**59.** $17\dfrac{3}{8} + 7\dfrac{7}{20}$ 　　　**60.** $14\dfrac{7}{12} + 29\dfrac{13}{21}$

61. $5\dfrac{7}{8} + 27\dfrac{5}{12}$

62. $7\dfrac{5}{6} + 3\dfrac{5}{9}$

63. $7\dfrac{5}{9} + 2\dfrac{7}{12}$

64. $3\dfrac{1}{2} + 2\dfrac{3}{4} + 1\dfrac{5}{6}$

65. $2\dfrac{1}{2} + 3\dfrac{2}{3} + 4\dfrac{1}{4}$

66. $3\dfrac{1}{3} + 7\dfrac{1}{5} + 2\dfrac{1}{7}$

67. $3\dfrac{1}{2} + 3\dfrac{1}{5} + 8\dfrac{1}{9}$

68. $6\dfrac{5}{9} + 6\dfrac{5}{12} + 2\dfrac{5}{18}$

69. $2\dfrac{3}{8} + 4\dfrac{7}{12} + 3\dfrac{5}{16}$

70. Find the sum of $2\dfrac{4}{9}$ and $5\dfrac{7}{12}$.

71. Find $5\dfrac{5}{6}$ more than $3\dfrac{3}{8}$.

72. What is $4\dfrac{3}{4}$ added to $9\dfrac{1}{3}$?

73. What is $4\dfrac{8}{9}$ added to $9\dfrac{1}{6}$?

74. Find the total of 2, $4\dfrac{5}{8}$, and $2\dfrac{2}{9}$.

75. Find the total of $1\dfrac{5}{8}$, 3, and $7\dfrac{7}{24}$.

 For Exercises 76 and 77, state whether the given sum can be a whole number. Answer *yes* or *no*.

76. The sum of two mixed numbers

77. The sum of a mixed number and a whole number

OBJECTIVE D　　**To solve application problems**

78. **Mechanics** Find the length of the shaft.

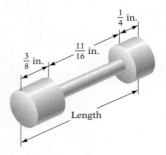

79. **Mechanics** Find the length of the shaft.

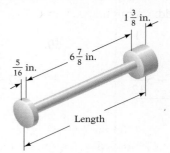

80. **Carpentry** A table 30 inches high has a top that is $1\dfrac{1}{8}$ inches thick. Find the total thickness of the table top after a $\dfrac{3}{16}$-inch veneer is applied.

 81. For the table pictured at the right, what does the sum $30 + 1\dfrac{1}{8} + \dfrac{3}{16}$ represent?

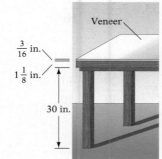

82. Wages You are working a part-time job that pays $11 an hour. You worked 5, $3\frac{3}{4}$, $2\frac{1}{3}$, $1\frac{1}{4}$, and $7\frac{2}{3}$ hours during the last five days.
  a. Find the total number of hours you worked during the last five days.
  b. Find your total wages for the five days.

83. Sports The course of a yachting race is in the shape of a triangle with sides that measure $4\frac{3}{10}$ miles, $3\frac{7}{10}$ miles, and $2\frac{1}{2}$ miles. Find the total length of the course.

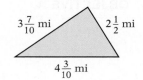
$3\frac{7}{10}$ mi · $2\frac{1}{2}$ mi · $4\frac{3}{10}$ mi

Construction The size of an interior door frame is determined by the width of the wall into which it is installed. The width of the wall is determined by the width of the stud in the wall and the thickness of the sheets of dry wall installed on each side of the wall. A 2 × 4 stud is $3\frac{5}{8}$ inches thick. A 2 × 6 stud is $5\frac{5}{8}$ inches thick. Use this information for Exercises 84 to 86.

84. Find the thickness of a wall constructed with 2 × 4 studs and dry wall that is $\frac{1}{2}$ inch thick.

85. Find the thickness of a wall constructed with 2 × 6 studs and dry wall that is $\frac{1}{2}$ inch thick.

86. A fire wall is a physical barrier in a building designed to limit the spread of fire. Suppose a fire wall is built between the garage and the kitchen of a house. Find the width of the fire wall if it is constructed using 2 × 4 studs and dry wall that is $\frac{5}{8}$ inch thick.

87. Construction Two pieces of wood must be bolted together. One piece of wood is $\frac{1}{2}$ inch thick. The second piece is $\frac{5}{8}$ inch thick. A washer will be placed on each of the outer sides of the two pieces of wood. Each washer is $\frac{1}{16}$ inch thick. The nut is $\frac{3}{16}$ inch thick. Find the minimum length of bolt needed to bolt the two pieces of wood together.

Applying the Concepts

88. What is a unit fraction? Find the sum of the three largest unit fractions. Is there a smallest unit fraction? If so, write it down. If not, explain why.

89. A survey was conducted to determine people's favorite color from among blue, green, red, purple, and other. The surveyor claims that $\frac{1}{3}$ of the people responded blue, $\frac{1}{6}$ responded green, $\frac{1}{8}$ responded red, $\frac{1}{12}$ responded purple, and $\frac{2}{5}$ responded some other color. Is this possible? Explain your answer.

SECTION

2.5   Subtraction of Fractions and Mixed Numbers

OBJECTIVE A   **To subtract fractions with the same denominator**

Fractions with the same denominator are subtracted by subtracting the numerators and placing the difference over the common denominator. After subtracting, write the fraction in simplest form.

HOW TO 1   Subtract: $\frac{5}{7} - \frac{3}{7}$

$$\begin{array}{r} \frac{5}{7} \\ -\frac{3}{7} \\ \hline \frac{2}{7} \end{array}$$

• Subtract the numerators and place the difference over the common denominator.

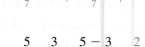

$$\frac{5}{7} - \frac{3}{7} = \frac{5-3}{7} = \frac{2}{7}$$

EXAMPLE • 1

Find $\frac{17}{30}$ less $\frac{11}{30}$.

Solution

$$\begin{array}{r} \frac{17}{30} \\ -\frac{11}{30} \\ \hline \frac{6}{30} = \frac{1}{5} \end{array}$$

• The denominators are the same. Subtract the numerators. Place the difference over the common denominator.

YOU TRY IT • 1

Subtract: $\frac{16}{27} - \frac{7}{27}$

Your solution

Solution on p. S5

OBJECTIVE B   **To subtract fractions with different denominators**

To subtract fractions with different denominators, first rewrite the fractions as equivalent fractions with a common denominator. As with adding fractions, the common denominator is the LCM of the denominators of the fractions.

HOW TO 2   Subtract: $\frac{5}{6} - \frac{1}{4}$

The common denominator is the LCM of 6 and 4. The LCM = 12.

Write equivalent fractions using the LCM.

$$\begin{array}{r} \frac{5}{6} = \frac{10}{12} \\ -\frac{1}{4} = \frac{3}{12} \end{array}$$

Subtract the fractions.

$$\begin{array}{r} \frac{5}{6} = \frac{10}{12} \\ -\frac{1}{4} = \frac{3}{12} \\ \hline \frac{7}{12} \end{array}$$

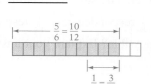

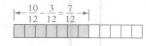

EXAMPLE · 2

Subtract: $\frac{11}{16} - \frac{5}{12}$

Solution

$$\frac{11}{16} = \frac{33}{48}$$
$$- \frac{5}{12} = \frac{20}{48}$$
$$\frac{13}{48}$$

• LCM = 48

YOU TRY IT · 2

Subtract: $\frac{13}{18} - \frac{7}{24}$

Your solution

Solution on p. S5

OBJECTIVE C **To subtract whole numbers, mixed numbers, and fractions**

To subtract mixed numbers without borrowing, subtract the fractional parts and then subtract the whole numbers.

HOW TO · 3 Subtract: $5\frac{5}{6} - 2\frac{3}{4}$

Subtract the fractional parts.

• The LCM of 6 and 4 is 12.

$$5\frac{5}{6} = 5\frac{10}{12}$$
$$- 2\frac{3}{4} = 2\frac{9}{12}$$
$$\overline{}$$

Subtract the whole numbers.

$$5\frac{5}{6} = 5\frac{10}{12}$$
$$- 2\frac{3}{4} = 2\frac{9}{12}$$
$$\frac{1}{12}$$

Subtraction of mixed numbers sometimes involves borrowing.

HOW TO · 4 Subtract: $5 - 2\frac{5}{8}$

Borrow 1 from 5.

$$5 = \overset{4}{\cancel{5}}\,1$$
$$- 2\frac{5}{8} = 2\frac{5}{8}$$

Write 1 as a fraction so that the fractions have the same denominators.

$$5 = 4\frac{8}{8}$$
$$- 2\frac{5}{8} = 2\frac{5}{8}$$

Subtract the mixed numbers.

$$5 = 4\frac{8}{8}$$
$$- 2\frac{5}{8} = 2\frac{5}{8}$$
$$2\frac{3}{8}$$

HOW TO · 5 Subtract: $7\frac{1}{6} - 2\frac{5}{8}$

Write equivalent fractions using the LCM.

$$7\frac{1}{6} = 7\frac{4}{24}$$
$$- 2\frac{5}{8} = 2\frac{15}{24}$$

Borrow 1 from 7. Add the 1 to $\frac{4}{24}$. Write $1\frac{4}{24}$ as $\frac{28}{24}$.

$$7\frac{1}{6} = \overset{6}{\cancel{7}}1\frac{4}{24} = 6\frac{28}{24}$$
$$- 2\frac{5}{8} = 2\frac{15}{24} = 2\frac{15}{24}$$

Subtract the mixed numbers.

$$7\frac{1}{6} = 6\frac{28}{24}$$
$$- 2\frac{5}{8} = 2\frac{15}{24}$$
$$4\frac{13}{24}$$

EXAMPLE • 3

Subtract: $15\frac{7}{8} - 12\frac{2}{3}$

Solution

$$15\frac{7}{8} = 15\frac{21}{24} \qquad \bullet \text{ LCM} = 24$$

$$\underline{-\ 12\frac{2}{3} = 12\frac{16}{24}}$$

$$3\frac{5}{24}$$

YOU TRY IT • 3

Subtract: $17\frac{5}{9} - 11\frac{5}{12}$

Your solution

EXAMPLE • 4

Subtract: $9 - 4\frac{3}{11}$

Solution

$$9 \quad = 8\frac{11}{11} \qquad \bullet \text{ LCM} = 11$$

$$\underline{-\ 4\frac{3}{11} = 4\frac{3}{11}}$$

$$4\frac{8}{11}$$

YOU TRY IT • 4

Subtract: $8 - 2\frac{4}{13} =$

Your solution

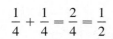

$7\frac{13}{13} =$

$-\ 2\frac{4}{13}$

$5\frac{9}{13}$

EXAMPLE • 5

Find $11\frac{5}{12}$ decreased by $2\frac{11}{16}$.

Solution

$$11\frac{5}{12} = 11\frac{20}{48} = 10\frac{68}{48} \qquad \bullet \text{ LCM} = 48$$

$$\underline{-\ 2\frac{11}{16} = 2\frac{33}{48} = 2\frac{33}{48}}$$

$$8\frac{35}{48}$$

YOU TRY IT • 5

What is $21\frac{7}{9}$ minus $7\frac{11}{12}$?

Your solution

Solutions on p. S6

OBJECTIVE D   To solve application problems

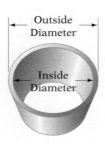

Outside Diameter

Inside Diameter

HOW TO • 6   The outside diameter of a bushing is $3\frac{3}{8}$ inches and the wall thickness is $\frac{1}{4}$ inch. Find the inside diameter of the bushing.

$$\frac{1}{4} + \frac{1}{4} = \frac{2}{4} = \frac{1}{2} \qquad \bullet \text{ Add } \frac{1}{4} \text{ and } \frac{1}{4} \text{ to find the total thickness of the two walls.}$$

$$3\frac{3}{8} = 3\frac{3}{8} = \ 2\frac{11}{8} \qquad \bullet \text{ Subtract the total thickness of the two walls from the outside diameter to find the inside diameter.}$$

$$\underline{-\ \frac{1}{2} = \frac{4}{8} = \quad \frac{4}{8}}$$

$$2\frac{7}{8}$$

The inside diameter of the bushing is $2\frac{7}{8}$ inches.

EXAMPLE • 6

A $2\frac{2}{3}$-inch piece is cut from a $6\frac{5}{8}$-inch board. How much of the board is left?

Strategy

To find the length remaining, subtract the length of the piece cut from the total length of the board.

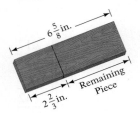

$6\frac{5}{8}$ in.

Remaining Piece

$2\frac{2}{3}$ in.

Solution

$$6\frac{5}{8} = 6\frac{15}{24} = 5\frac{39}{24}$$
$$-2\frac{2}{3} = 2\frac{16}{24} = 2\frac{16}{24}$$
$$\overline{\qquad\qquad\qquad 3\frac{23}{24}}$$

$3\frac{23}{24}$ inches of the board are left.

YOU TRY IT • 6

A flight from New York to Los Angeles takes $5\frac{1}{2}$ hours. After the plane has been in the air for $2\frac{3}{4}$ hours, how much flight time remains?

Your strategy

Your solution

EXAMPLE • 7

Two painters are staining a house. In 1 day one painter stained $\frac{1}{3}$ of the house, and the other stained $\frac{1}{4}$ of the house. How much of the job remains to be done?

Strategy

To find how much of the job remains:

• Find the total amount of the house already stained $\left(\frac{1}{3} + \frac{1}{4}\right)$.

• Subtract the amount already stained from 1, which represents the complete job.

Solution

$$\frac{1}{3} = \frac{4}{12} \qquad\qquad 1 = \frac{12}{12}$$
$$+\frac{1}{4} = \frac{3}{12} \qquad\qquad -\frac{7}{12} = \frac{7}{12}$$
$$\overline{\quad\frac{7}{12}\quad} \qquad\qquad \overline{\quad\frac{5}{12}\quad}$$

$\frac{5}{12}$ of the house remains to be stained.

YOU TRY IT • 7

A patient is put on a diet to lose 24 pounds in 3 months. The patient lost $7\frac{1}{2}$ pounds the first month and $5\frac{3}{4}$ pounds the second month. How much weight must be lost the third month to achieve the goal?

Your strategy

Your solution

Solutions on p. S6

2.5 EXERCISES

OBJECTIVE A   To subtract fractions with the same denominator

For Exercises 1 to 10, subtract.

1. $\dfrac{9}{17}$
    $-\dfrac{7}{17}$

2. $\dfrac{11}{15}$
    $-\dfrac{3}{15}$

3. $\dfrac{11}{12}$
    $-\dfrac{7}{12}$

4. $\dfrac{13}{15}$
    $-\dfrac{4}{15}$

5. $\dfrac{9}{20}$
    $-\dfrac{7}{20}$

6. $\dfrac{48}{55}$
    $-\dfrac{13}{55}$

7. $\dfrac{42}{65}$
    $-\dfrac{17}{65}$

8. $\dfrac{11}{24}$
    $-\dfrac{5}{24}$

9. $\dfrac{23}{30}$
    $-\dfrac{13}{30}$

10. $\dfrac{17}{42}$
     $-\dfrac{5}{42}$

11. What is $\dfrac{5}{14}$ less than $\dfrac{13}{14}$?

12. Find the difference between $\dfrac{7}{8}$ and $\dfrac{5}{8}$.

13. Find $\dfrac{17}{24}$ decreased by $\dfrac{11}{24}$.

14. What is $\dfrac{19}{30}$ minus $\dfrac{11}{30}$?

For Exercises 15 and 16, each statement describes the difference between a pair of fractions that have the same denominator. State whether the difference of the fractions will need to be rewritten in order to be in simplest form. Answer *yes* or *no*.

15. The difference between the numerators is a factor of the denominator.

16. The difference between the numerators is 1.

OBJECTIVE B   To subtract fractions with different denominators

For Exercises 17 to 26, subtract.

17. $\dfrac{2}{3}$
     $-\dfrac{1}{6}$

18. $\dfrac{7}{8}$
     $-\dfrac{5}{16}$

19. $\dfrac{5}{8}$
     $-\dfrac{2}{7}$

20. $\dfrac{5}{6}$
     $-\dfrac{3}{7}$

21. $\dfrac{5}{7}$
     $-\dfrac{3}{14}$

22. $\dfrac{5}{9}$
     $-\dfrac{7}{15}$

23. $\dfrac{8}{15}$
     $-\dfrac{7}{20}$

24. $\dfrac{7}{9}$
     $-\dfrac{1}{6}$

25. $\dfrac{9}{16}$
     $-\dfrac{17}{32}$

26. $\dfrac{29}{60}$
     $-\dfrac{3}{40}$

27. What is $\frac{3}{5}$ less than $\frac{11}{12}$?

28. What is $\frac{5}{9}$ less than $\frac{11}{15}$?

29. Find the difference between $\frac{11}{24}$ and $\frac{7}{18}$.

30. Find the difference between $\frac{9}{14}$ and $\frac{5}{42}$.

31. Find $\frac{11}{12}$ decreased by $\frac{11}{15}$.

32. Find $\frac{17}{20}$ decreased by $\frac{7}{15}$.

33. What is $\frac{13}{20}$ minus $\frac{1}{6}$?

34. What is $\frac{5}{6}$ minus $\frac{7}{9}$?

35. Which statement describes a pair of fractions for which the least common denominator is one of the denominators?

  (i) The denominator of one fraction is a factor of the denominator of the second fraction.

  (ii) The denominators of the two fractions have no common factors.

OBJECTIVE C   To subtract whole numbers, mixed numbers, and fractions

For Exercises 36 to 50, subtract.

36. $\quad 5\frac{7}{12}$
$\quad -2\frac{5}{12}$

37. $\quad 16\frac{11}{15}$
$\quad -11\frac{8}{15}$

38. $\quad 6\frac{1}{3}$
$\quad -1\frac{2}{3}$

39. $\quad 5\frac{7}{8}$
$\quad -1$

40. $\quad 10$
$\quad -6\frac{1}{3}$

41. $\quad 3$
$\quad -2\frac{5}{21}$

42. $\quad 6\frac{2}{5}$
$\quad -4\frac{4}{5}$

43. $\quad 16\frac{3}{8}$
$\quad -10\frac{7}{8}$

44. $\quad 25\frac{4}{9}$
$\quad -16\frac{7}{9}$

45. $\quad 8\frac{3}{7}$
$\quad -2\frac{6}{7}$

46. $\quad 16\frac{2}{5}$
$\quad -8\frac{4}{9}$

47. $\quad 23\frac{7}{8}$
$\quad -16\frac{2}{3}$

48. $\quad 82\frac{4}{33}$
$\quad -16\frac{5}{22}$

49. $\quad 6$
$\quad -4\frac{3}{5}$

50. $\quad 17$
$\quad -7\frac{8}{13}$

51. What is $7\frac{3}{5}$ less than $23\frac{3}{20}$?

52. Find the difference between $12\frac{3}{8}$ and $7\frac{5}{12}$.

53. What is $10\frac{5}{9}$ minus $5\frac{11}{15}$?

54. Find $6\frac{1}{3}$ decreased by $3\frac{3}{5}$.

55. Can the difference between a whole number and a mixed number ever be a whole number?

OBJECTIVE D   **To solve application problems**

56. **Mechanics** Find the missing dimension.

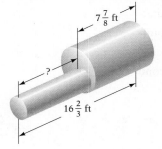

57. **Mechanics** Find the missing dimension.

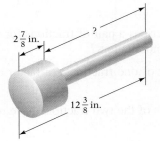

58. **Sports** In the Kentucky Derby the horses run $1\frac{1}{4}$ miles. In the Belmont Stakes they run $1\frac{1}{2}$ miles, and in the Preakness Stakes they run $1\frac{3}{16}$ miles. How much farther do the horses run in the Kentucky Derby than in the Preakness Stakes? How much farther do they run in the Belmont Stakes than in the Preakness Stakes?

© Reuters/Corbis

59. **Sports** In the running high jump in the 1948 Summer Olympic Games, Alice Coachman's distance was $66\frac{1}{8}$ inches. In the same event in the 1972 Summer Olympics, Urika Meyfarth jumped $75\frac{1}{2}$ inches, and in the 1996 Olympic Games, Stefka Kostadinova jumped $80\frac{3}{4}$ inches. Find the difference between Meyfarth's distance and Coachman's distance. Find the difference between Kostadinova's distance and Meyfarth's distance.

60. **Fundraising** A 12-mile walkathon has three checkpoints. The first checkpoint is $3\frac{3}{8}$ miles from the starting point. The second checkpoint is $4\frac{1}{3}$ miles from the first.
a. How many miles is it from the starting point to the second checkpoint?
b. How many miles is it from the second checkpoint to the finish line?

61. Hiking Two hikers plan a 3-day, $27\frac{1}{2}$-mile backpack trip carrying a total of 80 pounds. The hikers plan to travel $7\frac{3}{8}$ miles the first day and $10\frac{1}{3}$ miles the second day.

a. How many total miles do the hikers plan to travel the first two days?

b. How many miles will be left to travel on the third day?

For Exercises 62 and 63, refer to Exercise 61. Describe what each difference represents.

62. $27\frac{1}{2} - 7\frac{3}{8}$

63. $10\frac{1}{3} - 7\frac{3}{8}$

64. Health A patient with high blood pressure who weighs 225 pounds is put on a diet to lose 25 pounds in 3 months. The patient loses $8\frac{3}{4}$ pounds the first month and $11\frac{5}{8}$ pounds the second month. How much weight must be lost the third month for the goal to be achieved?

65. Sports A wrestler is entered in the 172-pound weight class in the conference finals coming up in 3 weeks. The wrestler needs to lose $12\frac{3}{4}$ pounds. The wrestler loses $5\frac{1}{4}$ pounds the first week and $4\frac{1}{4}$ pounds the second week.

a. Without doing the calculations, determine whether the wrestler can reach his weight class by losing less in the third week than was lost in the second week.

b. How many pounds must be lost in the third week for the desired weight to be reached?

66. Construction Find the difference in thickness between a fire wall constructed with 2×6 studs and dry wall that is $\frac{1}{2}$ inch thick and a fire wall constructed with 2×4 studs and dry wall that is $\frac{5}{8}$ inch thick. See Exercises 84 to 86 on page 83.

67. Finances If $\frac{4}{15}$ of an electrician's income is spent for housing, what fraction of the electrician's income is not spent for housing?

Applying the Concepts

68. Fill in the square to produce a true statement: $5\frac{1}{3} - \square = 2\frac{1}{2}$

69. Fill in the square to produce a true statement: $\square - 4\frac{1}{2} = 1\frac{5}{8}$

70. Fill in the blank squares at the right so that the sum of the numbers is the same along any row, column, or diagonal. The resulting square is called a magic square.

		$\frac{3}{4}$
1	$\frac{5}{8}$	
$\frac{1}{2}$		$\frac{7}{8}$

SECTION

2.6

Multiplication of Fractions and Mixed Numbers

OBJECTIVE A   **To multiply fractions**

The product of two fractions is the product of the numerators over the product of the denominators.

HOW TO · 1   Multiply: $\frac{2}{3} \times \frac{4}{5}$

$$\frac{2}{3} \times \frac{4}{5} = \frac{2 \cdot 4}{3 \cdot 5} = \frac{8}{15}$$

- **Multiply the numerators.**
- **Multiply the denominators.**

The product $\frac{2}{3} \times \frac{4}{5}$ can be read "$\frac{2}{3}$ times $\frac{4}{5}$," or "$\frac{2}{3}$ of $\frac{4}{5}$."

Reading the times sign as "of" is useful in application problems.

$\frac{4}{5}$ of the bar is shaded.

Shade $\frac{2}{3}$ of the $\frac{4}{5}$ already shaded.

$\frac{8}{15}$ of the bar is then shaded light yellow.

$\frac{2}{3}$ of $\frac{4}{5} = \frac{2}{3} \times \frac{4}{5} = \frac{8}{15}$

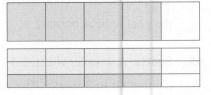

After multiplying two fractions, write the product in simplest form.

HOW TO · 2

Multiply: $\frac{3}{4} \times \frac{14}{15}$

$$\frac{3}{4} \times \frac{14}{15} = \frac{3 \cdot 14}{4 \cdot 15}$$

- **Multiply the numerators.**
- **Multiply the denominators.**

$$= \frac{3 \cdot 2 \cdot 7}{2 \cdot 2 \cdot 3 \cdot 5}$$

- **Write the prime factorization of each number.**

$$= \frac{\overset{1}{\cancel{3}} \cdot \overset{1}{\cancel{2}} \cdot 7}{\cancel{2} \cdot 2 \cdot \cancel{3} \cdot 5} = \frac{7}{10}$$

- **Eliminate the common factors. Then multiply the remaining factors in the numerator and denominator.**

This example could also be worked by using the GCF.

$$\frac{3}{4} \times \frac{14}{15} = \frac{42}{60}$$

- **Multiply the numerators.**
- **Multiply the denominators.**

$$= \frac{6 \cdot 7}{6 \cdot 10}$$

- **The GCF of 42 and 60 is 6. Factor 6 from 42 and 60.**

$$= \frac{\overset{1}{\cancel{6}} \cdot 7}{\cancel{6} \cdot 10} = \frac{7}{10}$$

- **Eliminate the GCF.**

EXAMPLE · 1

Multiply $\frac{4}{15}$ and $\frac{5}{28}$.

Solution

$$\frac{4}{15} \times \frac{5}{28} = \frac{4 \cdot 5}{15 \cdot 28} = \frac{\overset{1}{\cancel{2}} \cdot \overset{1}{\cancel{2}} \cdot \overset{1}{\cancel{5}}}{3 \cdot \underset{1}{\cancel{5}} \cdot \underset{1}{\cancel{2}} \cdot \underset{1}{\cancel{2}} \cdot 7} = \frac{1}{21}$$

YOU TRY IT · 1

Multiply $\frac{4}{21}$ and $\frac{7}{44}$.

Your solution

EXAMPLE · 2

Find the product of $\frac{9}{20}$ and $\frac{33}{35}$.

Solution

$$\frac{9}{20} \times \frac{33}{35} = \frac{9 \cdot 33}{20 \cdot 35} = \frac{3 \cdot 3 \cdot 3 \cdot 11}{2 \cdot 2 \cdot 5 \cdot 5 \cdot 7} = \frac{297}{700}$$

YOU TRY IT · 2

Find the product of $\frac{2}{21}$ and $\frac{10}{33}$.

Your solution

EXAMPLE · 3

What is $\frac{14}{9}$ times $\frac{12}{7}$?

Solution

$$\frac{14}{9} \times \frac{12}{7} = \frac{14 \cdot 12}{9 \cdot 7} = \frac{2 \cdot \overset{1}{\cancel{7}} \cdot 2 \cdot 2 \cdot \overset{1}{\cancel{3}}}{3 \cdot \underset{1}{\cancel{3}} \cdot \underset{1}{\cancel{7}}} = \frac{8}{3} = 2\frac{2}{3}$$

YOU TRY IT · 3

What is $\frac{16}{5}$ times $\frac{15}{24}$?

Your solution

Solutions on p. S6

OBJECTIVE B   **To multiply whole numbers, mixed numbers, and fractions**

To multiply a whole number by a fraction or a mixed number, first write the whole number as a fraction with a denominator of 1.

HOW TO · 3   Multiply: $4 \times \frac{3}{7}$

$$4 \times \frac{3}{7} = \frac{4}{1} \times \frac{3}{7} = \frac{4 \cdot 3}{1 \cdot 7} = \frac{2 \cdot 2 \cdot 3}{7} = \frac{12}{7} = 1\frac{5}{7}$$

• Write 4 with a denominator of 1; then multiply the fractions.

When one or more of the factors in a product is a mixed number, write the mixed number as an improper fraction before multiplying.

HOW TO · 4   Multiply: $2\frac{1}{3} \times \frac{3}{14}$

$$2\frac{1}{3} \times \frac{3}{14} = \frac{7}{3} \times \frac{3}{14} = \frac{7 \cdot 3}{3 \cdot 14} = \frac{\overset{1}{\cancel{7}} \cdot \overset{1}{\cancel{3}}}{\underset{1}{\cancel{3}} \cdot 2 \cdot \underset{1}{\cancel{7}}} = \frac{1}{2}$$

• Write $2\frac{1}{3}$ as an improper fraction; then multiply the fractions.

EXAMPLE · 4

Multiply: $4\frac{5}{6} \times \frac{12}{13}$

Solution

$$4\frac{5}{6} \times \frac{12}{13} = \frac{29}{6} \times \frac{12}{13} = \frac{29 \cdot 12}{6 \cdot 13}$$

$$= \frac{29 \cdot \overset{1}{\cancel{2}} \cdot 2 \cdot \overset{1}{\cancel{3}}}{\underset{1}{\cancel{2}} \cdot \underset{1}{\cancel{3}} \cdot 13} = \frac{58}{13} = 4\frac{6}{13}$$

YOU TRY IT · 4

Multiply: $5\frac{2}{5} \times \frac{5}{9}$

Your solution

EXAMPLE · 5

Find $5\frac{2}{3}$ times $4\frac{1}{2}$.

Solution

$$5\frac{2}{3} \times 4\frac{1}{2} = \frac{17}{3} \times \frac{9}{2} = \frac{17 \cdot 9}{3 \cdot 2}$$

$$= \frac{17 \cdot \overset{1}{\cancel{3}} \cdot 3}{\underset{1}{\cancel{3}} \cdot 2} = \frac{51}{2} = 25\frac{1}{2}$$

YOU TRY IT · 5

Multiply: $3\frac{2}{5} \times 6\frac{1}{4}$

Your solution

EXAMPLE · 6

Multiply: $4\frac{2}{5} \times 7$

Solution

$$4\frac{2}{5} \times 7 = \frac{22}{5} \times \frac{7}{1} = \frac{22 \cdot 7}{5 \cdot 1}$$

$$= \frac{2 \cdot 11 \cdot 7}{5} = \frac{154}{5} = 30\frac{4}{5}$$

YOU TRY IT · 6

Multiply: $3\frac{2}{7} \times 6$

Your solution

Solutions on p. S6

OBJECTIVE C　　**To solve application problems**

Length (ft)	Weight (lb/ft)
$6\frac{1}{2}$	$\frac{3}{8}$
$8\frac{5}{8}$	$1\frac{1}{4}$
$10\frac{3}{4}$	$2\frac{1}{2}$
$12\frac{7}{12}$	$4\frac{1}{3}$

The table at the left lists the lengths of steel rods and their corresponding weight per foot. The weight per foot is measured in pounds for each foot of rod and is abbreviated as lb/ft.

HOW TO · 5　　Find the weight of the steel bar that is $10\frac{3}{4}$ feet long.

Strategy
To find the weight of the steel bar, multiply its length by the weight per foot.

Solution

$$10\frac{3}{4} \times 2\frac{1}{2} = \frac{43}{4} \times \frac{5}{2} = \frac{43 \cdot 5}{4 \cdot 2} = \frac{215}{8} = 26\frac{7}{8}$$

The weight of the $10\frac{3}{4}$-foot rod is $26\frac{7}{8}$ pounds.

EXAMPLE · 7

An electrician earns $206 for each day worked. What are the electrician's earnings for working $4\frac{1}{2}$ days?

Strategy

To find the electrician's total earnings, multiply the daily earnings (206) by the number of days worked $\left(4\frac{1}{2}\right)$.

Solution

$$206 \times 4\frac{1}{2} = \frac{206}{1} \times \frac{9}{2}$$
$$= \frac{206 \cdot 9}{1 \cdot 2}$$
$$= 927$$

The electrician's earnings are $927.

YOU TRY IT · 7

Over the last 10 years, a house increased in value by $2\frac{1}{2}$ times. The price of the house 10 years ago was $170,000. What is the value of the house today?

Your strategy

Your solution

EXAMPLE · 8

The value of a small office building and the land on which it is built is $290,000. The value of the land is $\frac{1}{4}$ the total value. What is the dollar value of the building?

Strategy

To find the value of the building:

• Find the value of the land $\left(\frac{1}{4} \times 290,000\right)$.

• Subtract the value of the land from the total value (290,000).

Solution

$$\frac{1}{4} \times 290,000 = \frac{290,000}{4}$$
$$= 72,500 \quad \text{• Value of the land}$$
$$290,000 - 72,500 = 217,500$$

The value of the building is $217,500.

YOU TRY IT · 8

A paint company bought a drying chamber and an air compressor for spray painting. The total cost of the two items was $160,000. The drying chamber's cost was $\frac{4}{5}$ of the total cost. What was the cost of the air compressor?

Your strategy

Your solution

Solutions on pp. S6–S7

2.6 EXERCISES

OBJECTIVE A   **To multiply fractions**

For Exercises 1 to 32, multiply.

1. $\dfrac{2}{3} \times \dfrac{7}{8}$

2. $\dfrac{1}{2} \times \dfrac{2}{3}$

3. $\dfrac{5}{16} \times \dfrac{7}{15}$

4. $\dfrac{3}{8} \times \dfrac{6}{7}$

5. $\dfrac{1}{6} \times \dfrac{1}{8}$

6. $\dfrac{2}{5} \times \dfrac{5}{6}$

7. $\dfrac{11}{12} \times \dfrac{6}{7}$

8. $\dfrac{11}{12} \times \dfrac{3}{5}$

9. $\dfrac{8}{9} \times \dfrac{27}{4}$

10. $\dfrac{3}{5} \times \dfrac{3}{10}$

11. $\dfrac{5}{6} \times \dfrac{1}{2}$

12. $\dfrac{3}{8} \times \dfrac{5}{12}$

13. $\dfrac{16}{9} \times \dfrac{27}{8}$

14. $\dfrac{5}{8} \times \dfrac{16}{15}$

15. $\dfrac{3}{2} \times \dfrac{4}{9}$

16. $\dfrac{5}{3} \times \dfrac{3}{7}$

17. $\dfrac{7}{8} \times \dfrac{3}{14}$

18. $\dfrac{2}{9} \times \dfrac{1}{5}$

19. $\dfrac{1}{10} \times \dfrac{3}{8}$

20. $\dfrac{5}{12} \times \dfrac{6}{7}$

21. $\dfrac{15}{8} \times \dfrac{16}{3}$

22. $\dfrac{5}{6} \times \dfrac{4}{15}$

23. $\dfrac{1}{2} \times \dfrac{2}{15}$

24. $\dfrac{3}{8} \times \dfrac{5}{16}$

25. $\dfrac{5}{7} \times \dfrac{14}{15}$

26. $\dfrac{3}{8} \times \dfrac{15}{41}$

27. $\dfrac{5}{12} \times \dfrac{42}{65}$

28. $\dfrac{16}{33} \times \dfrac{55}{72}$

29. $\dfrac{12}{5} \times \dfrac{5}{3}$

30. $\dfrac{17}{9} \times \dfrac{81}{17}$

31. $\dfrac{16}{85} \times \dfrac{125}{84}$

32. $\dfrac{19}{64} \times \dfrac{48}{95}$

33. Give an example of a proper and an improper fraction whose product is 1.

34. Multiply $\frac{7}{12}$ and $\frac{15}{42}$.

35. Multiply $\frac{32}{9}$ and $\frac{3}{8}$.

36. Find the product of $\frac{5}{9}$ and $\frac{3}{20}$.

37. Find the product of $\frac{7}{3}$ and $\frac{15}{14}$.

38. What is $\frac{1}{2}$ times $\frac{8}{15}$?

39. What is $\frac{3}{8}$ times $\frac{12}{17}$?

OBJECTIVE B   To multiply whole numbers, mixed numbers, and fractions

For Exercises 40 to 71, multiply.

40. $4 \times \frac{3}{8}$

41. $14 \times \frac{5}{7}$

42. $\frac{2}{3} \times 6$

43. $\frac{5}{12} \times 40$

44. $\frac{1}{3} \times 1\frac{1}{3}$

45. $\frac{2}{5} \times 2\frac{1}{2}$

46. $1\frac{7}{8} \times \frac{4}{15}$

47. $2\frac{1}{5} \times \frac{5}{22}$

48. $4 \times 2\frac{1}{2}$

49. $9 \times 3\frac{1}{3}$

50. $2\frac{1}{7} \times 3$

51. $5\frac{1}{4} \times 8$

52. $3\frac{2}{3} \times 5$

53. $4\frac{2}{9} \times 3$

54. $\frac{1}{2} \times 3\frac{3}{7}$

55. $\frac{3}{8} \times 4\frac{4}{5}$

56. $6\frac{1}{8} \times \frac{4}{7}$

57. $5\frac{1}{3} \times \frac{5}{16}$

58. $\frac{3}{8} \times 4\frac{1}{2}$

59. $\frac{5}{7} \times 2\frac{1}{3}$

60. $0 \times 2\frac{2}{3}$

61. $6\frac{1}{8} \times 0$

62. $2\frac{5}{8} \times 3\frac{2}{5}$

63. $5\frac{3}{16} \times 5\frac{1}{3}$

64. $3\frac{1}{7} \times 2\frac{1}{8}$

65. $16\frac{5}{8} \times 1\frac{1}{16}$

66. $2\frac{2}{5} \times 3\frac{1}{12}$

67. $2\frac{2}{3} \times \frac{3}{20}$

68. $5\frac{1}{5} \times 3\frac{1}{13}$

69. $3\frac{3}{4} \times 2\frac{3}{20}$

70. $12\frac{3}{5} \times 1\frac{3}{7}$

71. $6\frac{1}{2} \times 1\frac{3}{13}$

72. True or false? If the product of a whole number and a fraction is a whole number, then the denominator of the fraction is a factor of the original whole number.

73. Multiply $2\frac{1}{2}$ and $3\frac{3}{5}$.

74. Multiply $4\frac{3}{8}$ and $3\frac{3}{5}$.

75. Find the product of $2\frac{1}{8}$ and $\frac{5}{17}$.

76. Find the product of $12\frac{2}{5}$ and $3\frac{7}{31}$. $\dfrac{62}{5} \times \dfrac{100}{31} = \dfrac{620}{155}$

Convert

OBJECTIVE C　　　To solve application problems　_Read all ? Carefully_

For Exercises 79 and 80, give your answer without actually doing a calculation.

79. Read Exercise 81. Will the requested cost be greater than or less than $12?

80. Read Exercise 83. Will the requested length be greater than or less than 4 feet?

81. **Consumerism** Salmon costs $4 per pound. Find the cost of $2\frac{3}{4}$ pounds of salmon.

82. **Exercise** Maria Rivera can walk $3\frac{1}{2}$ miles in 1 hour. At this rate, how far can Maria walk in $\frac{1}{3}$ hour?

83. **Carpentry** A board that costs $6 is $9\frac{1}{4}$ feet long. One-third of the board is cut off. What is the length of the piece cut off?

84. **Geometry** The perimeter of a square is equal to four times the length of a side of the square. Find the perimeter of a square whose side measures $16\frac{3}{4}$ inches.

$16\frac{3}{4}$ in.

85. **Geometry** To find the area of a square, multiply the length of one side of the square times itself. What is the area of a square whose side measures $5\frac{1}{4}$ feet? The area of the square will be in square feet.

86. **Geometry** The area of a rectangle is equal to the product of the length of the rectangle times its width. Find the area of a rectangle that has a length of $4\frac{2}{5}$ miles and a width of $3\frac{3}{10}$ miles. The area will be in square miles.

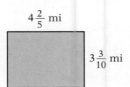

$4\frac{2}{5}$ mi

$3\frac{3}{10}$ mi

87. **Biofuels** See the news clipping at the right. How many bushels of corn produced each year are turned into ethanol?

Measurement The table at the right below shows the lengths of steel rods and their corresponding weights per foot. Use this table for Exercises 88 to 90.

88. Find the weight of the $6\frac{1}{2}$-foot steel rod.

89. Find the weight of the $12\frac{7}{12}$-foot steel rod.

90. Find the total weight of the $8\frac{5}{8}$-foot and the $10\frac{3}{4}$-foot steel rods.

91. **Sewing** The Booster Club is making 22 capes for the members of the high school marching band. Each cape is made from $1\frac{3}{8}$ yards of material at a cost of $12 per yard. Find the total cost of the material.

92. **Construction** On an architectural drawing of a kitchen, the front face of the cabinet below the sink is $23\frac{1}{2}$ inches from the back wall. Before the cabinet is installed, a plumber must install a drain in the floor halfway between the wall and the front face of the cabinet. Find the required distance from the wall to the center of the drain.

Applying the Concepts

93. The product of 1 and a number is $\frac{1}{2}$. Find the number.

94. **Time** Our calendar is based on the solar year, which is $365\frac{1}{4}$ days. Use this fact to explain leap years.

95. Which of the labeled points on the number line at the right could be the graph of the product of *B* and *C*?

$$\begin{array}{ccccccccc} \vdash & \bullet & \bullet & \bullet & + & \bullet & + & \bullet & \dashv \\ 0 & A & B & C & 1 & D & 2 & E & 3 \end{array}$$

96. Fill in the circles on the square at the right with the fractions $\frac{1}{6}, \frac{5}{18}, \frac{4}{9}, \frac{5}{9}, \frac{2}{3}, \frac{3}{4}, 1\frac{1}{9}, 1\frac{1}{2}$, and $2\frac{1}{4}$ so that the product of any row is equal to $\frac{5}{18}$. (*Note:* There is more than one possible answer.)

Length (ft)	Weight (lb/ft)
$6\frac{1}{2}$	$\frac{3}{8}$
$8\frac{5}{8}$	$1\frac{1}{4}$
$10\frac{3}{4}$	$2\frac{1}{2}$
$12\frac{7}{12}$	$4\frac{1}{3}$

© iStockphoto.com/Janice Richard

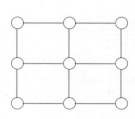

SECTION

2.7   Division of Fractions and Mixed Numbers

OBJECTIVE A   To divide fractions

The **reciprocal of a fraction** is the fraction with the numerator and denominator interchanged.

The reciprocal of $\frac{2}{3}$ is $\frac{3}{2}$.

The process of interchanging the numerator and denominator is called **inverting a fraction.**

To find the reciprocal of a whole number, first write the whole number as a fraction with a denominator of 1. Then find the reciprocal of that fraction.

The reciprocal of 5 is $\frac{1}{5}$. $\left(\text{Think } 5 = \frac{5}{1}.\right)$

Reciprocals are used to rewrite division problems as related multiplication problems. Look at the following two problems:

$$8 \div 2 = 4 \qquad\qquad 8 \times \frac{1}{2} = 4$$

8 divided by 2 is 4.   8 times the reciprocal of 2 is 4.

"Divided by" means the same as "times the reciprocal of." Thus "÷ 2" can be replaced with "$\times \frac{1}{2}$," and the answer will be the same. Fractions are divided by making this replacement.

HOW TO • 1   Divide: $\frac{2}{3} \div \frac{3}{4}$

$$\frac{2}{3} \div \frac{3}{4} = \frac{2}{3} \times \frac{4}{3} = \frac{2 \cdot 4}{3 \cdot 3} = \frac{2 \cdot 2 \cdot 2}{3 \cdot 3} = \frac{8}{9}$$

• Multiply the first fraction by the reciprocal of the second fraction.

EXAMPLE • 1

Divide: $\frac{5}{8} \div \frac{4}{9}$

Solution $\frac{5}{8} \div \frac{4}{9} = \frac{5}{8} \times \frac{9}{4} = \frac{5 \cdot 9}{8 \cdot 4}$

$$= \frac{5 \cdot 3 \cdot 3}{2 \cdot 2 \cdot 2 \cdot 2 \cdot 2} = \frac{45}{32} = 1\frac{13}{32}$$

YOU TRY IT • 1

Divide: $\frac{3}{7} \div \frac{2}{3}$

Your solution

EXAMPLE • 2

Divide: $\frac{3}{5} \div \frac{12}{25}$

Solution $\frac{3}{5} \div \frac{12}{25} = \frac{3}{5} \times \frac{25}{12} = \frac{3 \cdot 25}{5 \cdot 12}$

$$= \frac{\overset{1}{\cancel{3}} \cdot \overset{1}{\cancel{5}} \cdot 5}{\underset{1}{\cancel{5}} \cdot 2 \cdot 2 \cdot \underset{1}{\cancel{3}}} = \frac{5}{4} = 1\frac{1}{4}$$

YOU TRY IT • 2

Divide: $\frac{3}{4} \div \frac{9}{10}$

Your solution

Solutions on p. S7

OBJECTIVE B   To divide whole numbers, mixed numbers, and fractions

To divide a fraction and a whole number, first write the whole number as a fraction with a denominator of 1.

> **HOW TO · 2**   Divide: $\frac{3}{7} \div 5$
>
> $$\frac{3}{7} \div \boxed{5} = \frac{3}{7} \div \boxed{\frac{5}{1}} = \frac{3}{7} \times \frac{1}{5} = \frac{3 \cdot 1}{7 \cdot 5} = \frac{3}{35}$$ • Write 5 with a denominator of 1. Then divide the fractions.

When a number in a quotient is a mixed number, write the mixed number as an improper fraction before dividing.

> **HOW TO · 3**   Divide: $1\frac{13}{15} \div 4\frac{4}{5}$
>
> Write the mixed numbers as improper fractions. Then divide the fractions.
>
> $$1\frac{13}{15} \div 4\frac{4}{5} = \frac{28}{15} \div \frac{24}{5} = \frac{28}{15} \times \frac{5}{24} = \frac{28 \cdot 5}{15 \cdot 24} = \frac{\overset{1}{\cancel{2}} \cdot \overset{1}{\cancel{2}} \cdot 7 \cdot \overset{1}{\cancel{5}}}{3 \cdot \cancel{5} \cdot \cancel{2} \cdot \cancel{2} \cdot 2 \cdot 3} = \frac{7}{18}$$

EXAMPLE · 3

Divide $\frac{4}{9}$ by 5.

Solution

$$\frac{4}{9} \div 5 = \frac{4}{9} \div \frac{5}{1} = \frac{4}{9} \times \frac{1}{5}$$

$$= \frac{4 \cdot 1}{9 \cdot 5} = \frac{2 \cdot 2}{3 \cdot 3 \cdot 5} = \frac{4}{45}$$

• $5 = \frac{5}{1}$. The reciprocal of $\frac{5}{1}$ is $\frac{1}{5}$.

YOU TRY IT · 3

Divide $\frac{5}{7}$ by 6.

Your solution

EXAMPLE · 4

Find the quotient of $\frac{3}{8}$ and $2\frac{1}{10}$.

Solution

$$\frac{3}{8} \div 2\frac{1}{10} = \frac{3}{8} \div \frac{21}{10} = \frac{3}{8} \times \frac{10}{21}$$

$$= \frac{3 \cdot 10}{8 \cdot 21} = \frac{\overset{1}{\cancel{3}} \cdot \overset{1}{\cancel{2}} \cdot 5}{\underset{1}{\cancel{2}} \cdot 2 \cdot 2 \cdot \underset{1}{\cancel{3}} \cdot 7} = \frac{5}{28}$$

YOU TRY IT · 4

Find the quotient of $12\frac{3}{5}$ and 7.

Your solution

EXAMPLE · 5

Divide: $2\frac{3}{4} \div 1\frac{5}{7}$

Solution

$$2\frac{3}{4} \div 1\frac{5}{7} = \frac{11}{4} \div \frac{12}{7} = \frac{11}{4} \times \frac{7}{12} = \frac{11 \cdot 7}{4 \cdot 12}$$

$$= \frac{11 \cdot 7}{2 \cdot 2 \cdot 2 \cdot 2 \cdot 3} = \frac{77}{48} = 1\frac{29}{48}$$

YOU TRY IT · 5

Divide: $3\frac{2}{3} \div 2\frac{2}{5}$

Your solution

Solutions on p. S7

EXAMPLE • 6

Divide: $1\frac{13}{15} \div 4\frac{1}{5}$

Solution

$$1\frac{13}{15} \div 4\frac{1}{5} = \frac{28}{15} \div \frac{21}{5} = \frac{28}{15} \times \frac{5}{21} = \frac{28 \cdot 5}{15 \cdot 21}$$

$$= \frac{2 \cdot 2 \cdot \overset{1}{\cancel{7}} \cdot \overset{1}{\cancel{5}}}{3 \cdot \cancel{5} \cdot 3 \cdot \cancel{7}} = \frac{4}{9}$$

YOU TRY IT • 6

Divide: $2\frac{5}{6} \div 8\frac{1}{2}$

Your solution

EXAMPLE • 7

Divide: $4\frac{3}{8} \div 7$

Solution

$$4\frac{3}{8} \div 7 = \frac{35}{8} \div \frac{7}{1} = \frac{35}{8} \times \frac{1}{7}$$

$$= \frac{35 \cdot 1}{8 \cdot 7} = \frac{5 \cdot \overset{1}{\cancel{7}}}{2 \cdot 2 \cdot 2 \cdot \underset{1}{\cancel{7}}} = \frac{5}{8}$$

YOU TRY IT • 7

Divide: $6\frac{2}{5} \div 4$

Your solution

Solutions on p. S7

OBJECTIVE C　　**To solve application problems**

EXAMPLE • 8

A car used $15\frac{1}{2}$ gallons of gasoline on a 310-mile trip. How many miles can this car travel on 1 gallon of gasoline?

Strategy

To find the number of miles, divide the number of miles traveled by the number of gallons of gasoline used.

Solution

$$310 \div 15\frac{1}{2} = \frac{310}{1} \div \frac{31}{2}$$

$$= \frac{310}{1} \times \frac{2}{31} = \frac{310 \cdot 2}{1 \cdot 31}$$

$$= \frac{2 \cdot 5 \cdot \overset{1}{\cancel{31}} \cdot 2}{1 \cdot \underset{1}{\cancel{31}}} = \frac{20}{1} = 20$$

The car travels 20 miles on 1 gallon of gasoline.

YOU TRY IT • 8

A factory worker can assemble a product in $7\frac{1}{2}$ minutes. How many products can the worker assemble in 1 hour?

Your strategy

Your solution

Solutions on p. S7

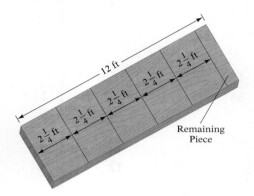

EXAMPLE · 9

A 12-foot board is cut into pieces $2\frac{1}{4}$ feet long for use as bookshelves. What is the length of the remaining piece after as many shelves as possible have been cut?

YOU TRY IT · 9

A 16-foot board is cut into pieces $3\frac{1}{3}$ feet long for shelves for a bookcase. What is the length of the remaining piece after as many shelves as possible have been cut?

Strategy

To find the length of the remaining piece:

• Divide the total length of the board (12) by the length of each shelf $\left(2\frac{1}{4}\right)$. This will give you the number of shelves cut, with a certain fraction of a shelf left over.
• Multiply the fractional part of the result in step 1 by the length of one shelf to determine the length of the remaining piece.

Your strategy

Solution

$$12 \div 2\frac{1}{4} = \frac{12}{1} \div \frac{9}{4} = \frac{12}{1} \times \frac{4}{9}$$

$$= \frac{12 \cdot 4}{1 \cdot 9} = \frac{16}{3} = 5\frac{1}{3}$$

There are 5 pieces that are each $2\frac{1}{4}$ feet long.

There is 1 piece that is $\frac{1}{3}$ of $2\frac{1}{4}$ feet long.

$$\frac{1}{3} \times 2\frac{1}{4} = \frac{1}{3} \times \frac{9}{4} = \frac{1 \cdot 9}{3 \cdot 4} = \frac{3}{4}$$

The length of the piece remaining is $\frac{3}{4}$ foot.

Your solution

Solution on p. S7

2.7 EXERCISES

OBJECTIVE A   To divide fractions

For Exercises 1 to 28, divide.

1. $\dfrac{1}{3} \div \dfrac{2}{5}$

2. $\dfrac{3}{7} \div \dfrac{3}{2}$

3. $\dfrac{3}{7} \div \dfrac{3}{7}$

4. $0 \div \dfrac{1}{2}$

5. $0 \div \dfrac{3}{4}$

6. $\dfrac{16}{33} \div \dfrac{4}{11}$

7. $\dfrac{5}{24} \div \dfrac{15}{36}$

8. $\dfrac{11}{15} \div \dfrac{1}{12}$

9. $\dfrac{1}{9} \div \dfrac{2}{3}$

10. $\dfrac{10}{21} \div \dfrac{5}{7}$

11. $\dfrac{2}{5} \div \dfrac{4}{7}$

12. $\dfrac{3}{8} \div \dfrac{5}{12}$

13. $\dfrac{1}{2} \div \dfrac{1}{4}$

14. $\dfrac{1}{3} \div \dfrac{1}{9}$

15. $\dfrac{1}{5} \div \dfrac{1}{10}$

16. $\dfrac{4}{15} \div \dfrac{2}{5}$

17. $\dfrac{7}{15} \div \dfrac{14}{5}$

18. $\dfrac{5}{8} \div \dfrac{15}{2}$

19. $\dfrac{14}{3} \div \dfrac{7}{9}$

20. $\dfrac{7}{4} \div \dfrac{9}{2}$

21. $\dfrac{5}{9} \div \dfrac{25}{3}$

22. $\dfrac{5}{16} \div \dfrac{3}{8}$

23. $\dfrac{2}{3} \div \dfrac{1}{3}$

24. $\dfrac{4}{9} \div \dfrac{1}{9}$

25. $\dfrac{5}{7} \div \dfrac{2}{7}$

26. $\dfrac{5}{6} \div \dfrac{1}{9}$

27. $\dfrac{2}{3} \div \dfrac{2}{9}$

28. $\dfrac{5}{12} \div \dfrac{5}{6}$

29. Divide $\dfrac{7}{8}$ by $\dfrac{3}{4}$.

30. Divide $\dfrac{7}{12}$ by $\dfrac{3}{4}$.

31. Find the quotient of $\dfrac{5}{7}$ and $\dfrac{3}{14}$.

32. Find the quotient of $\dfrac{6}{11}$ and $\dfrac{9}{32}$.

 33. True or false? If a fraction has a numerator of 1, then the reciprocal of the fraction is a whole number.

 34. True or false? The reciprocal of an improper fraction that is not equal to 1 is a proper fraction.

OBJECTIVE B   To divide whole numbers, mixed numbers, and fractions

For Exercises 35 to 73, divide.

35. $4 \div \dfrac{2}{3}$

36. $\dfrac{2}{3} \div 4$

37. $\dfrac{3}{2} \div 3$

38. $3 \div \dfrac{3}{2}$

39. $\dfrac{5}{6} \div 25$

40. $22 \div \dfrac{3}{11}$

41. $6 \div 3\dfrac{1}{3}$

42. $5\dfrac{1}{2} \div 11$

43. $6\dfrac{1}{2} \div \dfrac{1}{2}$

44. $\dfrac{3}{8} \div 2\dfrac{1}{4}$

45. $8\dfrac{1}{4} \div 2\dfrac{3}{4}$

46. $3\dfrac{5}{9} \div 32$

47. $4\dfrac{1}{5} \div 21$

48. $6\dfrac{8}{9} \div \dfrac{31}{36}$

49. $\dfrac{11}{12} \div 2\dfrac{1}{3}$

50. $\dfrac{7}{8} \div 3\dfrac{1}{4}$

51. $35 \div \dfrac{7}{24}$

52. $\dfrac{3}{8} \div 2\dfrac{3}{4}$

53. $\dfrac{11}{18} \div 2\dfrac{2}{9}$

54. $\dfrac{21}{40} \div 3\dfrac{3}{10}$

55. $2\dfrac{1}{16} \div 2\dfrac{1}{2}$

56. $7\dfrac{3}{5} \div 1\dfrac{7}{12}$

57. $1\dfrac{2}{3} \div \dfrac{3}{8}$

58. $16 \div \dfrac{2}{3}$

59. $1\dfrac{5}{8} \div 4$

60. $13\dfrac{3}{8} \div \dfrac{1}{4}$

61. $16 \div 1\dfrac{1}{2}$

62. $9 \div \dfrac{7}{8}$

63. $1\dfrac{1}{3} \div 5\dfrac{8}{9}$

64. $13\dfrac{2}{3} \div 0$

65. $82\dfrac{3}{5} \div 19\dfrac{1}{10}$

66. $45\dfrac{3}{5} \div 15$

67. $102 \div 1\dfrac{1}{2}$

68. $0 \div 3\dfrac{1}{2}$

69. $8\dfrac{2}{7} \div 1$

70. $6\dfrac{9}{16} \div 1\dfrac{3}{32}$

71. $8\dfrac{8}{9} \div 2\dfrac{13}{18}$

72. $10\dfrac{1}{5} \div 1\dfrac{7}{10}$

73. $7\dfrac{3}{8} \div 1\dfrac{27}{32}$

74. Divide $7\dfrac{7}{9}$ by $5\dfrac{5}{6}$.

75. Divide $2\dfrac{3}{4}$ by $1\dfrac{23}{32}$.

76. Find the quotient of $8\dfrac{1}{4}$ and $1\dfrac{5}{11}$.

77. Find the quotient of $\dfrac{14}{17}$ and $3\dfrac{1}{9}$.

  78. True or false? The reciprocal of a mixed number is an improper fraction.

  79. True or false? A fraction divided by its reciprocal is 1.

OBJECTIVE C   To solve application problems

  For Exercises 80 and 81, give your answer without actually doing a calculation.

80. Read Exercise 82. Will the requested number of boxes be greater than or less than 600?

81. Read Exercise 83. Will the requested number of servings be greater than or less than 16?

82. **Consumerism** Individual cereal boxes contain $\dfrac{3}{4}$ ounce of cereal. How many boxes can be filled with 600 ounces of cereal?

83. **Consumerism** A box of Post's Great Grains cereal costing \$4 contains 16 ounces of cereal. How many $1\dfrac{1}{3}$-ounce servings are in this box?

84. **Gemology** A $\dfrac{5}{8}$-karat diamond was purchased for \$1200. What would a similar diamond weighing 1 karat cost?

85. **Real Estate** The Inverness Investor Group bought $8\dfrac{1}{3}$ acres of land for \$200,000. What was the cost of each acre?

86. **Fuel Efficiency** A car used $12\dfrac{1}{2}$ gallons of gasoline on a 275-mile trip. How many miles can the car travel on 1 gallon of gasoline?

87. **Mechanics** A nut moves $\dfrac{5}{32}$ inch for each turn. Find the number of turns it will take for the nut to move $1\dfrac{7}{8}$ inches.

88. **Real Estate** The Hammond Company purchased $9\frac{3}{4}$ acres of land for a housing project. One and one-half acres were set aside for a park.

   a. How many acres are available for housing?

   b. How many $\frac{1}{4}$-acre parcels of land can be sold after the land for the park is set aside?

89. **The Food Industry** A chef purchased a roast that weighed $10\frac{3}{4}$ pounds. After the fat was trimmed and the bone removed, the roast weighed $9\frac{1}{3}$ pounds.

   a. What was the total weight of the fat and bone?

   b. How many $\frac{1}{3}$-pound servings can be cut from the trimmed roast?

90. **Carpentry** A 15-foot board is cut into pieces $3\frac{1}{2}$ feet long for a bookcase. What is the length of the piece remaining after as many shelves as possible have been cut?

91. **Construction** The railing of a stairway extends onto a landing. The distance between the end posts of the railing on the landing is $22\frac{3}{4}$ inches. Five posts are to be inserted, evenly spaced, between the end posts. Each post has a square base that measures $1\frac{1}{4}$ inches. Find the distance between each pair of posts.

92. **Construction** The railing of a stairway extends onto a landing. The distance between the end posts of the railing on the landing is $42\frac{1}{2}$ inches. Ten posts are to be inserted, evenly spaced, between the end posts. Each post has a square base that measures $1\frac{1}{2}$ inches. Find the distance between each pair of posts.

Applying the Concepts

Loans The figure at the right shows how the money borrowed on home equity loans is spent. Use this graph for Exercises 93 and 94.

93. What fractional part of the money borrowed on home equity loans is spent on debt consolidation and home improvement?

94. What fractional part of the money borrowed on home equity loans is spent on home improvement, cars, and tuition?

How Money Borrowed on Home Equity Loans Is Spent
Source: Consumer Bankers Association

95. **Puzzles** You completed $\frac{1}{3}$ of a jigsaw puzzle yesterday and $\frac{1}{2}$ of the puzzle today. What fraction of the puzzle is left to complete?

96. Finances A bank recommends that the maximum monthly payment for a home be $\frac{1}{3}$ of your total monthly income. Your monthly income is $4500. What would the bank recommend as your maximum monthly house payment?

97. Sports During the second half of the 1900s, greenskeepers mowed the grass on golf putting surfaces progressively lower. The table at the right shows the average grass height by decade. What was the difference between the average height of the grass in the 1980s and its average height in the 1950s?

Average Height of Grass on Golf Putting Surfaces	
Decade	**Height (in inches)**
1950s	$\frac{1}{4}$
1960s	$\frac{7}{32}$
1970s	$\frac{3}{16}$
1980s	$\frac{5}{32}$
1990s	$\frac{1}{8}$

Source: Golf Course Superintendents Association

98. Wages You have a part-time job that pays $9 an hour. You worked 5 hours, $3\frac{3}{4}$ hours, $1\frac{1}{4}$ hours, and $2\frac{1}{3}$ hours during the four days you worked last week. Find your total earnings for last week's work.

99. Board Games A wooden travel game board has hinges that allow the board to be folded in half. If the dimensions of the open board are 14 inches by 14 inches by $\frac{7}{8}$ inch, what are the dimensions of the board when it is closed?

Nutrition According to the Center for Science in the Public Interest, the average teenage boy drinks $3\frac{1}{3}$ cans of soda per day. The average teenage girl drinks $2\frac{1}{3}$ cans of soda per day. Use this information for Exercises 100 and 101.

100. If a can of soda contains 150 calories, how many calories does the average teenage boy consume each week in soda?

101. How many more cans of soda per week does the average teenage boy drink than the average teenage girl?

Bill Aron/PhotoEdit, Inc.

102. Maps On a map, two cities are $4\frac{5}{8}$ inches apart. If $\frac{3}{8}$ inch on the map represents 60 miles, what is the number of miles between the two cities?

103. Fill in the box to make a true statement.

a. $\frac{3}{4} \cdot \square = \frac{1}{2}$ **b.** $\frac{2}{3} \cdot \square = 1\frac{3}{4}$

104. Publishing A page of type in a certain textbook is $7\frac{1}{2}$ inches wide. If the page is divided into three equal columns, with $\frac{3}{8}$ inch between columns, how wide is each column?

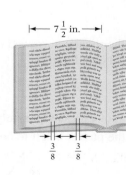

$\longleftarrow 7\frac{1}{2}$ in. $\longrightarrow$

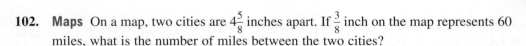

$\frac{3}{8}$ $\frac{3}{8}$

SECTION

2.8

Order, Exponents, and the Order of Operations Agreement

OBJECTIVE A
To identify the order relation between two fractions

Recall that whole numbers can be graphed as points on the number line. Fractions can also be graphed as points on the number line.

The graph of $\frac{3}{4}$ on the number line

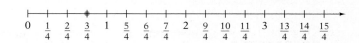

The number line can be used to determine the order relation between two fractions. A fraction that appears to the left of a given fraction is less than the given fraction. A fraction that appears to the right of a given fraction is greater than the given fraction.

$$\frac{1}{8} < \frac{3}{8} \qquad \frac{6}{8} > \frac{3}{8}$$

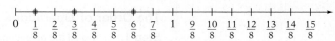

To find the order relation between two fractions with the same denominator, compare the numerators. The fraction that has the smaller numerator is the smaller fraction. When the denominators are different, begin by writing equivalent fractions with a common denominator; then compare the numerators.

> **HOW TO • 1** Find the order relation between $\frac{11}{18}$ and $\frac{5}{8}$.
>
> The LCM of 18 and 8 is 72.
>
> $\frac{11}{18} = \frac{44}{72}$ ← Smaller numerator $\qquad \frac{11}{18} < \frac{5}{8}$ or $\frac{5}{8} > \frac{11}{18}$
>
> $\frac{5}{8} = \frac{45}{72}$ ← Larger numerator

EXAMPLE • 1

Place the correct symbol, $<$ or $>$, between the two numbers.

$\frac{5}{12} \qquad \frac{7}{18}$

Solution $\quad \frac{5}{12} = \frac{15}{36} \qquad \frac{7}{18} = \frac{14}{36}$

$\qquad \frac{5}{12} > \frac{7}{18}$

YOU TRY IT • 1

Place the correct symbol, $<$ or $>$, between the two numbers.

$\frac{9}{14} \qquad \frac{13}{21}$

Your solution

Solution on p. S8

OBJECTIVE B
To simplify expressions containing exponents

Repeated multiplication of the same fraction can be written in two ways:

$$\frac{1}{2} \cdot \frac{1}{2} \cdot \frac{1}{2} \cdot \frac{1}{2} \quad \text{or} \quad \left(\frac{1}{2}\right)^4 \leftarrow \text{Exponent}$$

The exponent indicates how many times the fraction occurs as a factor in the multiplication. The expression $\left(\frac{1}{2}\right)^4$ is in exponential notation.

EXAMPLE • 2

Simplify: $\left(\frac{5}{6}\right)^3 \cdot \left(\frac{3}{5}\right)^2$

Solution

$$\left(\frac{5}{6}\right)^3 \cdot \left(\frac{3}{5}\right)^2 = \left(\frac{5}{6} \cdot \frac{5}{6} \cdot \frac{5}{6}\right) \cdot \left(\frac{3}{5} \cdot \frac{3}{5}\right)$$

$$= \frac{\overset{1}{\cancel{5}} \cdot \overset{1}{\cancel{5}} \cdot 5 \cdot \overset{1}{\cancel{3}} \cdot \overset{1}{\cancel{3}}}{2 \cdot \underset{1}{\cancel{3}} \cdot 2 \cdot \underset{1}{\cancel{3}} \cdot 2 \cdot 3 \cdot \underset{1}{\cancel{5}} \cdot \underset{1}{\cancel{5}}} = \frac{5}{24}$$

YOU TRY IT • 2

Simplify: $\left(\frac{7}{11}\right)^2 \cdot \left(\frac{2}{7}\right)$

Your solution

Solution on p. S8

OBJECTIVE C   **To use the Order of Operations Agreement to simplify expressions**

The Order of Operations Agreement is used for fractions as well as whole numbers.

> **The Order of Operations Agreement**
>
> **Step 1.** Do all the operations inside parentheses.
> **Step 2.** Simplify any number expressions containing exponents.
> **Step 3.** Do multiplications and divisions as they occur from left to right.
> **Step 4.** Do additions and subtractions as they occur from left to right.

HOW TO • 2   Simplify $\frac{14}{15} - \left(\frac{1}{2}\right)^2 \times \left(\frac{2}{3} + \frac{4}{5}\right)$.

$$\frac{14}{15} - \left(\frac{1}{2}\right)^2 \times \underbrace{\left(\frac{2}{3} + \frac{4}{5}\right)}$$

1. Perform operations in parentheses.

$$\frac{14}{15} - \underbrace{\left(\frac{1}{2}\right)^2} \times \frac{22}{15}$$

2. Simplify expressions with exponents.

$$\frac{14}{15} - \underbrace{\frac{1}{4} \times \frac{22}{15}}$$

3. Do multiplication and division as they occur from left to right.

$$\underbrace{\frac{14}{15} - \frac{11}{30}}$$

4. Do addition and subtraction as they occur from left to right.

$$\frac{17}{30}$$

One or more of the above steps may not be needed to simplify an expression. In that case, proceed to the next step in the Order of Operations Agreement.

EXAMPLE • 3

Simplify: $\left(\frac{3}{4}\right)^2 \div \left(\frac{3}{8} - \frac{1}{12}\right)$

Solution

$$\left(\frac{3}{4}\right)^2 \div \left(\frac{3}{8} - \frac{1}{12}\right)$$

$$= \left(\frac{3}{4}\right)^2 \div \left(\frac{7}{24}\right) = \frac{9}{16} \div \frac{7}{24}$$

$$= \frac{9}{16} \cdot \frac{24}{7} = \frac{27}{14} = 1\frac{13}{14}$$

YOU TRY IT • 3

Simplify: $\left(\frac{1}{13}\right)^2 \cdot \left(\frac{1}{4} + \frac{1}{6}\right) \div \frac{5}{13}$

Your solution

Solution on p. S8

2.8 EXERCISES

| OBJECTIVE A | To identify the order relation between two fractions |

For Exercises 1 to 12, place the correct symbol, < or >, between the two numbers.

1. $\dfrac{11}{40}$ $\dfrac{19}{40}$ 2. $\dfrac{92}{103}$ $\dfrac{19}{103}$ 3. $\dfrac{2}{3}$ $\dfrac{5}{7}$ 4. $\dfrac{2}{5}$ $\dfrac{3}{8}$

5. $\dfrac{5}{8}$ $\dfrac{7}{12}$ 6. $\dfrac{11}{16}$ $\dfrac{17}{24}$ 7. $\dfrac{7}{9}$ $\dfrac{11}{12}$ 8. $\dfrac{5}{12}$ $\dfrac{7}{15}$

9. $\dfrac{13}{14}$ $\dfrac{19}{21}$ 10. $\dfrac{13}{18}$ $\dfrac{7}{12}$ 11. $\dfrac{7}{24}$ $\dfrac{11}{30}$ 12. $\dfrac{13}{36}$ $\dfrac{19}{48}$

 13. Without writing the fractions $\dfrac{4}{5}$ and $\dfrac{1}{7}$ with a common denominator, decide which fraction is larger.

| OBJECTIVE B | To simplify expressions containing exponents |

For Exercises 14 to 29, simplify.

14. $\left(\dfrac{3}{8}\right)^2$ 15. $\left(\dfrac{5}{12}\right)^2$ 16. $\left(\dfrac{2}{9}\right)^3$ 17. $\left(\dfrac{1}{2}\right) \cdot \left(\dfrac{2}{3}\right)^2$

18. $\left(\dfrac{2}{3}\right) \cdot \left(\dfrac{1}{2}\right)^4$ 19. $\left(\dfrac{1}{3}\right)^2 \cdot \left(\dfrac{3}{5}\right)^3$ 20. $\left(\dfrac{2}{5}\right)^3 \cdot \left(\dfrac{5}{7}\right)^2$ 21. $\left(\dfrac{5}{9}\right)^3 \cdot \left(\dfrac{18}{25}\right)^2$

22. $\left(\dfrac{1}{3}\right)^4 \cdot \left(\dfrac{9}{11}\right)^2$ 23. $\left(\dfrac{1}{2}\right)^6 \cdot \left(\dfrac{32}{35}\right)^2$ 24. $\left(\dfrac{2}{3}\right)^4 \cdot \left(\dfrac{81}{100}\right)^2$ 25. $\left(\dfrac{1}{6}\right) \cdot \left(\dfrac{6}{7}\right)^2 \cdot \left(\dfrac{2}{3}\right)$

26. $\left(\dfrac{2}{7}\right) \cdot \left(\dfrac{7}{8}\right)^2 \cdot \left(\dfrac{8}{9}\right)$ 27. $3 \cdot \left(\dfrac{3}{5}\right)^3 \cdot \left(\dfrac{1}{3}\right)^2$ 28. $4 \cdot \left(\dfrac{3}{4}\right)^3 \cdot \left(\dfrac{4}{7}\right)^2$ 29. $11 \cdot \left(\dfrac{3}{8}\right)^3 \cdot \left(\dfrac{8}{11}\right)^2$

 30. True or false? When simplified, the expression $\left(\dfrac{1}{2}\right)^{24} \cdot \left(\dfrac{1}{3}\right)^{35}$ is a fraction with a numerator of 1.

| OBJECTIVE C | To use the Order of Operations Agreement to simplify expressions |

For Exercises 31 to 49, simplify.

31. $\dfrac{1}{2} - \dfrac{1}{3} + \dfrac{2}{3}$

32. $\dfrac{2}{5} + \dfrac{3}{10} - \dfrac{2}{3}$

33. $\dfrac{1}{3} \div \dfrac{1}{2} + \dfrac{3}{4}$

34. $\dfrac{4}{5} + \dfrac{3}{7} \cdot \dfrac{14}{15}$

35. $\left(\dfrac{3}{4}\right)^2 - \dfrac{5}{12}$

36. $\left(\dfrac{3}{5}\right)^3 - \dfrac{3}{25}$

37. $\dfrac{5}{6} \cdot \left(\dfrac{2}{3} - \dfrac{1}{6}\right) + \dfrac{7}{18}$

38. $\dfrac{3}{4} \cdot \left(\dfrac{11}{12} - \dfrac{7}{8}\right) + \dfrac{5}{16}$

39. $\dfrac{7}{12} - \left(\dfrac{2}{3}\right)^2 + \dfrac{5}{8}$

40. $\dfrac{11}{16} - \left(\dfrac{3}{4}\right)^2 + \dfrac{7}{12}$

41. $\dfrac{3}{4} \cdot \left(\dfrac{4}{9}\right)^2 + \dfrac{1}{2}$

42. $\dfrac{9}{10} \cdot \left(\dfrac{2}{3}\right)^3 + \dfrac{2}{3}$

43. $\left(\dfrac{1}{2} + \dfrac{3}{4}\right) \div \dfrac{5}{8}$

44. $\left(\dfrac{2}{3} + \dfrac{5}{6}\right) \div \dfrac{5}{9}$

45. $\dfrac{3}{8} \div \left(\dfrac{5}{12} + \dfrac{3}{8}\right)$

46. $\dfrac{7}{12} \div \left(\dfrac{2}{3} + \dfrac{5}{9}\right)$

47. $\left(\dfrac{3}{8}\right)^2 \div \left(\dfrac{3}{7} + \dfrac{3}{14}\right)$

48. $\left(\dfrac{5}{6}\right)^2 \div \left(\dfrac{5}{12} + \dfrac{2}{3}\right)$

49. $\dfrac{2}{5} \div \dfrac{3}{8} \cdot \dfrac{4}{5}$

50. Insert parentheses into the expression $\dfrac{2}{9} \cdot \dfrac{5}{6} + \dfrac{3}{4} \div \dfrac{3}{5}$ so that **a.** the first operation to be performed is addition and **b.** the first operation to be performed is division.

Applying the Concepts

51. **The Food Industry** The table at the right shows the results of a survey that asked fast-food patrons their criteria for choosing where to go for fast food. For example, 3 out of every 25 people surveyed said that the speed of the service was most important.

  a. According to the survey, do more people choose a fast-food restaurant on the basis of its location or the quality of the food?

  b. Which criterion was cited by the most people?

Fast-Food Patrons' Top Criteria for Fast-Food Restaurants	
Food quality	$\dfrac{1}{4}$
Location	$\dfrac{13}{50}$
Menu	$\dfrac{4}{25}$
Price	$\dfrac{2}{25}$
Speed	$\dfrac{3}{25}$
Other	$\dfrac{13}{100}$

Source: Maritz Marketing Research, Inc.

FOCUS ON PROBLEM SOLVING

Common Knowledge   An application problem may not provide all the information that is needed to solve the problem. Sometimes, however, the necessary information is common knowledge.

> **HOW TO · 1**   You are traveling by bus from Boston to New York. The trip is 4 hours long. If the bus leaves Boston at 10 A.M., what time should you arrive in New York?
>
> What other information do you need to solve this problem?
>
> You need to know that, using a 12-hour clock, the hours run
>
> 10 A.M.
> 11 A.M.
> 12 P.M.
> 1 P.M.
> 2 P.M.
>
> Four hours after 10 A.M. is 2 P.M.
>
> You should arrive in New York at 2 P.M.

> **HOW TO · 2**   You purchase a 44¢ stamp at the Post Office and hand the clerk a one-dollar bill. How much change do you receive?
>
> What information do you need to solve this problem?
>
> You need to know that there are 100¢ in one dollar.
>
> Your change is 100¢ − 44¢.
>
> 100 − 44 = 56
>
> You receive 56¢ in change.

What information do you need to know to solve each of the following problems?

1. You sell a dozen tickets to a fundraiser. Each ticket costs $10. How much money do you collect?

2. The weekly lab period for your science course is 1 hour and 20 minutes long. Find the length of the science lab period in minutes.

3. An employee's monthly salary is $3750. Find the employee's annual salary.

4. A survey revealed that eighth graders spend an average of 3 hours each day watching television. Find the total time an eighth grader spends watching TV each week.

5. You want to buy a carpet for a room that is 15 feet wide and 18 feet long. Find the amount of carpet that you need.

PROJECTS AND GROUP ACTIVITIES

Music In musical notation, notes are printed on a **staff,** which is a set of five horizontal lines and the spaces between them. The notes of a musical composition are grouped into **measures,** or **bars.** Vertical lines separate measures on a staff. The shape of a note indicates how long it should be held. The whole note has the longest time value of any note. Each time value is divided by 2 in order to find the next smallest time value.

The **time signature** is a fraction that appears at the beginning of a piece of music. The numerator of the fraction indicates the number of beats in a measure. The denominator indicates what kind of note receives 1 beat. For example, music written in $\frac{2}{4}$ time has 2 beats to a measure, and a quarter note receives 1 beat. One measure in $\frac{2}{4}$ time may have 1 half note, 2 quarter notes, 4 eighth notes, or any other combination of notes totaling 2 beats. Other common time signatures are $\frac{4}{4}$, $\frac{3}{4}$, and $\frac{6}{8}$.

1. Explain the meaning of the 6 and the 8 in the time signature $\frac{6}{8}$.

2. Give some possible combinations of notes in one measure of a piece written in $\frac{4}{4}$ time.

3. What does a dot at the right of a note indicate? What is the effect of a dot at the right of a half note? At the right of a quarter note? At the right of an eighth note?

4. Symbols called rests are used to indicate periods of silence in a piece of music. What symbols are used to indicate the different time values of rests?

5. Find some examples of musical compositions written in different time signatures. Use a few measures from each to show that the sum of the time values of the notes and rests in each measure equals the numerator of the time signature.

Construction Suppose you are involved in building your own home. Design a stairway from the first floor of the house to the second floor. Some of the questions you will need to answer follow.

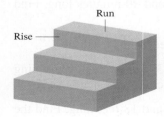

What is the distance from the floor of the first story to the floor of the second story?

Typically, what is the number of steps in a stairway?

What is a reasonable length for the run of each step?

What is the width of the wood being used to build the staircase?

In designing the stairway, remember that each riser should be the same height, that each run should be the same length, and that the width of the wood used for the steps will have to be incorporated into the calculation.

Fractions of Diagrams   The diagram that follows has been broken up into nine areas separated by heavy lines. Eight of the areas have been labeled *A* through *H*. The ninth area is shaded. Determine which lettered areas would have to be shaded so that half of the entire diagram is shaded and half is not shaded. Write down the strategy that you or your group used to arrive at the solution. Compare your strategy with that of other individual students or groups.

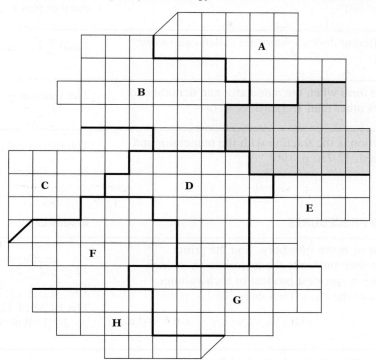

Tips for Success
Three important features of this text that can be used to prepare for a test are the
• Chapter Summary
• Chapter Review Exercises
• Chapter Test
See *AIM for Success* at the front of the book.

CHAPTER 2

SUMMARY

KEY WORDS	EXAMPLES
A number that is a multiple of two or more numbers is a *common multiple* of those numbers. The *least common multiple (LCM)* is the smallest common multiple of two or more numbers. [2.1A, p. 64]	12, 24, 36, 48, . . . are common multiples of 4 and 6. The LCM of 4 and 6 is 12.
A number that is a factor of two or more numbers is a *common factor* of those numbers. The *greatest common factor (GCF)* is the largest common factor of two or more numbers. [2.1B, p. 65]	The common factors of 12 and 16 are 1, 2, and 4. The GCF of 12 and 16 is 4.
A *fraction* can represent the number of equal parts of a whole. In a fraction, the *fraction bar* separates the *numerator* and the *denominator*. [2.2A, p. 68]	In the fraction $\frac{3}{4}$, the numerator is 3 and the denominator is 4.

In a *proper fraction,* the numerator is smaller than the denominator; a proper fraction is a number less than 1. In an *improper fraction,* the numerator is greater than or equal to the denominator; an improper fraction is a number greater than or equal to 1. A *mixed number* is a number greater than 1 with a whole-number part and a fractional part. [2.2A, p. 68]

$\frac{2}{5}$ is proper fraction.

$\frac{7}{6}$ is an improper fraction.

$4\frac{1}{10}$ is a mixed number; 4 is the whole-number part and $\frac{1}{10}$ is the fractional part.

Equal fractions with different denominators are called *equivalent fractions.* [2.3A, p. 72]

$\frac{3}{4}$ and $\frac{6}{8}$ are equivalent fractions.

A fraction is in *simplest form* when the numerator and denominator have no common factors other than 1. [2.3B, p. 73]

The fraction $\frac{11}{12}$ is in simplest form.

The *reciprocal* of a fraction is the fraction with the numerator and denominator interchanged. [2.7A, p. 100]

The reciprocal of $\frac{3}{8}$ is $\frac{8}{3}$.

The reciprocal of 5 is $\frac{1}{5}$.

ESSENTIAL RULES AND PROCEDURES

EXAMPLES

To find the LCM of two or more numbers, find the prime factorization of each number and write the factorization of each number in a table. Circle the greatest product in each column. The LCM is the product of the circled numbers. [2.1A, p. 64]

	2	3
12 =	②·②	3
18 =	2	③·③

The LCM of 12 and 18 is $2 \cdot 2 \cdot 3 \cdot 3 = 36$.

To find the GCF of two or more numbers, find the prime factorization of each number and write the factorization of each number in a table. Circle the least product in each column that does not have a blank. The GCF is the product of the circled numbers. [2.1B, p. 65]

	2	3
12 =	2·2	③
18 =	②	3·3

The GCF of 12 and 18 is $2 \cdot 3 = 6$.

To write an improper fraction as a mixed number or a whole number, divide the numerator by the denominator. [2.2B, p. 69]

$\frac{29}{6} = 29 \div 6 = 4\frac{5}{6}$

To write a mixed number as an improper fraction, multiply the denominator of the fractional part of the mixed number by the whole-number part. Add this product and the numerator of the fractional part. The sum is the numerator of the improper fraction. The denominator remains the same. [2.2B, p. 69]

$3\frac{2}{5} = \frac{5 \times 3 + 2}{5} = \frac{17}{5}$

To find equivalent fractions by raising to higher terms, multiply the numerator and denominator of the fraction by the same number. [2.3A, p. 72]

$\frac{3}{4} = \frac{3 \cdot 5}{4 \cdot 5} = \frac{15}{20}$

$\frac{3}{4}$ and $\frac{15}{20}$ are equivalent fractions.

To write a fraction in simplest form, factor the numerator and denominator of the fraction; then eliminate the common factors. [2.3B, p. 73]

$\frac{30}{45} = \frac{2 \cdot \overset{1}{\cancel{3}} \cdot \overset{1}{\cancel{5}}}{\underset{1}{\cancel{3}} \cdot 3 \cdot \underset{1}{\cancel{5}}} = \frac{2}{3}$

To add fractions with the same denominator, add the numerators and place the sum over the common denominator. [2.4A, p. 76]

$$\frac{5}{12} + \frac{11}{12} = \frac{16}{12} = 1\frac{4}{12} = 1\frac{1}{3}$$

To add fractions with different denominators, first rewrite the fractions as equivalent fractions with a common denominator. (The common denominator is the LCM of the denominators of the fractions.) Then add the fractions. [2.4B, p. 76]

$$\frac{1}{4} + \frac{2}{5} = \frac{5}{20} + \frac{8}{20} = \frac{13}{20}$$

To subtract fractions with the same denominator, subtract the numerators and place the difference over the common denominator. [2.5A, p. 84]

$$\frac{9}{16} - \frac{5}{16} = \frac{4}{16} = \frac{1}{4}$$

To subtract fractions with different denominators, first rewrite the fractions as equivalent fractions with a common denominator. (The common denominator is the LCM of the denominators of the fractions.) Then subtract the fractions. [2.5B, p. 84]

$$\frac{2}{3} - \frac{7}{16} = \frac{32}{48} - \frac{21}{48} = \frac{11}{48}$$

To multiply two fractions, multiply the numerators; this is the numerator of the product. Multiply the denominators; this is the denominator of the product. [2.6A, p. 92]

$$\frac{3}{4} \cdot \frac{2}{9} = \frac{3 \cdot 2}{4 \cdot 9} = \frac{\overset{1}{\cancel{3}} \cdot \overset{1}{\cancel{2}}}{\underset{1}{\cancel{2}} \cdot 2 \cdot \underset{1}{\cancel{3}} \cdot 3} = \frac{1}{6}$$

To divide two fractions, multiply the first fraction by the reciprocal of the second fraction. [2.7A, p. 100]

$$\frac{8}{15} \div \frac{4}{5} = \frac{8}{15} \cdot \frac{5}{4} = \frac{8 \cdot 5}{15 \cdot 4}$$
$$= \frac{\overset{1}{\cancel{2}} \cdot \overset{1}{\cancel{2}} \cdot 2 \cdot \overset{1}{\cancel{5}}}{3 \cdot \underset{1}{\cancel{5}} \cdot \underset{1}{\cancel{2}} \cdot \underset{1}{\cancel{2}}} = \frac{2}{3}$$

The find the order relation between two fractions with the same denominator, compare the numerators. The fraction that has the smaller numerator is the smaller fraction. [2.8A, p. 109]

$$\frac{17}{25} \leftarrow \text{Smaller numerator}$$
$$\frac{19}{25} \leftarrow \text{Larger numerator}$$
$$\frac{17}{25} < \frac{19}{25}$$

To find the order relation between two fractions with different denominators, first rewrite the fractions with a common denominator. The fraction that has the smaller numerator is the smaller fraction. [2.8A, p. 109]

$$\frac{3}{5} = \frac{24}{40} \quad \frac{5}{8} = \frac{25}{40}$$
$$\frac{24}{40} < \frac{25}{40}$$
$$\frac{3}{5} < \frac{5}{8}$$

Order of Operations Agreement [2.8C, p. 110]

Step 1 Do all the operations inside parentheses.

Step 2 Simplify any numerical expressions containing exponents.

Step 3 Do multiplication and division as they occur from left to right.

Step 4 Do addition and subtraction as they occur from left to right.

$$\left(\frac{1}{3}\right)^2 + \left(\frac{5}{6} - \frac{7}{12}\right) \cdot (4)$$
$$= \left(\frac{1}{3}\right)^2 + \left(\frac{1}{4}\right) \cdot (4)$$
$$= \frac{1}{9} + \left(\frac{1}{4}\right) \cdot (4)$$
$$= \frac{1}{9} + 1 = 1\frac{1}{9}$$

CHAPTER 2

CONCEPT REVIEW

Test your knowledge of the concepts presented in this chapter. Answer each question.
Then check your answers against the ones provided in the Answer Section.

1. How do you find the LCM of 75, 30, and 50?

2. How do you find the GCF of 42, 14, and 21?

3. How do you write an improper fraction as a mixed number?

4. When is a fraction in simplest form?

5. When adding fractions, why do you have to convert to equivalent fractions with a common denominator?

6. How do you add mixed numbers?

7. If you are subtracting a mixed number from a whole number, why do you need to borrow?

8. When multiplying two fractions, why is it better to eliminate the common factors before multiplying the remaining factors in the numerator and denominator?

9. When multiplying two fractions that are less than 1, will the product be greater than 1, less than the smaller number, or between the smaller number and the bigger number?

10. How are reciprocals used when dividing fractions?

11. When a fraction is divided by a whole number, why do we write the whole number as a fraction before dividing?

12. When comparing two fractions, why is it important to look at both the numerators and denominators to determine which is larger?

13. In the expression $\left(\frac{5}{6}\right)^2 - \left(\frac{3}{4} - \frac{2}{3}\right) \div \frac{1}{2}$, in what order should the operations be performed?

CHAPTER 2

REVIEW EXERCISES

1. Write $\frac{30}{45}$ in simplest form.

2. Simplify: $\left(\frac{3}{4}\right)^3 \cdot \frac{20}{27}$

3. Express the shaded portion of the circles as an improper fraction.

4. Find the total of $\frac{2}{3}$, $\frac{5}{6}$, and $\frac{2}{9}$.

5. Place the correct symbol, $<$ or $>$, between the two numbers.
$\frac{11}{18}$ $\frac{17}{24}$

6. Subtract: $\begin{array}{r} 18\frac{1}{6} \\ -3\frac{5}{7} \\ \hline \end{array}$

7. Simplify: $\frac{2}{7}\left(\frac{5}{8} - \frac{1}{3}\right) \div \frac{3}{5}$

8. Multiply: $2\frac{1}{3} \times 3\frac{7}{8}$

9. Divide: $1\frac{1}{3} \div \frac{2}{3}$

10. Find $\frac{17}{24}$ decreased by $\frac{3}{16}$.

11. Divide: $8\frac{2}{3} \div 2\frac{3}{5}$

12. Find the GCF of 20 and 48.

13. Write an equivalent fraction with the given denominator.
$\frac{2}{3} = \frac{}{36}$

14. What is $\frac{15}{28}$ divided by $\frac{5}{7}$?

15. Write an equivalent fraction with the given denominator.
$\frac{8}{11} = \frac{}{44}$

16. Multiply: $2\frac{1}{4} \times 7\frac{1}{3}$

17. Find the LCM of 18 and 12.

18. Write $\frac{16}{44}$ in simplest form.

19. Add: $\frac{3}{8} + \frac{5}{8} + \frac{1}{8}$

20. Subtract: $\begin{array}{r} 16 \\ -\,5\frac{7}{8} \\ \hline \end{array}$

21. Add: $4\frac{4}{9} + 2\frac{1}{6} + 11\frac{17}{27}$

22. Find the GCF of 15 and 25.

23. Write $\frac{17}{5}$ as a mixed number.

24. Simplify: $\left(\frac{4}{5} - \frac{2}{3}\right)^2 \div \frac{4}{15}$

25. Add: $\frac{3}{8} + 1\frac{2}{3} + 3\frac{5}{6}$

26. Find the LCM of 18 and 27.

27. Subtract: $\frac{11}{18} - \frac{5}{18}$

28. Write $2\frac{5}{7}$ as an improper fraction.

29. Divide: $\frac{5}{6} \div \frac{5}{12}$

30. Multiply: $\frac{5}{12} \times \frac{4}{25}$

31. What is $\frac{11}{50}$ multiplied by $\frac{25}{44}$?

32. Express the shaded portion of the circles as a mixed number.

33. **Meteorology** During 3 months of the rainy season, $5\frac{7}{8}$, $6\frac{2}{3}$, and $8\frac{3}{4}$ inches of rain fell. Find the total rainfall for the 3 months.

34. **Real Estate** A home building contractor bought $4\frac{2}{3}$ acres of land for $168,000. What was the cost of each acre?

35. **Sports** A 15-mile race has three checkpoints. The first checkpoint is $4\frac{1}{2}$ miles from the starting point. The second checkpoint is $5\frac{3}{4}$ miles from the first checkpoint. How many miles is the second checkpoint from the finish line?

36. **Fuel Efficiency** A compact car gets 36 miles on each gallon of gasoline. How many miles can the car travel on $6\frac{3}{4}$ gallons of gasoline?

CHAPTER 2

TEST

1. Multiply: $\frac{9}{11} \times \frac{44}{81}$

2. Find the GCF of 24 and 80.

3. Divide: $\frac{5}{9} \div \frac{7}{18}$

4. Simplify: $\left(\frac{3}{4}\right)^2 \div \left(\frac{2}{3} + \frac{5}{6}\right) - \frac{1}{12}$

5. Write $9\frac{4}{5}$ as an improper fraction.

6. What is $5\frac{2}{3}$ multiplied by $1\frac{7}{17}$?

7. Write $\frac{40}{64}$ in simplest form.

8. Place the correct symbol, $<$ or $>$, between the two numbers.

$\frac{3}{8}$ $\frac{5}{12}$

9. Simplify: $\left(\frac{1}{4}\right)^3 \div \left(\frac{1}{8}\right)^2 - \frac{1}{6}$

10. Find the LCM of 24 and 40.

11. Subtract: $\frac{17}{24} - \frac{11}{24}$

12. Write $\frac{18}{5}$ as a mixed number.

13. Find the quotient of $6\frac{2}{3}$ and $3\frac{1}{6}$.

14. Write an equivalent fraction with the given denominator.

15. Add: $\dfrac{5}{6}$

$\dfrac{7}{9}$

$+\dfrac{1}{15}$

16. Subtract: $23\dfrac{1}{8}$

$-9\dfrac{9}{44}$

17. What is $\dfrac{9}{16}$ minus $\dfrac{5}{12}$?

18. Simplify: $\left(\dfrac{2}{3}\right)^4 \cdot \dfrac{27}{32}$

19. Add: $\dfrac{7}{12} + \dfrac{11}{12} + \dfrac{5}{12}$

20. What is $12\dfrac{5}{12}$ more than $9\dfrac{17}{20}$?

21. Express the shaded portion of the circles as an improper fraction.

22. **Compensation** An electrician earns $240 for each day worked. What is the total of the electrician's earnings for working $3\dfrac{1}{2}$ days?

23. **Real Estate** Grant Miura bought $7\dfrac{1}{4}$ acres of land for a housing project. One and three-fourths acres were set aside for a park, and the remaining land was developed into $\dfrac{1}{2}$-acre lots. How many lots were available for sale?

24. **Architecture** A scale of $\dfrac{1}{2}$ inch to 1 foot is used to draw the plans for a house. The scale measurements for three walls are given in the table at the right. Complete the table to determine the actual wall lengths for the three walls a, b, and c.

Wall	Scale	Actual Wall Length
a	$6\dfrac{1}{4}$ in.	?
b	9 in.	?
c	$7\dfrac{7}{8}$ in.	?

25. **Meteorology** In 3 successive months, the rainfall measured $11\dfrac{1}{2}$ inches, $7\dfrac{5}{8}$ inches, and $2\dfrac{1}{3}$ inches. Find the total rainfall for the 3 months.

CUMULATIVE REVIEW EXERCISES

1. Round 290,496 to the nearest thousand.

2. Subtract: $\begin{array}{r} 390,047 \\ -\ 98,769 \\ \hline \end{array}$

3. Find the product of 926 and 79.

4. Divide: $57\overline{)30,792}$

5. Simplify: $4 \cdot (6 - 3) \div 6 - 1$

6. Find the prime factorization of 44.

7. Find the LCM of 30 and 42.

8. Find the GCF of 60 and 80.

9. Write $7\frac{2}{3}$ as an improper fraction.

10. Write $\frac{25}{4}$ as a mixed number.

11. Write an equivalent fraction with the given denominator.

$$\frac{5}{16} = \frac{}{48}$$

12. Write $\frac{24}{60}$ in simplest form.

13. What is $\frac{9}{16}$ more than $\frac{7}{12}$?

14. Add: $\begin{array}{r} 3\frac{7}{8} \\ 7\frac{5}{12} \\ +\ 2\frac{15}{16} \\ \hline \end{array}$

15. Find $\frac{3}{8}$ less than $\frac{11}{12}$.

16. Subtract: $\begin{array}{r} 5\frac{1}{6} \\ -\ 3\frac{7}{18} \\ \hline \end{array}$

17. Multiply: $\frac{3}{8} \times \frac{14}{15}$

18. Multiply: $3\frac{1}{8} \times 2\frac{2}{5}$

19. Divide: $\frac{7}{16} \div \frac{5}{12}$

20. Find the quotient of $6\frac{1}{8}$ and $2\frac{1}{3}$.

21. Simplify: $\left(\frac{1}{2}\right)^3 \cdot \frac{8}{9}$

22. Simplify: $\left(\frac{1}{2} + \frac{1}{3}\right) \div \left(\frac{2}{5}\right)^2$

23. Banking Molly O'Brien had $1359 in a checking account. During the week, Molly wrote checks for $128, $54, and $315. Find the amount in the checking account at the end of the week.

24. Entertainment The tickets for a movie were $10 for an adult and $4 for a student. Find the total income from the sale of 87 adult tickets and 135 student tickets.

Kevin Lee/Getty Images

25. Measurement Find the total weight of three packages that weigh $1\frac{1}{2}$ pounds, $7\frac{7}{8}$ pounds, and $2\frac{2}{3}$ pounds.

26. Carpentry A board $2\frac{5}{8}$ feet long is cut from a board $7\frac{1}{3}$ feet long. What is the length of the remaining piece?

27. Fuel Efficiency A car travels 27 miles on each gallon of gasoline. How many miles can the car travel on $8\frac{1}{3}$ gallons of gasoline?

28. Real Estate Jimmy Santos purchased $10\frac{1}{3}$ acres of land to build a housing development. Jimmy donated 2 acres for a park. How many $\frac{1}{3}$-acre parcels can be sold from the remaining land?

Decimals

OBJECTIVES

SECTION 3.1
A To write decimals in standard form and in words
B To round a decimal to a given place value

SECTION 3.2
A To add decimals
B To solve application problems

SECTION 3.3
A To subtract decimals
B To solve application problems

SECTION 3.4
A To multiply decimals
B To solve application problems

SECTION 3.5
A To divide decimals
B To solve application problems

SECTION 3.6
A To convert fractions to decimals
B To convert decimals to fractions
C To identify the order relation between two decimals or between a decimal and a fraction

ARE YOU READY?

Take the Chapter 3 Prep Test to find out if you are ready to learn to:

- Round decimals
- Add, subtract, multiply, and divide decimals
- Convert between fractions and decimals
- Compare decimals and fractions

PREP TEST

Do these exercises to prepare for Chapter 3.

1. Express the shaded portion of the rectangle as a fraction.

2. Round 36,852 to the nearest hundred.

3. Write 4791 in words.

4. Write six thousand eight hundred forty-two in standard form.

For Exercises 5 to 8, add, subtract, multiply, or divide.

5. $37 + 8892 + 465$

6. $2403 - 765$

7. 844×91

8. $23\overline{)6412}$

SECTION 3.1 Introduction to Decimals

OBJECTIVE A To write decimals in standard form and in words

The price tag on a sweater reads $61.88. The number 61.88 is in **decimal notation.** A number written in decimal notation is often called simply a **decimal.**

A number written in decimal notation has three parts.

61	.	88
Whole-number part	**Decimal point**	**Decimal part**

The decimal part of the number represents a number less than 1. For example, $.88 is less than $1. The decimal point (.) separates the whole-number part from the decimal part.

The position of a digit in a decimal determines the digit's place value. The place-value chart is extended to the right to show the place value of digits to the right of a decimal point.

In the decimal 458.302719, the position of the digit 7 determines that its place value is ten-thousandths.

Note the relationship between fractions and numbers written in decimal notation.

Seven tenths	Seven hundredths	Seven thousandths
$\dfrac{7}{10} = 0.7$	$\dfrac{7}{100} = 0.07$	$\dfrac{7}{1000} = 0.007$
1 zero in 10	2 zeros in 100	3 zeros in 1000
1 decimal place in 0.7	2 decimal places in 0.07	3 decimal places in 0.007

To write a decimal in words, write the decimal part of the number as though it were a whole number, and then name the place value of the last digit.

0.9684   Nine thousand six hundred eighty-four ten-thousandths

The decimal point in a decimal is read as "and."

372.516   Three hundred seventy-two and five hundred sixteen thousandths

$$\frac{3}{4} \cdot \frac{3}{4} = \frac{9}{12}$$

$$\frac{16}{16} \over 32$$

$$\frac{11}{16} - \frac{9}{16} + \frac{7}{12} =$$

Reduce $\frac{2}{16} + \frac{7}{12} = 9$ (24) Common factor.

Evaluate = ÷

q = X,

$\frac{124}{2}$ → $2\overline{)50}$

$1 = \frac{2}{2}$

$\frac{27}{3}$ → $8\overline{)1}$

$\frac{11}{20} - \frac{5}{12} =$

$\frac{3}{60} \times \frac{6}{7} = \frac{3}{35}$

$\frac{x}{60}$

$2L$

$24\frac{1}{4} = \frac{25}{20}$

$- 6\frac{3}{10} = \frac{6}{20}$

$\frac{19}{20}$

$7\frac{5}{7} \times 5\frac{4}{9} =$

$\frac{54}{7} \times \frac{49}{9} = \frac{42}{1} = 42$

$\frac{3}{1} \times \frac{521}{1} = \frac{1,563}{1} = 1,563$

$\frac{2}{9} \div \frac{5}{24} =$

$\frac{2}{9} \times \frac{24}{5} = \frac{6}{5} = 1\frac{1}{5}$

$\frac{2}{9} = \frac{1}{51}$

$9\frac{3}{5} \div 1\frac{3}{5} =$

$\frac{48}{5} \div \frac{8}{5} =$

$\frac{48}{5} \times \frac{5}{8} = \frac{6}{1} = 6$

$\frac{168}{-42}$ → $\frac{126}{1} \div 5\frac{1}{2} =$

$\frac{23}{1} \times \frac{1}{1} = 23$ lots

$\frac{1}{3} \cdot \frac{1}{3} \cdot \frac{1}{3} = \frac{1}{27} \times \frac{9}{5} \cdot \frac{9}{5} = \frac{81}{25}$

$\frac{1}{27}$

$\frac{1}{3} \cdot \frac{1}{3} \cdot \frac{1}{3} \cdot \frac{1}{3} = \frac{1}{81}$

$\frac{1}{81} \times \frac{81}{25} = \frac{1}{25} =$

$\frac{5}{7} < \frac{6}{7}$ = Cross Multiplying

$\begin{array}{cc} 54 & 55 \\ 9/11 & < 5/6 \end{array}$ =

Comparison.

The Bigger fraction   Sec. 2.8 (A) - pg. 111, 1-13

$\overset{55}{5/9} > \overset{54}{6/11}$

Left to Right

$\times \begin{smallmatrix}16\\4\\\hline 64\end{smallmatrix}$ $\begin{smallmatrix}2,6\\\times 4\\\hline 64\end{smallmatrix}$ $\begin{smallmatrix}5\text{te }4\text{?}\\\times 7\\\hline 343\end{smallmatrix}$

Math

Ex.

Practice test on turd $7/5 \div 3/8$ $4/5 = 2/5 \div \frac{3}{10}$

$(2\frac{1}{3})^2 = \frac{7}{3} \cdot \frac{7}{3} = \frac{7}{3}$

$\frac{2}{5} \div \frac{3}{10} =$

$\frac{2}{5} \times \frac{10}{3} = \frac{4}{3} = 1\frac{1}{3}$

1.) $\left(\frac{4}{7}\right)^2 = \frac{4}{7} \times \frac{4}{7} = \frac{16}{49} =$

$2/5 \cdot 8/3 \cdot 4/5 = \frac{64}{75}$

$3/9 \times 4/5 = \frac{12}{45} =$

2.) $\left(\frac{2}{3}\right)^3 \cdot \left(1\frac{1}{8}\right)^2 =$ 328 $\frac{420}{1000}$

$\frac{16}{27} \cdot \left(\frac{9}{8} \cdot \frac{9}{8}\right) = \frac{81}{64} =$

Always Work
Common Factor

$\boxed{2/3 \cdot 2/3} \cdot 2/3 \times 9/8 \cdot 9/8 =$

Whole #

Homework pg. 112

31-50

$\left(\frac{1}{3} + \frac{5}{24}\right) \div \frac{5}{9} = \frac{4}{6} + \frac{5}{6} = $ $9/6 \div \frac{5}{9} =$

$5/7 \div 9/7$ $9/42 \times 9/5 = \frac{27}{10} = 2\frac{7}{10}$

Add & Sub = Rank
X & ÷ = Equal Rank

42.03 =
forty-two & three hundredths

208.023 =
two hundred, eight and twenty three ~~hundredths~~ Thousandths

405,000. 026 = 206

400,005. 0206

The decimal point did not make its appearance until the early 1600s. Stevin's notation used subscripts with circles around them after each digit: 0 for ones, 1 for tenths (which he called "primes"), 2 for hundredths (called "seconds"), 3 for thousandths ("thirds"), and so on. For example, 1.375 would have been written

1 3 7 5
⓪ ① ② ③

To write a decimal in standard form when it is written in words, write the whole-number part, replace the word *and* with a decimal point, and write the decimal part so that the last digit is in the given place-value position.

Four and twenty-three <u>hundredths</u>
3 is in the hundredths place.

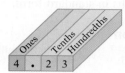

When writing a decimal in standard form, you may need to insert zeros after the decimal point so that the last digit is in the given place-value position.

Ninety-one and eight <u>thousandths</u>
8 is in the thousandths place.
Insert two zeros so that the 8 is in the thousandths place.

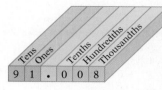

Sixty-five <u>ten-thousandths</u>
5 is in the ten-thousandths place.
Insert two zeros so that the 5 is in the ten-thousandths place.

EXAMPLE · 1

Name the place value of the digit 8 in the number 45.687.

Solution
The digit 8 is in the hundredths place.

YOU TRY IT · 1

Name the place value of the digit 4 in the number 907.1342.

Your solution

EXAMPLE · 2

Write $\dfrac{43}{100}$ as a decimal.

Solution
$\dfrac{43}{100} = 0.43$ • **Forty-three hundredths**

YOU TRY IT · 2

Write $\dfrac{501}{1000}$ as a decimal.

Your solution

EXAMPLE · 3

Write 0.289 as a fraction.

Solution
$0.289 = \dfrac{289}{1000}$ • **289 thousandths**

YOU TRY IT · 3

Write 0.67 as a fraction.

Your solution

EXAMPLE · 4

Write 293.50816 in words.

Solution
Two hundred ninety-three and fifty thousand eight hundred sixteen hundred-thousandths

YOU TRY IT · 4

Write 55.6083 in words.

Your solution

Solutions on p. S8

EXAMPLE · 5

Write twenty-three and two hundred forty-seven millionths in standard form.

Solution

23.000247   • **7 is in the millionths place.**

YOU TRY IT · 5

Write eight hundred six and four hundred ninety-one hundred-thousandths in standard form.

Your solution

Solution on p. S8

OBJECTIVE B   **To round a decimal to a given place value**

 Tips for Success

Have you considered joining a study group? Getting together regularly with other students in the class to go over material and quiz each other can be very beneficial. See *AIM for Success* at the front of the book.

In general, rounding decimals is similar to rounding whole numbers except that the digits to the right of the given place value are dropped instead of being replaced by zeros.

If the digit to the right of the given place value is less than 5, that digit and all digits to the right are dropped.

Round 6.9237 to the nearest hundredth.

      ┌─── Given place value (hundredths)

6.9237

      └── 3 < 5   Drop the digits 3 and 7.

6.9237 rounded to the nearest hundredth is 6.92.

If the digit to the right of the given place value is greater than or equal to 5, increase the digit in the given place value by 1, and drop all digits to its right.

Round 12.385 to the nearest tenth.

      ┌─── Given place value (tenths)

12.385

      └── 8 > 5   Increase 3 by 1 and drop all digits to the right of 3.

12.385 rounded to the nearest tenth is 12.4.

 Take Note

In the example at the right, the zero in the given place value is not dropped. This indicates that the number is rounded to the nearest thousandth. If we dropped the zero and wrote 0.47, it would indicate that the number was rounded to the nearest hundredth.

HOW TO · 1   Round 0.46972 to the nearest thousandth.

      ┌─── Given place value (thousandths)

0.46972

      └── 7 > 5 Round up by adding 1 to the 9 (9 + 1 = 10). Carry the 1 to the hundredths place (6 + 1 = 7).

0.46972 rounded to the nearest thousandth is 0.470.

EXAMPLE • 6

Round 0.9375 to the nearest thousandth.

Solution

┌── Given place value

0.9375

└── 5 = 5

0.9375 rounded to the nearest thousandth is 0.938.

YOU TRY IT • 6

Round 3.675849 to the nearest ten-thousandth.

Your solution

EXAMPLE • 7

Round 2.5963 to the nearest hundredth.

Solution

┌── Given place value

2.5963

└── 6 > 5

2.5963 rounded to the nearest hundredth is 2.60.

YOU TRY IT • 7

Round 48.907 to the nearest tenth.

Your solution

EXAMPLE • 8

Round 72.416 to the nearest whole number.

Solution

┌── Given place value

72.416

└── 4 < 5

72.416 rounded to the nearest whole number is 72.

YOU TRY IT • 8

Round 31.8652 to the nearest whole number.

Your solution

EXAMPLE • 9

On average, an American goes to the movies 4.56 times per year. To the nearest whole number, how many times per year does an American go to the movies?

Solution

4.56 rounded to the nearest whole number is 5.
An American goes to the movies about 5 times per year.

YOU TRY IT • 9

One of the driest cities in the Southwest is Yuma, Arizona, with an average annual precipitation of 2.65 inches. To the nearest inch, what is the average annual precipitation in Yuma?

Your solution

Solutions on p. S8

3.1 EXERCISES

OBJECTIVE A　　To write decimals in standard form and in words

For Exercises 1 to 6, name the place value of the digit 5.

1.　76.31587　　　　　**2.**　291.508　　　　　**3.**　432.09157

4.　0.0006512　　　　**5.**　38.2591　　　　　**6.**　0.0000853

For Exercises 7 to 12, write the fraction as a decimal.

7. $\dfrac{3}{10}$　　**8.** $\dfrac{9}{10}$　　**9.** $\dfrac{21}{100}$　　**10.** $\dfrac{87}{100}$　　**11.** $\dfrac{461}{1000}$　　**12.** $\dfrac{853}{1000}$

For Exercises 13 to 18, write the decimal as a fraction.

13.　0.1　　**14.**　0.3　　**15.**　0.47　　**16.**　0.59　　**17.**　0.289　　**18.**　0.601

For Exercises 19 to 27, write the number in words.

19.　0.37　　　　　　**20.**　25.6　　　　　　**21.**　9.4

22.　1.004　　　　　**23.**　0.0053　　　　　**24.**　41.108

25.　0.045　　　　　**26.**　3.157　　　　　**27.**　26.04

For Exercises 28 to 35, write the number in standard form.

28.　Six hundred seventy-two thousandths

29.　Three and eight hundred six ten-thousandths

30.　Nine and four hundred seven
ten-thousandths

31.　Four hundred seven and three hundredths

32.　Six hundred twelve and seven hundred
four thousandths

33.　Two hundred forty-six and twenty-four
thousandths

34.　Two thousand sixty-seven and nine thousand
two ten-thousandths

35.　Seventy-three and two thousand six hundred
eighty-four hundred-thousandths

36. Suppose the first nonzero digit to the right of the decimal point in a decimal number is in the hundredths place. If the number has three consecutive nonzero digits to the right of the decimal point, and all other digits are zero, what place value names the number?

OBJECTIVE B To round a decimal to a given place value

For Exercises 37 to 51, round the number to the given place value.

37. 6.249   Tenths

38. 5.398   Tenths

39. 21.007   Tenths

40. 30.0092   Tenths

41. 18.40937   Hundredths

42. 413.5972   Hundredths

43. 72.4983   Hundredths

44. 6.061745   Thousandths

45. 936.2905   Thousandths

46. 96.8027   Whole number

47. 47.3192   Whole number

48. 5439.83   Whole number

49. 7014.96   Whole number

50. 0.023591   Ten-thousandths

51. 2.975268   Hundred-thousandths

52. **Measurement** A nickel weighs about 0.1763668 ounce. Find the weight of a nickel to the nearest hundredth of an ounce.

53. **Sports** Runners in the Boston Marathon run a distance of 26.21875 miles. To the nearest tenth of a mile, find the distance that an entrant who completes the Boston Marathon runs.

AFP/Getty Images

For Exercises 54 and 55, give an example of a decimal number that satisfies the given condition.

54. The number rounded to the nearest tenth is greater than the number rounded to the nearest hundredth.

55. The number rounded to the nearest hundredth is equal to the number rounded to the nearest thousandth.

Applying the Concepts

56. Indicate which digits of the number, if any, need not be entered on a calculator.
   a. 1.500   **b.** 0.908   **c.** 60.07   **d.** 0.0032

57. **a.** Find a number between 0.1 and 0.2.   **b.** Find a number between 1 and 1.1.
   c. Find a number between 0 and 0.005.

SECTION

3.2　Addition of Decimals

OBJECTIVE A　　To add decimals

To add decimals, write the numbers so that the decimal points are on a vertical line. Add as for whole numbers, and write the decimal point in the sum directly below the decimal points in the addends.

HOW TO • 1　　Add: 0.237 + 4.9 + 27.32

Note that by placing the decimal points on a vertical line, we make sure that digits of the same place value are added.

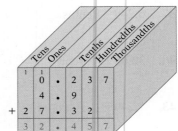

EXAMPLE • 1

Find the sum of 42.3, 162.903, and 65.0729.

Solution

$$
\begin{array}{r}
\overset{1\,1\,1}{}42.3 \\
162.903 \\
+\ \ 65.0729 \\
\hline
270.2759
\end{array}
$$

• Place the decimal points on a vertical line.

EXAMPLE • 2

Add: 0.83 + 7.942 + 15

Solution

$$
\begin{array}{r}
\overset{1\,1}{}0.83 \\
7.942 \\
+15. \\
\hline
23.772
\end{array}
$$

YOU TRY IT • 1

Find the sum of 4.62, 27.9, and 0.62054.

Your solution

YOU TRY IT • 2

Add: 6.05 + 12 + 0.374

Your solution

Solutions on p. S8

ESTIMATION

Estimating the Sum of Two or More Decimals

Calculate 23.037 + 16.7892. Then use estimation to determine whether the sum is reasonable.

Add to find the exact sum.　　23.037 **+** 16.7892 **=** 39.8262

To estimate the sum, round each number to the same place value. Here we have rounded to the nearest whole number. Then add. The estimated answer is 40, which is very close to the exact sum, 39.8262.

$$
\begin{array}{r}
23.037 \approx\ \ 23 \\
+16.7892 \approx +17 \\
\hline
40
\end{array}
$$

OBJECTIVE B   To solve application problems

The graph at the right shows the breakdown by age group of Americans who are hearing-impaired. Use this graph for Example 3 and You Try It 3.

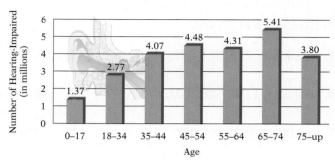

Breakdown by Age Group of Americans Who Are Hearing-Impaired
Source: American Speech-Language-Hearing Association

EXAMPLE · 3

Determine the number of Americans under the age of 45 who are hearing-impaired.

Strategy
To determine the number, add the numbers of hearing impaired ages 0 to 17, 18 to 34, and 35 to 44.

Solution

```
  1.37
  2.77
+ 4.07
------
  8.21
```

8.21 million Americans under the age of 45 are hearing-impaired.

YOU TRY IT · 3

Determine the number of Americans ages 45 and older who are hearing-impaired.

Your strategy

Your solution

EXAMPLE · 4

Dan Burhoe earned a salary of $210.48 for working 3 days this week as a food server. He also received $82.75, $75.80, and $99.25 in tips during the 3 days. Find his total income for the 3 days of work.

Strategy
To find the total income, add the tips (82.75, 75.80, and 99.25) to the salary (210.48).

Solution
210.48 + 82.75 + 75.80 + 99.25 = 468.28

Dan's total income for the 3 days of work was $468.28.

YOU TRY IT · 4

Anita Khavari, an insurance executive, earns a salary of $875 every 4 weeks. During the past 4-week period, she received commissions of $985.80, $791.46, $829.75, and $635.42. Find her total income for the past 4-week period.

Your strategy

Your solution

Solutions on p. S8

3.2 EXERCISES

OBJECTIVE A   To add decimals

For Exercises 1 to 17, add.

1. 16.008 + 2.0385 + 132.06

2. 17.32 + 1.0579 + 16.5

3. 1.792 + 67 + 27.0526

4. 8.772 + 1.09 + 26.5027

5. 3.02 + 62.7 + 3.924

6. 9.06 + 4.976 + 59.6

7. 82.006 + 9.95 + 0.927

8. 0.826 + 8.76 + 79.005

9. 4.307 + 99.82 + 9.078

10.
    0.3
  + 0.07

11.
    0.29
  + 0.4

12.
    1.007
  + 2.1

13.
    7.3
  + 9.005

14.
    4.9257
    27.05
  + 9.0063

15.
    8.72
    99.073
  + 2.9736

16.
    62.4
    9.827
  + 692.44

17.
    8
    89.43
  + 7.0659

 For Exercises 18 to 21, use a calculator to add. Then round the numbers to the nearest whole number and use estimation to determine whether the sum you calculated is reasonable.

18.
    342.42
    89.625
  + 176.2

19.
    219.9
    0.872
  + 13.42

20.
    823.9
    82.65
  + 46.923

21.
    678.92
    97.6
  + 5.423

 22. For a certain decimal addition problem, each addend rounded to the nearest whole number is greater than the addend itself. Must the sum of the rounded numbers be greater than the exact sum?

 23. If none of the addends of a decimal addition problem is a whole number, is it possible for the sum to be a whole number?

OBJECTIVE B   To solve application problems

24. Mechanics Find the length of the shaft.

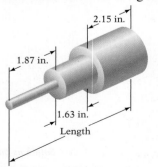

2.15 in.
1.87 in.
1.63 in.
Length

25. Mechanics Find the length of the shaft.

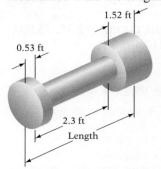

1.52 ft
0.53 ft
2.3 ft
Length

26. **Banking** You have $2143.57 in your checking account. You make deposits of $210.98, $45.32, $1236.34, and $27.99. Find the amount in your checking account after you have made the deposits if no money has been withdrawn.

27. **Geometry** The perimeter of a triangle is the sum of the lengths of the three sides of the triangle. Find the perimeter of a triangle that has sides that measure 4.9 meters, 6.1 meters, and 7.5 meters.

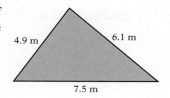

28. **Demography** The world's population in 2050 is expected to be 8.9 billion people. It is projected that in that year, Asia's population will be 5.3 billion and Africa's population will be 1.8 billion. What are the combined populations of Asia and Africa expected to be in 2050? (*Source:* United Nations Population Division, World Population Prospects)

29. **TV Viewership** The table at the right shows the numbers of viewers, in millions, of three network evening news programs for the week of January 28 to February 1, 2008. Calculate the total number of people who watched these three news programs that week.

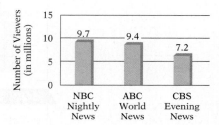

30. **The Stock Market** On May 1, 2008, the Dow Jones Industrial Average climbed 189.87 points after starting the day at 12,820.13. The Nasdaq Composite started the day at 2412.80 and rose 67.91 points during the day. The Standard & Poor 500 Index began the day at 1385.59 and ended the day 23.75 points higher. Find the values of **a.** the Dow Jones Industrial Average, **b.** the Nasdaq Composite, and **c.** the Standard & Poor 500 Index at the end of the trading day on May 1, 2008.

31. **Measurement** Can a piece of rope 4 feet long be wrapped around the box shown at the right?

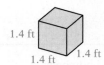

Applying the Concepts

Consumerism The table at the right gives the prices for selected products in a grocery store. Use this table for Exercises 32 and 33.

32. Does a customer with $10 have enough money to purchase raisin bran, bread, milk, and butter?

33. Name three items that would cost more than $8 but less than $9. (There is more than one answer.)

Product	Cost
Raisin bran	$3.29
Butter	$2.79
Bread	$1.99
Popcorn	$2.19
Potatoes	$3.49
Cola (6-pack)	$2.99
Mayonnaise	$3.99
Lunch meat	$3.39
Milk	$2.59
Toothpaste	$2.69

3.3　Subtraction of Decimals

OBJECTIVE A　　To subtract decimals

To subtract decimals, write the numbers so that the decimal points are on a vertical line. Subtract as for whole numbers, and write the decimal point in the difference directly below the decimal point in the subtrahend.

HOW TO · 1　Subtract $21.532 - 9.875$ and check.

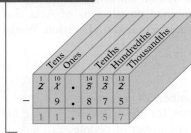

Placing the decimal points on a vertical line ensures that digits of the same place value are subtracted.

Check:

Subtrahend	9.875
+ Difference	+ 11.657
= Minuend	21.532

HOW TO · 2　Subtract $4.3 - 1.7942$ and check.

$$\begin{array}{r} \overset{3\;\;12\;9\;9\;10}{4.3000} \\ -\;1.7942 \\ \hline 2.5058 \end{array}$$

If necessary, insert zeros in the minuend before subtracting.

Check:

$$\begin{array}{r} 1.7942 \\ + 2.5058 \\ \hline 4.3000 \end{array}$$

EXAMPLE · 1

Subtract $39.047 - 7.96$ and check.

Solution

$$\begin{array}{r} \overset{8\;\;9\;14}{3\,9.0\,4\,7} \\ -\;\;\;7.9\,6 \\ \hline 3\,1.0\,8\,7 \end{array}$$

Check:

$$\begin{array}{r} 7.96 \\ + 31.087 \\ \hline 39.047 \end{array}$$

YOU TRY IT · 1

Subtract $72.039 - 8.47$ and check.

Your solution

EXAMPLE · 2

Find 9.23 less than 29 and check.

Solution

$$\begin{array}{r} \overset{1\;18\;\;9\;10}{2\,9.0\,0} \\ -\;\;9.2\,3 \\ \hline 1\,9.7\,7 \end{array}$$

Check:

$$\begin{array}{r} 9.23 \\ + 19.77 \\ \hline 29.00 \end{array}$$

YOU TRY IT · 2

Subtract $35 - 9.67$ and check.

Your solution

EXAMPLE · 3

Subtract $1.2 - 0.8235$ and check.

Solution

$$\begin{array}{r} \overset{0\;\;11\;9\;9\;10}{1.2\,0\,0\,0} \\ -\;0.8\,2\,3\,5 \\ \hline 0.3\,7\,6\,5 \end{array}$$

Check:

$$\begin{array}{r} 0.8235 \\ + 0.3765 \\ \hline 1.2000 \end{array}$$

YOU TRY IT · 3

Subtract $3.7 - 1.9715$ and check.

Your solution

Solutions on pp. S8–S9

ESTIMATION

Estimating the Difference Between Two Decimals

Calculate $820.23 - 475.748$. Then use estimation to determine whether the difference is reasonable.

Subtract to find the exact difference.

$$820.23 \; \boxed{-} \; 475.748 \; \boxed{=} \; 344.482$$

To estimate the difference, round each number to the same place value. Here we have rounded to the nearest ten. Then subtract. The estimated answer is 340, which is very close to the exact difference, 344.482.

$$\begin{array}{r} 820.23 \approx 820 \\ -475.748 \approx -480 \\ \hline 340 \end{array}$$

OBJECTIVE B   To solve application problems

EXAMPLE · 4

You bought a book for $15.87. How much change did you receive from a $20.00 bill?

Strategy

To find the amount of change, subtract the cost of the book (15.87) from $20.00.

Solution

$$\begin{array}{r} 20.00 \\ -15.87 \\ \hline 4.13 \end{array}$$

You received $4.13 in change.

YOU TRY IT · 4

Your breakfast cost $6.85. How much change did you receive from a $10.00 bill?

Your strategy

Your solution

EXAMPLE · 5

You had a balance of $87.93 on your student debit card. You then used the card, deducting $15.99 for a CD, $6.85 for lunch, and $28.50 for a ticket to the football game. What is your new student debit card balance?

Strategy

To find your new debit card balance:
• Add to find the total of the three deductions (15.99 + 6.85 + 28.50).
• Subtract the total of the three deductions from the old balance (87.93).

Solution

$$\begin{array}{r} 15.99 \\ 6.85 \\ +28.50 \\ \hline 51.34 \end{array} \text{ total of deductions} \qquad \begin{array}{r} 87.93 \\ -51.34 \\ \hline 36.59 \end{array}$$

Your new debit card balance is $36.59.

YOU TRY IT · 5

You had a balance of $2472.69 in your checking account. You then wrote checks for $1025.60, $79.85, and $162.47. Find the new balance in your checking account.

Your strategy

Your solution

Solutions on p. S9

3.3 EXERCISES

OBJECTIVE A   To subtract decimals

For Exercises 1 to 24, subtract and check.

1. 24.037 − 18.41

2. 26.029 − 19.31

3. 123.07 − 9.4273

4. 214 − 7.143

5. 16.5 − 9.7902

6. 13.2 − 8.6205

7. 235.79 − 20.093

8. 463.27 − 40.095

9. 63.005 − 9.1274

10. 23.004 − 7.2175

11. 92 − 19.2909

12. 41.2405 − 25.2709

13.   0.32
      − 0.0058

14.   0.78
      − 0.0073

15.   3.005
      − 1.982

16.   6.007
      − 2.734

17.   352.16
      − 90.994

18.   872
      − 80.753

19.   724.32
      − 69

20.   625.46
      − 77.509

21.   362.394
      − 19.4672

22.   421.385
      − 17.5293

23.   19
      − 10.372

24.   23.4
      − 0.921

For Exercises 25 to 27, use the relationship between addition and subtraction to write the subtraction problem you would use to find the missing addend.

25. _____ + 2.325 = 7.01   **26.** 5.392 + _____ = 8.07   **27.** _____ + 8.967 = 19.35

For Exercises 28 to 31, use a calculator to subtract. Then round the numbers to the nearest whole number and use estimation to determine whether the difference you calculated is reasonable.

28.   93.079256
      − 66.09249

29.   3.7529
      − 1.00784

30.   76.53902
      − 45.73005

31.   9.07325
      − 1.924

OBJECTIVE B   To solve application problems

32. Mechanics Find the missing dimension.

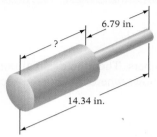

6.79 in.

?

14.34 in.

PPT

33. Mechanics Find the missing dimension.

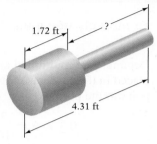

1.72 ft

?

4.31 ft

PPT

34. Business The manager of the Edgewater Cafe takes a reading of the cash register tape each hour. At 1:00 P.M. the tape read $967.54. At 2:00 P.M. the tape read $1437.15. Find the amount of sales between 1:00 P.M. and 2:00 P.M.

35. Moviegoing The graph at the right shows the average annual numbers of theater tickets sold each decade. Find the difference between the average annual number of theater tickets sold in the 1990s and in the 1970s.

36. Coal In a recent year, 1.163 billion tons of coal were produced in the United States. In the same year, U.S. consumption of coal was 1.112 billion tons. (*Source:* Department of Energy) How many more million tons of coal were produced than were consumed that year?

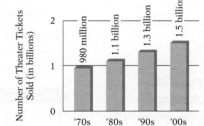

Number of Theater Tickets Sold (in billions)

980 million   1.1 billion   1.3 billion   1.5 billion

'70s   '80s   '90s   '00s

Average Annual Number of Theater Tickets Sold Each Decade

Source: National Association of Theater Owners

37. Super Bowl Super Bowl XLII was watched on the Fox network by 97.4 million people. On the same network, 63.9 million people watched the Super Bowl post-game show. (*Source:* Nielsen Network Research) How many more people watched Super Bowl XLII than watched the Super Bowl post-game show?

38. You have $30 to spend, and you make purchases that cost $6.74 and $13.68. Which expressions correctly represent the amount of money you have left?
(i) 30 − 6.74 + 13.68   (ii) (6.74 + 13.68) − 30
(iii) 30 − (6.74 + 13.68)   (iv) 30 − 6.74 − 13.68

Paul Spinelli/Getty Images

Applying the Concepts

39. Find the largest amount by which the estimate of the sum of two decimals rounded to the given place value could differ from the exact sum.
a. Tenths   **b.** Hundredths   **c.** Thousandths

SECTION

3.4   Multiplication of Decimals

OBJECTIVE A   To multiply decimals

Point of Interest

Benjamin Banneker (1731–1806) was the first African American to earn distinction as a mathematician and scientist. He was on the survey team that determined the boundaries of Washington, D.C. The mathematics of surveying requires extensive use of decimals.

Decimals are multiplied as though they were whole numbers. Then the decimal point is placed in the product. Writing the decimals as fractions shows where to write the decimal point in the product.

$$0.\underline{3} \times 5 = \frac{3}{10} \times \frac{5}{1} = \frac{15}{10} = 1.\underline{5}$$

1 decimal place   1 decimal place

$$0.\underline{3} \times 0.\underline{5} = \frac{3}{10} \times \frac{5}{10} = \frac{15}{100} = 0.\underline{15}$$

1 decimal place   1 decimal place   2 decimal places

$$0.\underline{3} \times 0.\underline{05} = \frac{3}{10} \times \frac{5}{100} = \frac{15}{1000} = 0.\underline{015}$$

1 decimal place   2 decimal places   3 decimal places

To multiply decimals, multiply the numbers as with whole numbers. Write the decimal point in the product so that the number of decimal places in the product is the sum of the decimal places in the factors.

Integrating Technology

Scientific calculators have a floating decimal point. This means that the decimal point is automatically placed in the answer. For example, for the product at the right, enter

21 · 4 **x** 0 · 36 **=**

The display reads 7.704, with the decimal point in the correct position.

HOW TO 1   Multiply: 21.4×0.36

$$\begin{array}{r} 21.4 \\ \times\ 0.36 \\ \hline 1284 \\ 642\ \ \\ \hline 7.704 \end{array}$$

21.4 — 1 decimal place
× 0.36 — 2 decimal places
7.704 — 3 decimal places

HOW TO 2   Multiply: 0.037×0.08

$$\begin{array}{r} 0.037 \\ \times\ \ 0.08 \\ \hline 0.00296 \end{array}$$

0.037 — 3 decimal places
× 0.08 — 2 decimal places
0.00296 — 5 decimal places

• **Two zeros must be inserted between the 2 and the decimal point so that there are 5 decimal places in the product.**

To multiply a decimal by a power of 10 (10, 100, 1000, . . .), move the decimal point to the right the same number of places as there are zeros in the power of 10.

$3.8925 \times 1\underline{0}\ = 38.925$

1 zero   1 decimal place

$3.8925 \times 1\underline{00}\ = 389.25$

2 zeros   2 decimal places

$3.8925 \times 1\underline{000}\ = 3892.5$

3 zeros   3 decimal places

$3.8925 \times 1\underline{0,000}\ = 38,925.$

4 zeros   4 decimal places

$3.8925 \times 1\underline{00,000}\ = 389,250.$

5 zeros   5 decimal places

• Note that a zero must be inserted before the decimal point.

Note that if the power of 10 is written in exponential notation, the exponent indicates how many places to move the decimal point.

$$3.8925 \times 10^1 = 38.925$$
1 decimal place

$$3.8925 \times 10^2 = 389.25$$
2 decimal places

$$3.8925 \times 10^3 = 3892.5$$
3 decimal places

$$3.8925 \times 10^4 = 38,925.$$
4 decimal places

$$3.8925 \times 10^5 = 389,250.$$
5 decimal places

EXAMPLE · 1

Multiply: 920×3.7

Solution

$$
\begin{array}{r}
920 \\
\times \quad 3.7 \\
\hline
644\ 0 \\
2760 \\
\hline
3404.0
\end{array}
$$

• 1 decimal place

• 1 decimal place

YOU TRY IT · 1

Multiply: 870×4.6

Your solution

EXAMPLE · 2

Find 0.00079 multiplied by 0.025.

Solution

$$
\begin{array}{r}
0.00079 \\
\times \quad 0.025 \\
\hline
395 \\
158 \\
\hline
0.00001975
\end{array}
$$

• 5 decimal places
• 3 decimal places

• 8 decimal places

YOU TRY IT · 2

Find 0.000086 multiplied by 0.057.

Your solution

EXAMPLE · 3

Find the product of 3.69 and 2.07.

Solution

$$
\begin{array}{r}
3.69 \\
\times \quad 2.07 \\
\hline
2583 \\
7380 \\
\hline
7.6383
\end{array}
$$

• 2 decimal places
• 2 decimal places

• 4 decimal places

YOU TRY IT · 3

Find the product of 4.68 and 6.03.

Your solution

EXAMPLE · 4

Multiply: $42.07 \times 10,000$

Solution

$42.07 \times 10,000 = 420,700$

YOU TRY IT · 4

Multiply: 6.9×1000

Your solution

EXAMPLE · 5

Find 3.01 times 10^3.

Solution

$3.01 \times 10^3 = 3010$

YOU TRY IT · 5

Find 4.0273 times 10^2.

Your solution

Solutions on p. S9

ESTIMATION

Estimating the Product of Two Decimals

Calculate 28.259 × 0.029. Then use estimation to determine whether the product is reasonable.

Multiply to find the exact product.

To estimate the product, round each number so that it contains one nonzero digit. Then multiply. The estimated answer is 0.90, which is very close to the exact product, 0.819511.

28.259 **x** 0.029 **=** 0.819511

$$28.259 \approx 30$$
$$\underline{\times 0.029 \approx \times 0.03}$$
$$0.90$$

OBJECTIVE B   To solve application problems

The tables that follow list water rates and meter fees for a city. These tables are used for Example 6 and You Try It 6.

Water Charges	
Commercial	$1.39/1000 gal
Comm Restaurant	$1.39/1000 gal
Industrial	$1.39/1000 gal
Institutional	$1.39/1000 gal
Res—No Sewer	
Residential—SF	
>0 <200 gal per day	$1.15/1000 gal
>200 <1500 gal per day	$1.39/1000 gal
>1500 gal per day	$1.54/1000 gal

Meter Charges	
Meter	**Meter Fee**
5/8" & 3/4"	$13.50
1"	$21.80
1-1/2"	$42.50
2"	$67.20
3"	$133.70
4"	$208.20
6"	$415.10
8"	$663.70

EXAMPLE · 6

Find the total bill for an industrial water user with a 6-inch meter that used 152,000 gallons of water for July and August.

Strategy

To find the total cost of water:
• Find the cost of water by multiplying the cost per 1000 gallons (1.39) by the number of 1000-gallon units used.
• Add the cost of the water to the meter fee (415.10).

Solution

$$\text{Cost of water} = \frac{152,000}{1000} \cdot 1.39 = 211.28$$

Total cost = 211.28 + 415.10 = 626.38

The total cost is $626.38.

YOU TRY IT · 6

Find the total bill for a commercial user that used 5000 gallons of water per day for July and August. The user has a 3-inch meter.

Your strategy

Your solution

Solution on p. S9

EXAMPLE • 7

It costs $.036 an hour to operate an electric motor. How much does it cost to operate the motor for 120 hours?

Strategy

To find the cost of running the motor for 120 hours, multiply the hourly cost (0.036) by the number of hours the motor is run (120).

Solution

$$
\begin{array}{r}
0.036 \\
\times\ \ \ 120 \\
\hline
720 \\
36\ \ \ \\
\hline
4.320
\end{array}
$$

The cost of running the motor for 120 hours is $4.32.

YOU TRY IT • 7

The cost of electricity to run a freezer for 1 hour is $.035. This month the freezer has run for 210 hours. Find the total cost of running the freezer this month.

Your strategy

Your solution

EXAMPLE • 8

Jason Ng earns a salary of $440 for a 40-hour workweek. This week he worked 12 hours of overtime at a rate of $16.50 for each hour of overtime worked. Find his total income for the week.

Strategy

To find Jason's total income for the week:

• Find the overtime pay by multiplying the hourly overtime rate (16.50) by the number of hours of overtime worked (12).
• Add the overtime pay to the weekly salary (440).

Solution

$$
\begin{array}{r}
16.50 \\
\times\ \ \ 12 \\
\hline
33\ 00 \\
165\ 0\ \ \\
\hline
198.00
\end{array}
\quad
\begin{array}{r}
440.00 \\
+\ 198.00 \\
\hline
638.00
\end{array}
$$

198.00 Overtime pay

Jason's total income for the week is $638.00.

YOU TRY IT • 8

You make a down payment of $175 on an electronic drum kit and agree to make payments of $37.18 a month for the next 18 months to repay the remaining balance. Find the total cost of the electronic drum kit.

Your strategy

Your solution

Solutions on p. S9

3.4 EXERCISES

OBJECTIVE A   To multiply decimals

For Exercises 1 to 73, multiply.

1.   0.9
     × 0.4

2.   0.7
     × 0.9

3.   0.5
     × 0.5

4.   0.7
     × 0.7

5.   7.7
     × 0.9

6.   3.4
     × 0.4

7.   9.2
     × 0.2

8.   2.6
     × 0.7

9.   7.4
     × 0.1

10.   3.8
      × 0.1

11.   7.9
      × 5

12.   9.3
      × 7

13.   0.68
      ×   4

14.   0.83
      ×   9

15.   0.67
      × 0.9

16.   0.84
      × 0.3

17.   2.5
      × 5.4

18.   3.9
      × 1.9

19.   0.83
      × 5.2

20.   0.24
      × 2.7

21.   1.47
      × 0.09

22.   6.37
      × 0.05

23.   8.92
      × 0.004

24.   6.75
      × 0.007

25.   0.49
      × 0.16

26.   0.38
      × 0.21

27.   7.6
      × 0.01

28.   5.1
      × 0.01

29.   8.62
      ×   4

30.   5.83
      ×   7

31.   64.5
      ×   9

32.   37.8
      ×   8

33.   2.19
      × 9.2

34.   1.25
      × 5.6

35.   1.85
      × 0.023

36.   37.8
      × 0.052

37.   0.478
      × 0.37

38.   0.526
      × 0.22

39.   48.3
      × 0.0041

40.   67.2
      × 0.0086

41. 0.413
       × 0.0016

42. 0.517
       × 0.0029

43. 8.005
       × 0.067

44. 9.032
       × 0.019

45. 4.29 × 0.1

46. 6.78 × 0.1

47. 5.29 × 0.4

48. 6.78 × 0.5

49. 0.68 × 0.7

50. 0.56 × 0.9

51. 1.4 × 0.73

52. 6.3 × 0.37

53. 5.2 × 7.3

54. 7.4 × 2.9

55. 3.8 × 0.61

56. 7.2 × 0.72

57. 0.32 × 10

58. 6.93 × 10

59. 0.065 × 100

60. 0.039 × 100

61. 6.2856 × 1000

62. 3.2954 × 1000

63. 3.2 × 1000

64. 0.006 × 10,000

65. 3.57 × 10,000

66. 8.52×10^1

67. 0.63×10^1

68. 82.9×10^2

69. 0.039×10^2

70. 6.8×10^3

71. 4.9×10^4

72. 6.83×10^4

73. 0.067×10^2

74. A number is rounded to the nearest thousandth. What is the smallest power of 10 the number must be multiplied by to give a product that is a whole number?

75. Find the product of 0.0035 and 3.45.

76. Find the product of 237 and 0.34.

77. Multiply 3.005 by 0.00392.

78. Multiply 20.34 by 1.008.

79. Multiply 1.348 by 0.23.

80. Multiply 0.000358 by 3.56.

81. Find the product of 23.67 and 0.0035.

82. Find the product of 0.00346 and 23.1.

83. Find the product of 5, 0.45, and 2.3.

84. Find the product of 0.03, 23, and 9.45.

For Exercises 85 to 96, use a calculator to multiply. Then use estimation to determine whether the product you calculated is reasonable.

85.
$$\begin{array}{r} 28.5 \\ \times\ \ 3.2 \\ \hline \end{array}$$

86.
$$\begin{array}{r} 86.3 \\ \times\ \ 4.4 \\ \hline \end{array}$$

87.
$$\begin{array}{r} 2.38 \\ \times\ 0.44 \\ \hline \end{array}$$

88.
$$\begin{array}{r} 9.82 \\ \times\ 0.77 \\ \hline \end{array}$$

89.
$$\begin{array}{r} 0.866 \\ \times\ \ \ 4.5 \\ \hline \end{array}$$

90.
$$\begin{array}{r} 0.239 \\ \times\ \ \ 8.2 \\ \hline \end{array}$$

91.
$$\begin{array}{r} 4.34 \\ \times\ 2.59 \\ \hline \end{array}$$

92.
$$\begin{array}{r} 6.87 \\ \times\ 9.98 \\ \hline \end{array}$$

93.
$$\begin{array}{r} 8.434 \\ \times\ 0.044 \\ \hline \end{array}$$

94.
$$\begin{array}{r} 7.037 \\ \times\ 0.094 \\ \hline \end{array}$$

95.
$$\begin{array}{r} 28.44 \\ \times\ \ 1.12 \\ \hline \end{array}$$

96.
$$\begin{array}{r} 86.57 \\ \times\ \ 7.33 \\ \hline \end{array}$$

OBJECTIVE B   **To solve application problems**

97. **Electricity** In the United States, a homeowner's average monthly bill for electricity is $95.66. (*Source:* Department of Energy) What is a U.S. homeowner's average annual cost of electricity?

98. **Transportation** A long-haul truck driver earns $.43 for each mile driven. How much will a truck driver earn for driving 1507 miles from Boston to Miami?

99. **Recycling** Four hundred empty soft drink cans weigh 18.75 pounds. A recycling center pays $.75 per pound for the cans. Find the amount received for the 400 cans. Round to the nearest cent.

100. **Recycling** A recycling center pays $.045 per pound for newspapers.
a. Estimate the payment for recycling 520 pounds of newspapers.
b. Find the actual amount received from recycling the newspapers.

101. **Taxes** The tax per gallon of gasoline in California is $.192. (*Source:* Tax Foundation) If you fill your gasoline tank with 12.5 gallons of gasoline in California, how much will you pay in taxes?

102. Geometry The perimeter of a square is equal to four times the length of a side of the square. Find the perimeter of a square whose side measures 2.8 meters.

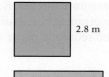

2.8 m

103. Geometry The area of a rectangle is equal to the product of the length of the rectangle times its width. Find the area of a rectangle that has a length of 6.75 feet and a width of 3.5 feet. The area will be in square feet.

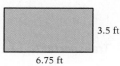

3.5 ft

6.75 ft

104. Finance You bought a car for $5000 down and made payments of $499.50 each month for 36 months.
a. Find the amount of the payments over the 36 months.
b. Find the total cost of the car.

105. Compensation A nurse earns a salary of $1396 for a 40-hour work week. This week the nurse worked 15 hours of overtime at a rate of $52.35 for each hour of overtime worked.
a. Find the nurse's overtime pay.
b. Find the nurse's total income for the week.

106. Consumerism Bay Area Rental Cars charges $25 a day and $.25 per mile for renting a car. You rented a car for 3 days and drove 235 miles. Find the total cost of renting the car.

107. Shipping The graph at the right shows United States Postal Service rates for express mail. How much would it cost a company to mail 25 express mail packages, each weighing $\frac{1}{4}$ pound, post office to addressee?

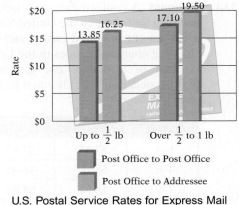

Rate

$20

$15

$10

$5

$0

13.85   16.25   17.10   19.50

Up to $\frac{1}{2}$ lb   Over $\frac{1}{2}$ to 1 lb

▉ Post Office to Post Office

▉ Post Office to Addressee

U.S. Postal Service Rates for Express Mail

PPT

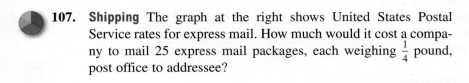

For Exercises 108 and 109, use the graph of Postal Service rates from Exercise 107. Write a verbal description of what each expression represents.

108. 17.10×15

109. $(16.25 \times 4) + (19.50 \times 9)$

110. Taxes For tax purposes, the standard deduction on tax returns for the business use of a car in 2007 was 48.5¢ per mile. In 2008, the standard deduction was 50.5¢ per mile. Find the amount deductible on a 2008 tax return for driving a business car 11,842 miles during the year.

111. Electronic Checks See the news clipping at the right. Find the added cost to the government for issuing paper checks to the 4 million Social Security recipients who request a paper check because they do not have a bank account.

In the News

Paper Checks Cost Government Millions

The federal government's cost to issue a paper check is $.89, while the cost for an electronic check is $.09.

Source: finance.yahoo.com

112. Transportation A taxi costs $2.50 and $.20 for each $\frac{1}{8}$ mile driven. Find the cost of hiring a taxi to get from the airport to the hotel, a distance of $5\frac{1}{2}$ miles.

113. Business The table at the right lists three pieces of steel required for a repair project.
   a. Find the total cost of grade 1.
   b. Find the total cost of grade 2.
   c. Find the total cost of grade 3.
   d. Find the total cost of the three pieces of steel.

Grade of Steel	Weight (pounds per foot)	Required Number of Feet	Cost per Pound
1	2.2	8	$3.20
2	3.4	6.5	$3.35
3	6.75	15.4	$3.94

114. Business A confectioner ships holiday packs of candy and nuts anywhere in the United States. At the right is a price list for nuts and candy, and below is a table of shipping charges to zones in the United States. For any fraction of a pound, use the next higher weight. Sixteen ounces (16 oz) is equal to 1 pound.

Code	Description	Price
112	Almonds 16 oz	$6.75
116	Cashews 8 oz	$5.90
117	Cashews 16 oz	$8.50
130	Macadamias 7 oz	$7.25
131	Macadamias 16 oz	$11.95
149	Pecan halves 8 oz	$8.25
155	Mixed nuts 8 oz	$6.80
160	Cashew brittle 8 oz	$5.95
182	Pecan roll 8 oz	$6.70
199	Chocolate peanuts 8 oz	$5.90

Pounds	Zone 1	Zone 2	Zone 3	Zone 4
1–3	$7.55	$7.85	$8.25	$8.75
4–6	$8.10	$8.40	$8.80	$9.30
7–9	$8.50	$8.80	$9.20	$9.70
10–12	$8.90	$9.20	$9.60	$10.10

Find the cost of sending the following orders to the given mail zone.

a. Code	Quantity	b. Code	Quantity	c. Code	Quantity
116	2	112	1	117	3
130	1	117	4	131	1
149	3	131	2	155	2
182	4	160	3	160	4
Mail to zone 4.		182	5	182	1
		Mail to zone 3.		199	3
				Mail to zone 2.	

Applying the Concepts

115. Show how the decimal is placed in the product of 1.3 × 2.31 by first writing each number as a fraction and then multiplying. Then change the product back to decimal notation.

116. **Automotive Repair** Chris works at B & W Garage as an auto mechanic and has just completed an engine overhaul for a customer. To determine the cost of the repair job, Chris keeps a list of times worked and parts used. A parts list and a list of the times worked are shown below.

Parts Used		Time Spent	
Item	*Quantity*	*Day*	*Hours*
Gasket set	1	Monday	7.0
Ring set	1	Tuesday	7.5
Valves	8	Wednesday	6.5
Wrist pins	8	Thursday	8.5
Valve springs	16	Friday	9.0
Rod bearings	8		
Main bearings	5		
Valve seals	16		
Timing chain	1		

Price List		
Item Number	*Description*	*Unit Price*
27345	Valve spring	$9.25
41257	Main bearing	$17.49
54678	Valve	$16.99
29753	Ring set	$169.99
45837	Gasket set	$174.90
23751	Timing chain	$50.49
23765	Fuel pump	$229.99
28632	Wrist pin	$23.55
34922	Rod bearing	$13.69
2871	Valve seal	$1.69

a. Organize a table of data showing the parts used, the unit price for each part, and the price of the quantity used. *Hint:* Use the following headings for the table.

Quantity   Item Number   Description   Unit Price   Total

b. Add up the numbers in the "Total" column to find the total cost of the parts.

c. If the charge for labor is $46.75 per hour, compute the cost of labor.

d. What is the total cost for parts and labor?

117. Explain how the decimal point is placed when a number is multiplied by 10, 100, 1000, 10,000, etc.

118. Explain how the decimal point is placed in the product of two decimals.

SECTION

3.5   Division of Decimals

OBJECTIVE A   To divide decimals

To divide decimals, move the decimal point in the divisor to the right to make the divisor a whole number. Move the decimal point in the dividend the same number of places to the right. Place the decimal point in the quotient directly over the decimal point in the dividend, and then divide as with whole numbers.

HOW TO 1   Divide: $3.25\overline{)15.275}$

$$3.\underset{\curvearrowright}{25.}\,)15.\underset{\curvearrowright}{27.}5$$

- Move the decimal point 2 places to the right in the divisor and then in the dividend. Place the decimal point in the quotient.

$$
\begin{array}{r}
4.7 \\
325.\,)\overline{1527.5} \\
-1300 \\
\hline
227\,5 \\
-227\,5 \\
\hline
0
\end{array}
$$

- Divide as with whole numbers.

Moving the decimal point the same number of decimal places in the divisor and dividend does not change the value of the quotient, because this process is the same as multiplying the numerator and denominator of a fraction by the same number. In the example above,

$$3.25\overline{)15.275} = \frac{15.275}{3.25} = \frac{15.275 \times 100}{3.25 \times 100} = \frac{1527.5}{325} = 325\overline{)1527.5}$$

When dividing decimals, we usually round the quotient off to a specified place value, rather than writing the quotient with a remainder.

HOW TO 2   Divide: $0.3\overline{)0.56}$
Round to the nearest hundredth.

$$
\begin{array}{r}
1.866 \approx 1.87 \\
0.\underset{\curvearrowright}{3.}\,)\,0.\underset{\curvearrowright}{5.}600 \\
-\ 3 \\
\hline
2\,6 \\
-2\,4 \\
\hline
20 \\
-18 \\
\hline
20 \\
-18 \\
\end{array}
$$

We must carry the division to the thousandths place to round the quotient to the nearest hundredth. Therefore, zeros must be inserted in the dividend so that the quotient has a digit in the thousandths place.

Integrating Technology

A calculator displays the quotient to the limit of the calculator's display. Enter

to determine the number of places your calculator displays.

HOW TO 3 Divide 57.93 by 3.24. Round to the nearest thousandth.

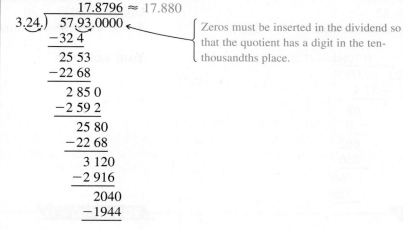

$$17.8796 \approx 17.880$$

Zeros must be inserted in the dividend so that the quotient has a digit in the ten-thousandths place.

To divide a decimal by a power of 10 (10, 100, 1000, . . .), move the decimal point to the left the same number of places as there are zeros in the power of 10.

$34.65 \div 1\underline{0} = 3.\underset{\smile}{4}65$
   1 zero   1 decimal place

$34.65 \div 1\underline{00} = 0.\underset{\smile}{3}465$
   2 zeros   2 decimal places

$34.65 \div 1\underline{000} = 0.\underset{\smile}{0}3465$
   3 zeros   3 decimal places

• Note that a zero must be inserted between the 3 and the decimal point.

$34.65 \div 1\underline{0,000} = 0.\underset{\smile}{0}03465$
   4 zeros   4 decimal places

• Note that two zeros must be inserted between the 3 and the decimal point.

If the power of 10 is written in exponential notation, the exponent indicates how many places to move the decimal point.

$34.65 \div 10^1 = 3.465$ 1 decimal place

$34.65 \div 10^2 = 0.3465$ 2 decimal places

$34.65 \div 10^3 = 0.03465$ 3 decimal places

$34.65 \div 10^4 = 0.003465$ 4 decimal places

EXAMPLE 1

Divide: $0.1344 \div 0.032$

Solution

$$0.032. \overline{)0.134.4}$$
$$\underline{-128}$$
$$\quad 64$$
$$\underline{-64}$$
$$\quad\; 0$$

quotient 4.2

• Move the decimal point 3 places to the right in the divisor and the dividend.

YOU TRY IT 1

Divide: $0.1404 \div 0.052$

Your solution

Solution on p. S9

EXAMPLE • 2

Divide: $58.092 \div 82$
Round to the nearest thousandth.

Solution

$$
\begin{array}{r}
0.7084 \approx 0.708 \\
82\overline{)\ 58.0920} \\
-57\ 4 \\
\hline
69 \\
-\ \ 0 \\
\hline
692 \\
-656 \\
\hline
360 \\
-328 \\
\hline
\end{array}
$$

EXAMPLE • 3

Divide: $420.9 \div 7.06$
Round to the nearest tenth.

Solution

$$
\begin{array}{r}
59.61 \approx 59.6 \\
7.06.\overline{)\ 420.90.00} \\
-353\ 0 \\
\hline
67\ 90 \\
-63\ 54 \\
\hline
4\ 36\ 0 \\
-4\ 23\ 6 \\
\hline
12\ 40 \\
-\ 7\ 06 \\
\hline
\end{array}
$$

EXAMPLE • 4

Divide: $402.75 \div 1000$

Solution
$402.75 \div 1000 = 0.40275$

EXAMPLE • 5

What is 0.625 divided by 10^2?

Solution
$0.625 \div 10^2 = 0.00625$

YOU TRY IT • 2

Divide: $37.042 \div 76$
Round to the nearest thousandth.

Your solution

YOU TRY IT • 3

Divide: $370.2 \div 5.09$
Round to the nearest tenth.

Your solution

YOU TRY IT • 4

Divide: $309.21 \div 10,000$

Your solution

YOU TRY IT • 5

What is 42.93 divided by 10^4?

Your solution

Solutions on p. S10

ESTIMATION

Estimating the Quotient of Two Decimals

Calculate $282.18 \div 0.48$. Then use estimation to determine whether the quotient is reasonable.

Divide to find the exact quotient.

$282.18\ \boxed{\div}\ 0.48\ \boxed{=}\ 587.875$

To estimate the quotient, round each number so that it contains one nonzero digit. Then divide. The estimated answer is 600, which is very close to the exact quotient, 587.875.

$282.18 \div 0.48 \approx 300 \div 0.5$
$\qquad\qquad\qquad = 600$

OBJECTIVE B To solve application problems

 The graph at the right shows average hourly earnings in the United States for selected years. Use this table for Example 6 and You Try It 6.

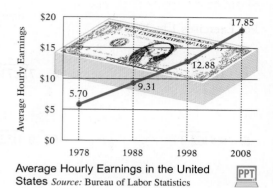

Average Hourly Earnings in the United States *Source:* Bureau of Labor Statistics

---

EXAMPLE · 6

How many times greater were the average hourly earnings in 2008 than in 1978? Round to the nearest whole number.

Strategy
To find how many times greater the average hourly earnings were, divide the 2008 average hourly earnings (17.85) by the 1978 average hourly earnings (5.70).

Solution
$17.85 \div 5.70 \approx 3$

The average hourly earnings in 2008 were about 3 times greater than in 1978.

YOU TRY IT · 6

How many times greater were the average hourly earnings in 1998 than in 1978? Round to the nearest tenth.

Your strategy

Your solution

EXAMPLE · 7

A 1-year subscription to a monthly magazine costs $90. The price of each issue at the newsstand is $9.80. How much would you save per issue by buying a year's subscription rather than buying each issue at the newsstand?

Strategy
To find the amount saved:

• Find the subscription price per issue by dividing the cost of the subscription (90) by the number of issues (12).
• Subtract the subscription price per issue from the newsstand price (9.80).

Solution
$90 \div 12 = 7.50$
$9.80 - 7.50 = 2.30$

The savings would be $2.30 per issue.

YOU TRY IT · 7

A Nielsen survey of the number of people (in millions) who watch television each day of the week is given in the table below.

Mon.	Tues.	Wed.	Thu.	Fri.	Sat.	Sun.
91.9	89.8	90.6	93.9	78.0	77.1	87.7

Find the average number of people who watch television per day.

Your strategy

Your solution

Solutions on p. S10

3.5 EXERCISES

OBJECTIVE A   **To divide decimals**

For Exercises 1 to 20, divide.

1. $3\overline{)2.46}$

2. $7\overline{)3.71}$

3. $0.8\overline{)3.84}$

4. $0.9\overline{)6.93}$

5. $0.7\overline{)62.3}$

6. $0.4\overline{)52.8}$

7. $0.4\overline{)24}$

8. $0.5\overline{)65}$

9. $0.7\overline{)59.01}$

10. $0.9\overline{)8.721}$

11. $0.5\overline{)16.15}$

12. $0.8\overline{)77.6}$

13. $0.7\overline{)3.542}$

14. $0.6\overline{)2.436}$

15. $6.3\overline{)8.19}$

16. $3.2\overline{)7.04}$

17. $3.6\overline{)0.396}$

18. $2.7\overline{)0.648}$

19. $6.9\overline{)26.22}$

20. $1.7\overline{)84.66}$

For Exercises 21 to 29, divide. Round to the nearest tenth.

21. $55.62 \div 8.8$

22. $25.43 \div 5.4$

23. $5.427 \div 9.5$

24. $1.837 \div 1.4$

25. $18.4 \div 7.3$

26. $52.9 \div 8.1$

27. $0.183 \div 0.17$

28. $0.381 \div 0.47$

29. $6.924 \div 0.053$

For Exercises 30 to 38, divide. Round to the nearest hundredth.

30. $4.817 \div 16$

31. $6.467 \div 8$

32. $0.0418 \div 0.53$

33. $0.0647 \div 0.72$

34. $7 \div 0.55$

35. $38.665 \div 0.95$

36. $13.97 \div 25.4$

37. $27.738 \div 60.8$

38. $3.171 \div 45.6$

For Exercises 39 to 47, divide. Round to the nearest thousandth.

39. $1.028 \div 54$

40. $6.729 \div 27$

41. $0.0437 \div 0.5$

42. $75.469 \div 77.8$

43. $34.31 \div 95.3$

44. $0.2695 \div 2.67$

45. $0.4871 \div 4.72$

46. $0.1142 \div 17.2$

47. $0.2307 \div 26.7$

For Exercises 48 to 56, divide. Round to the nearest whole number.

48. $16.5 \div 4$

49. $89.76 \div 90$

50. $1.94 \div 0.3$

51. $1.0478 \div 0.413$

52. $2.148 \div 0.519$

53. $0.79 \div 0.778$

54. $3.092 \div 0.075$

55. $392 \div 6.9$

56. $8.729 \div 0.075$

For Exercises 57 to 74, divide.

57. $4.07 \div 10$

58. $0.039 \div 10$

59. $42.67 \div 10$

60. $389.7 \div 100$

61. $1.037 \div 100$

62. $237.835 \div 100$

63. $8.295 \div 1000$

64. $82,547 \div 1000$

65. $825.37 \div 1000$

66. $8.35 \div 10^1$

67. $0.32 \div 10^1$

68. $87.65 \div 10^1$

69. $23.627 \div 10^2$

70. $2.954 \div 10^2$

71. $0.0053 \div 10^2$

72. $289.32 \div 10^3$

73. $1.8932 \div 10^3$

74. $0.139 \div 10^3$

75. Divide 44.208 by 2.4.

76. Divide 0.04664 by 0.44.

77. Find the quotient of 723.15 and 45.

78. Find the quotient of 3.3463 and 3.07.

79. Divide 13.5 by 10^3.

80. Divide 0.045 by 10^5.

81. Find the quotient of 23.678 and 1000.

82. Find the quotient of 7.005 and 10,000.

83. What is 0.0056 divided by 0.05?

84. What is 123.8 divided by 0.02?

 For Exercises 85 to 93, use a calculator to divide. Round to the nearest ten-thousandth. Then use estimation to determine whether the quotient you calculated is reasonable.

85. 42.42 ÷ 3.8

86. 69.8 ÷ 7.2

87. 389 ÷ 0.44

88. 642 ÷ 0.83

89. 6.394 ÷ 3.5

90. 8.429 ÷ 4.2

91. 1.235 ÷ 0.021

92. 7.456 ÷ 0.072

93. 95.443 ÷ 1.32

 94. A four-digit whole number is divided by 1000. Is the quotient less than 1 or greater than 1?

OBJECTIVE B   **To solve application problems**

 95. A 12-pack of bottled spring water sells for $3.85. State whether to use *multiplication* or *division* to find the specified amount.
  a. The cost of one bottle of spring water
  b. The cost of four 12-packs of spring water

 96. A city school district spends $8754 per student, and the school district in a nearby suburb spends $10,296 per student. Which expression represents how many times greater the amount spent per student by the suburban schools is than the amount spent per student by the city schools?
  (i) 10,296 − 8754   (ii) 10,296 ÷ 8754
  (iii) 8754 ÷ 10,296   (iv) 10,296 × 8754

97. **Sports** Ramon, a high school football player, gained 162 yards on 26 carries in a high school football game. Find the average number of yards gained per carry. Round to the nearest hundredth.

98. **Consumerism** A case of diet cola costs $6.79. If there are 24 cans in a case, find the cost per can. Round to the nearest cent.

99. Transportation A truck driver is paid by the number of miles driven. If a truck driver earns $.46 per mile, how many miles must the trucker drive in 1 hour to earn $16.00 per mile? Round to the nearest mile.

100. Tourism See the news clipping at the right. Find the average amount spent by each visitor to the United States. Round to the nearest cent.

101. Carpentry Anne is building bookcases that are 3.4 feet long. How many complete shelves can be cut from a 12-foot board?

102. Travel When the Massachusetts Turnpike opened, the toll for a passenger car that traveled the entire 136 miles of it was $5.60. Calculate the cost per mile. Round to the nearest cent.

Eric Fowke/PhotoEdit, Inc.

103. Investments An oil company has issued 3,541,221,500 shares of stock. The company paid $6,090,990,120 in dividends. Find the dividend for each share of stock. Round to the nearest cent.

104. Insurance Earl is 52 years old and is buying $70,000 of life insurance for an annual premium of $703.80. If he pays the annual premium in 12 equal installments, how much is each monthly payment?

105. Fuel Efficiency A car with an odometer reading of 17,814.2 is filled with 9.4 gallons of gas. At an odometer reading of 18,130.4, the tank is empty and the car is filled with 12.4 gallons of gas. How many miles does the car travel on 1 gallon of gasoline?

106. Carbon Footprint Depending on the efficiency of a power plant, 1 ton of coal can produce about 3000 kilowatt-hours of electricity. Suppose your family uses 25 kilowatt-hours of electricity per month. How many tons of coal will your family use in 1 year?

107. Carbon Footprint One barrel of oil produces approximately 800 kilowatt-hours of electricity. Suppose you use 27 kilowatt-hours of electricity per month. How many barrels of oil will you use in 1 year?

108. Going Green See the news clipping at the right.
   a. Find the reduction in solid waste per month if every U.S. household viewed and paid its bills online.
   b. Find the reduction in greenhouse gas emissions per month if every household viewed and paid its bills online.
   Write your answers in standard form, rounded to the nearest whole number.

Applying the Concepts

109. **Education** According to the National Center for Education Statistics, 10.03 million women and 7.46 million men were enrolled at institutions of higher learning in a recent year. How many more women than men were attending institutions of higher learning in that year?

The Military The table at the right shows the advertising budgets of four branches of the U.S. armed services in a recent year. Use this table for Exercises 110 to 112.

Service	Advertising Budget
Army	$85.3 million
Air Force	$41.1 million
Navy	$20.5 million
Marines	$15.9 million

Source: CMR/TNS Media Intelligence

110. Find the difference between the Army's advertising budget and the Marines' advertising budget.

111. How many times greater was the Army's advertising budget than the Navy's advertising budget? Round to the nearest tenth.

112. What was the total of the advertising budgets for the four branches of the service?

113. **Population Growth** The U.S. population of people ages 85 and over is expected to grow from 4.2 million in 2000 to 8.9 million in 2030. How many times greater is the population of this segment expected to be in 2030 than in 2000? Round to the nearest tenth.

114. Explain how the decimal point is moved when a number is divided by 10, 100, 1000, 10,000, etc.

115. **Sports** Explain how baseball batting averages are determined. Then find Detroit Tiger's right fielder Magglio Ordonez's batting average with 216 hits out of 595 at bats. Round to the nearest thousandth.

116. Explain how the decimal point is placed in the quotient when a number is divided by a decimal.

For Exercises 117 to 122, insert +, −, ×, or ÷ into the square so that the statement is true.

117. $3.45 \,\square\, 0.5 = 6.9$

118. $3.46 \,\square\, 0.24 = 0.8304$

119. $6.009 \,\square\, 4.68 = 1.329$

120. $0.064 \,\square\, 1.6 = 0.1024$

121. $9.876 \,\square\, 23.12 = 32.996$

122. $3.0381 \,\square\, 1.23 = 2.47$

For Exercises 123 to 125, fill in the square to make a true statement.

123. $6.47 - \square = 1.253$

124. $6.47 + \square = 9$

125. $0.009 \div \square = 0.36$

SECTION

3.6 Comparing and Converting Fractions and Decimals

OBJECTIVE A **To convert fractions to decimals**

Every fraction can be written as a decimal. To write a fraction as a decimal, divide the numerator of the fraction by the denominator. The quotient can be rounded to the desired place value.

HOW TO · 1   Convert $\frac{3}{7}$ to a decimal.

$$7\overline{)3.00000} \quad \begin{array}{r} 0.42857 \end{array}$$

$\frac{3}{7}$ rounded to the nearest hundredth is 0.43.

$\frac{3}{7}$ rounded to the nearest thousandth is 0.429.

$\frac{3}{7}$ rounded to the nearest ten-thousandth is 0.4286.

HOW TO · 2   Convert $3\frac{2}{9}$ to a decimal. Round to the nearest thousandth.

$$3\frac{2}{9} = \frac{29}{9} \qquad 9\overline{)29.0000}^{\,3.2222} \qquad 3\frac{2}{9} \text{ rounded to the nearest thousandth is } 3.222.$$

EXAMPLE · 1

Convert $\frac{3}{8}$ to a decimal.
Round to the nearest hundredth.

Solution $8\overline{)3.000}^{\,0.375} \approx 0.38$

YOU TRY IT · 1

Convert $\frac{9}{16}$ to a decimal.
Round to the nearest tenth.

Your solution

EXAMPLE · 2

Convert $2\frac{3}{4}$ to a decimal.
Round to the nearest tenth.

Solution $2\frac{3}{4} = \frac{11}{4}$ $4\overline{)11.00}^{\,2.75} \approx 2.8$

YOU TRY IT · 2

Convert $4\frac{1}{6}$ to a decimal.
Round to the nearest hundredth.

Your solution

Solutions on p. S10

OBJECTIVE B **To convert decimals to fractions**

To convert a decimal to a fraction, remove the decimal point and place the decimal part over a denominator equal to the place value of the last digit in the decimal.

$0.4\overset{\text{hundredths}}{7} = \dfrac{47}{100}$

$7.4\overset{\text{hundredths}}{5} = 7\dfrac{45}{100} = 7\dfrac{9}{20}$

$0.27\overset{\text{thousandths}}{5} = \dfrac{275}{1000} = \dfrac{11}{40}$

$0.16\overset{\text{hundredths}}{\tfrac{2}{3}} = \dfrac{16\frac{2}{3}}{100} = 16\frac{2}{3} \div 100 = \dfrac{50}{3} \times \dfrac{1}{100} = \dfrac{1}{6}$

EXAMPLE · 3

Convert 0.82 and 4.75 to fractions.

Solution

$$0.82 = \frac{82}{100} = \frac{41}{50}$$

$$4.75 = 4\frac{75}{100} = 4\frac{3}{4}$$

YOU TRY IT · 3

Convert 0.56 and 5.35 to fractions.

Your solution

EXAMPLE · 4

Convert $0.15\frac{2}{3}$ to a fraction.

Solution

$$0.15\frac{2}{3} = \frac{15\frac{2}{3}}{100} = 15\frac{2}{3} \div 100$$

$$= \frac{47}{3} \times \frac{1}{100} = \frac{47}{300}$$

YOU TRY IT · 4

Convert $0.12\frac{7}{8}$ to a fraction.

Your solution

Solutions on p. S10

OBJECTIVE C   **To identify the order relation between two decimals or between a decimal and a fraction**

Decimals, like whole numbers and fractions, can be graphed as points on the number line. The number line can be used to show the order of decimals. A decimal that appears to the right of a given number is greater than the given number. A decimal that appears to the left of a given number is less than the given number.

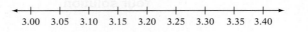

3.00 3.05 3.10 3.15 3.20 3.25 3.30 3.35 3.40

Note that 3, 3.0, and 3.00 represent the same number.

HOW TO · 3   Find the order relation between $\frac{3}{8}$ and 0.38.

$$\frac{3}{8} = 0.375 \qquad 0.38 = 0.380$$ • Convert the fraction $\frac{3}{8}$ to a decimal.

$$0.375 < 0.380$$ • Compare the two decimals.

$$\frac{3}{8} < 0.38$$ • Convert 0.375 back to a fraction.

EXAMPLE · 5

Place the correct symbol, < or >, between the numbers.

$$\frac{5}{16} \qquad 0.32$$

Solution

$$\frac{5}{16} \approx 0.313$$ • Convert $\frac{5}{16}$ to a decimal.

$$0.313 < 0.32$$ • Compare the two decimals.

$$\frac{5}{16} < 0.32$$ • Convert 0.313 back to a fraction.

YOU TRY IT · 5

Place the correct symbol, < or >, between the numbers.

$$0.63 \qquad \frac{5}{8}$$

Your solution

Solution on p. S10

3.6 EXERCISES

OBJECTIVE A To convert fractions to decimals

For Exercises 1 to 24, convert the fraction to a decimal.
Round to the nearest thousandth.

1. $\dfrac{5}{8}$ **2.** $\dfrac{7}{12}$ **3.** $\dfrac{2}{3}$ **4.** $\dfrac{5}{6}$ **5.** $\dfrac{1}{6}$ **6.** $\dfrac{7}{8}$

7. $\dfrac{5}{12}$ **8.** $\dfrac{9}{16}$ **9.** $\dfrac{7}{4}$ **10.** $\dfrac{5}{3}$ **11.** $1\dfrac{1}{2}$ **12.** $2\dfrac{1}{3}$

13. $\dfrac{16}{4}$ **14.** $\dfrac{36}{9}$ **15.** $\dfrac{3}{1000}$ **16.** $\dfrac{5}{10}$ **17.** $7\dfrac{2}{25}$ **18.** $16\dfrac{7}{9}$

19. $37\dfrac{1}{2}$ **20.** $\dfrac{5}{24}$ **21.** $\dfrac{4}{25}$ **22.** $3\dfrac{1}{3}$ **23.** $8\dfrac{2}{5}$ **24.** $5\dfrac{4}{9}$

 For Exercises 25 to 28, without actually doing any division, state whether the decimal
equivalent of the given fraction is greater than 1 or less than 1.

25. $\dfrac{54}{57}$ **26.** $\dfrac{176}{129}$ **27.** $\dfrac{88}{80}$ **28.** $\dfrac{2007}{2008}$

OBJECTIVE B To convert decimals to fractions

For Exercises 29 to 53, convert the decimal to a fraction.

29. 0.8   **30.** 0.4   **31.** 0.32   **32.** 0.48   **33.** 0.125

34. 0.485   **35.** 1.25   **36.** 3.75   **37.** 16.9   **38.** 17.5

39. 8.4   **40.** 10.7   **41.** 8.437   **42.** 9.279   **43.** 2.25

44. 7.75   **45.** $0.15\frac{1}{3}$ **46.** $0.17\frac{2}{3}$ **47.** $0.87\frac{7}{8}$ **48.** $0.12\frac{5}{9}$

49. 7.38   **50.** 0.33   **51.** 0.57   **52.** $0.33\frac{1}{3}$ **53.** $0.66\frac{2}{3}$

 54. Is $0.27\frac{4}{9}$ greater than 0.27 or less than 0.27?

OBJECTIVE C   **To identify the order relation between two decimals or between a decimal and a fraction**

For Exercises 55 to 74, place the correct symbol, $<$ or $>$, between the numbers.

55. 0.15   0.5   **56.** 0.6   0.45   **57.** 6.65   6.56   **58.** 3.89   3.98

59. 2.504   2.054   **60.** 0.025   0.105   **61.** $\frac{3}{8}$ 0.365   **62.** $\frac{4}{5}$ 0.802

63. $\frac{2}{3}$ 0.65   **64.** 0.85 $\frac{7}{8}$ **65.** $\frac{5}{9}$ 0.55   **66.** $\frac{7}{12}$ 0.58

67. 0.62 $\frac{7}{15}$ **68.** $\frac{11}{12}$ 0.92   **69.** 0.161 $\frac{1}{7}$ **70.** 0.623   0.6023

71. 0.86   0.855   **72.** 0.87   0.087   **73.** 1.005   0.5   **74.** 0.033   0.3

 75. Use the inequality symbol $<$ to rewrite the order relation expressed by the inequality $17.2 > 0.172$.

 76. Use the inequality symbol $>$ to rewrite the order relation expressed by the inequality $0.0098 < 0.98$.

Applying the Concepts

77. Air Pollution An emissions test for cars requires that of the total engine exhaust, less than 1 part per thousand $\left(\frac{1}{1000} = 0.001\right)$ be hydrocarbon emissions. Using this figure, determine which of the cars in the table at the right would fail the emissions test.

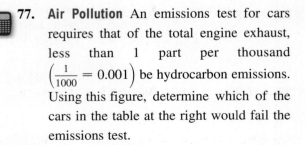

Car	Total Engine Exhaust	Hydrocarbon Emission
1	367,921	360
2	401,346	420
3	298,773	210
4	330,045	320
5	432,989	450

 78. Explain how terminating, repeating, and nonrepeating decimals differ. Give an example of each kind of decimal.

FOCUS ON PROBLEM SOLVING

Relevant Information

Problems in mathematics or real life involve a question or a need and information or circumstances related to that question or need. Solving problems in the sciences usually involves a question, an observation, and measurements of some kind.

One of the challenges of problem solving in the sciences is to separate the information that is relevant to the problem from other information. Following is an example from the physical sciences in which some relevant information was omitted.

Hooke's Law states that the distance that a weight will stretch a spring is directly proportional to the weight on the spring. That is, $d = kF$, where d is the distance the spring is stretched and F is the force. In an experiment to verify this law, some physics students were continually getting inconsistent results. Finally, the instructor discovered that the heat produced when the lights were turned on was affecting the experiment. In this case, relevant information was omitted—namely, that the temperature of the spring can affect the distance it will stretch.

A lawyer drove 8 miles to the train station. After a 35-minute ride of 18 miles, the lawyer walked 10 minutes to the office. Find the total time it took the lawyer to get to work.

From this situation, answer the following before reading on.

a. What is asked for?

b. Is there enough information to answer the question?

c. Is information given that is not needed?

Here are the answers.

a. We want the total time for the lawyer to get to work.

b. No. We do not know the time it takes the lawyer to get to the train station.

c. Yes. Neither the distance to the train station nor the distance of the train ride is necessary to answer the question.

For each of the following problems, answer the questions printed in red above.

1. A customer bought 6 boxes of strawberries and paid with a $20 bill. What was the change?

2. A board is cut into two pieces. One piece is 3 feet longer than the other piece. What is the length of the original board?

3. A family rented a car for their vacation and drove 680 miles. The cost of the rental car was $21 per day with 150 free miles per day and $.15 for each mile driven above the number of free miles allowed. How many miles did the family drive per day?

4. An investor bought 8 acres of land for $80,000. One and one-half acres were set aside for a park, and the remaining land was developed into one-half-acre lots. How many lots were available for sale?

5. You wrote checks of $43.67, $122.88, and $432.22 after making a deposit of $768.55. How much do you have left in your checking account?

PROJECTS AND GROUP ACTIVITIES

Fractions as Terminating or Repeating Decimals

The fraction $\frac{3}{4}$ is equivalent to 0.75. The decimal 0.75 is a **terminating decimal** because there is a remainder of zero when 3 is divided by 4. The fraction $\frac{1}{3}$ is equivalent to 0.333 . . . . The three dots mean the pattern continues on and on. 0.333 . . . is a **repeating decimal.** To determine whether a fraction can be written as a terminating decimal, first write the fraction in simplest form. Then look at the denominator of the fraction. If it contains prime factors of only 2s and/or 5s, then it can be expressed as a terminating decimal. If it contains prime factors other than 2s or 5s, it represents a repeating decimal.

✓ **Take Note**

If the denominator of a fraction in simplest form is 20, then it can be written as a terminating decimal because 20 = 2 · 2 · 5 (only prime factors of 2 and 5). If the denominator of a fraction in simplest form is 6, it represents a repeating decimal because it contains the prime factor 3 (a number other than 2 or 5).

1. Assume that each of the following numbers is the denominator of a fraction written in simplest form. Does the fraction represent a terminating or repeating decimal?
    a. 4   **b.** 5   **c.** 7   **d.** 9   **e.** 10   **f.** 12   **g.** 15
    h. 16   **i.** 18   **j.** 21   **k.** 24   **l.** 25   **m.** 28   **n.** 40

2. Write two other numbers that, as denominators of fractions in simplest form, represent terminating decimals, and write two other numbers that, as denominators of fractions in simplest form, represent repeating decimals.

CHAPTER 3

SUMMARY

KEY WORDS

A number written in *decimal notation* has three parts: a *whole-number part*, a *decimal point*, and a *decimal part*. The decimal part of a number represents a number less than 1. A number written in decimal notation is often simply called a *decimal*. [3.1A, p. 126]

EXAMPLES

For the decimal 31.25, 31 is the whole-number part and 25 is the decimal part.

ESSENTIAL RULES AND PROCEDURES

To write a decimal in words, write the decimal part as if it were a whole number. Then name the place value of the last digit. The decimal point is read as "and." [3.1A, p. 126]

EXAMPLES

The decimal 12.875 is written in words as twelve and eight hundred seventy-five thousandths.

To write a decimal in standard form when it is written in words, write the whole-number part, replace the word *and* with a decimal point, and write the decimal part so that the last digit is in the given place-value position. [3.1A, p. 127]

The decimal forty-nine and sixty-three thousandths is written in standard form as 49.063.

To round a decimal to a given place value, use the same rules used with whole numbers, except drop the digits to the right of the given place value instead of replacing them with zeros. [3.1B, p. 128]

2.7134 rounded to the nearest tenth is 2.7.

0.4687 rounded to the nearest hundredth is 0.47.

To add decimals, write the decimals so that the decimal points are on a vertical line. Add as you would with whole numbers. Then write the decimal point in the sum directly below the decimal points in the addends. [3.2A, p. 132]

$$\begin{array}{r} \overset{1\ \ 1}{1.35} \\ 20.8 \\ +\ 0.76 \\ \hline 22.91 \end{array}$$

To subtract decimals, write the decimals so that the decimal points are on a vertical line. Subtract as you would with whole numbers. Then write the decimal point in the difference directly below the decimal point in the subtrahend. [3.3A, p. 136]

$$\begin{array}{r} \overset{2\ \ 15\ \quad 6\ \ 10}{3\cancel{5}.8\cancel{7}\cancel{0}} \\ -\ 9.641 \\ \hline 26.229 \end{array}$$

To multiply decimals, multiply the numbers as you would whole numbers. Then write the decimal point in the product so that the number of decimal places in the product is the sum of the decimal places in the factors. [3.4A, p. 140]

$$\begin{array}{rl} 26.83 & \text{2 decimal places} \\ \times\quad 0.45 & \text{2 decimal places} \\ \hline 13415 & \\ 10732\quad\ & \\ \hline 12.0735 & \text{4 decimal places} \end{array}$$

To multiply a decimal by a power of 10, move the decimal point to the right the same number of places as there are zeros in the power of 10. If the power of 10 is written in exponential notation, the exponent indicates how many places to move the decimal point. [3.4A, pp. 140, 141]

$$3.97 \cdot 10{,}000 = 39{,}700$$
$$0.641 \cdot 10^5 = 64{,}100$$

To divide decimals, move the decimal point in the divisor to the right so that it is a whole number. Move the decimal point in the dividend the same number of places to the right. Place the decimal point in the quotient directly above the decimal point in the dividend. Then divide as you would with whole numbers. [3.5A, p. 150]

$$\begin{array}{r} 6.2 \\ 0.39.\overline{)\,2.41.8} \\ -2\,34\ \\ \hline 7\,8 \\ -7\,8 \\ \hline 0 \end{array}$$

To divide a decimal by a power of 10, move the decimal point to the left the same number of places as there are zeros in the power of 10. If the power of 10 is written in exponential notation, the exponent indicates how many places to move the decimal point. [3.5A, p. 151]

$$972.8 \div 1000 = 0.9728$$
$$61.305 \div 10^4 = 0.0061305$$

To convert a fraction to a decimal, divide the numerator of the fraction by the denominator. [3.6A, p. 159]

$$\frac{7}{8} = 7 \div 8 = 0.875$$

To convert a decimal to a fraction, remove the decimal point and place the decimal part over a denominator equal to the place value of the last digit in the decimal. [3.6B, p. 159]

0.85 is eighty-five <u>hundredths</u>.

$$0.85 = \frac{85}{100} = \frac{17}{20}$$

To find the order relation between a decimal and a fraction, first rewrite the fraction as a decimal. Then compare the two decimals. [3.6C, p. 160]

Because $\frac{3}{11} \approx 0.273$, and

$$0.273 > 0.26, \frac{3}{11} > 0.26.$$

CHAPTER 3

CONCEPT REVIEW

Test your knowledge of the concepts presented in this chapter. Answer each question. Then check your answers against the ones provided in the Answer Section.

1. How do you round a decimal to the nearest tenth?

2. How do you write the decimal 0.37 as a fraction?

3. How do you write the fraction $\dfrac{173}{10,000}$ as a decimal?

4. When adding decimals of different place values, what do you do with the decimal points?

5. Where do you put the decimal point in the product of two decimals?

6. How do you estimate the product of two decimals?

7. What do you do with the decimal point when dividing decimals?

8. Which is greater, the decimal 0.63 or the fraction $\dfrac{5}{8}$?

9. How many zeros must be inserted when dividing 0.763 by 0.6 and rounding to the nearest hundredth?

10. How do you subtract a decimal from a whole number that has no decimal point?

CHAPTER 3

REVIEW EXERCISES

1. Find the quotient of 3.6515 and 0.067.

2. Find the sum of 369.41, 88.3, 9.774, and 366.474.

3. Place the correct symbol, $<$ or $>$, between the two numbers.
0.055   0.1

4. Write 22.0092 in words.

5. Round 0.05678235 to the nearest hundred-thousandth.

6. Convert $2\frac{1}{3}$ to a decimal. Round to the nearest hundredth.

7. Convert 0.375 to a fraction.

8. Add: $3.42 + 0.794 + 32.5$

9. Write thirty-four and twenty-five thousandths in standard form.

10. Place the correct symbol, $<$ or $>$, between the two numbers.
$\frac{5}{8}$ 0.62

11. Convert $\frac{7}{9}$ to a decimal. Round to the nearest thousandth.

12. Convert 0.66 to a fraction.

13. Subtract: $27.31 - 4.4465$

14. Round 7.93704 to the nearest hundredth.

15. Find the product of 3.08 and 2.9.

16. Write 342.37 in words.

17. Write three and six thousand seven hundred fifty-three hundred-thousandths in standard form.

18. Multiply: 34.79
          $\times$ 0.74

19. Divide: $0.053\overline{)0.349482}$

20. What is 7.796 decreased by 2.9175?

21. **Banking** You had a balance of $895.68 in your checking account. You then wrote checks for $145.72 and $88.45. Find the new balance in your checking account.

For Exercises 22 and 23, use the news clipping at the right.

22. **Fuel Consumption** Find the difference between the amount United expects to pay per gallon of fuel and the amount Southwest expects to pay per gallon of fuel.

23. **Fuel Consumption** What is Northwest's cost per gallon of fuel? Round to the nearest cent. Is Northwest's cost per gallon of fuel greater than or less than United's cost per gallon?

24. **Travel** In a recent year, 30.6 million Americans drove to their destinations over Thanksgiving, and 4.8 million Americans traveled by plane. (*Source:* AAA) How many times greater is the number who drove than the number who flew? Round to the nearest tenth.

25. **Nutrition** According to the American School Food Service Association, 1.9 million gallons of milk are served in school cafeterias every day. How many gallons of milk are served in school cafeterias during a 5-day school week?

In the News

A Few Extra Minutes Can Save Millions

Drivers know that they can get more miles per gallon of gasoline by reducing their speed on expressways. The same is true for airplanes. Southwest Airlines expects to save $42 million in jet fuel costs this year by adding only a few more minutes to the time of each flight. On a Northwest Airlines flight between Minneapolis and Paris, 160 gallons of fuel was saved by flying more slowly and adding only 8 minutes to the flight. It saved Northwest $535. This year, Southwest Airlines expects to pay $2.35 per gallon for fuel, while United Airlines expects to pay $3.31 per gallon.

Source: John Wilen, AP Business Writer; Yahoo! News, May 1, 2008

© Ariel Skelley/Corbis

CHAPTER 3

TEST

1. Place the correct symbol, < or >, between the two numbers.
   0.66   0.666

2. Subtract: 13.027
                − 8.94

3. Write 45.0302 in words.

4. Convert $\frac{9}{13}$ to a decimal. Round to the nearest thousandth.

5. Convert 0.825 to a fraction.

6. Round 0.07395 to the nearest ten-thousandth.

7. Find 0.0569 divided by 0.037. Round to the nearest thousandth.

8. Find 9.23674 less than 37.003.

9. Round 7.0954625 to the nearest thousandth.

10. Divide: 0.006)‾1.392

11. Add: 270.93
             97.
              1.976
          + 88.675

12. **Mechanics** Find the missing dimension.

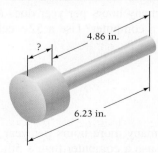

4.86 in.

?

6.23 in.

13. Multiply: 1.37
                    × 0.004

14. What is the total of 62.3, 4.007, and 189.65?

15. Write two hundred nine and seven thousand eighty-six hundred-thousandths in standard form.

16. **Finances** A car was bought for $16,734.40, with a down payment of $2500. The balance was paid in 36 monthly payments. Find the amount of each monthly payment.

17. **Compensation** You received a salary of $727.50, a commission of $1909.64, and a bonus of $450. Find your total income.

18. **Consumerism** A long-distance telephone call costs $.85 for the first 3 minutes and $.42 for each additional minute. Find the cost of a 12-minute long-distance telephone call.

Computers The table at the right shows the average number of hours per week that students use a computer. Use this table for Exercises 19 and 20.

19. On average, how many hours per year does a 10th-grade student use a computer? Use a 52-week year.

Grade Level	Average Number of Hours of Computer Use per Week
Prekindergarten–kindergarten	3.9
1st – 3rd	4.9
4th – 6th	4.2
7th – 8th	6.9
9th – 12th	6.7

Source: Find/SVP American Learning Household Survey

20. On average, how many more hours per year does a 2nd-grade student use a computer than a 5th-grade student? Use a 52-week year.

CUMULATIVE REVIEW EXERCISES

1. Divide: $89\overline{)20{,}932}$

2. Simplify: $2^3 \cdot 4^2$

3. Simplify: $2^2 - (7 - 3) \div 2 + 1$

4. Find the LCM of 9, 12, and 24.

5. Write $\frac{22}{5}$ as a mixed number.

6. Write $4\frac{5}{8}$ as an improper fraction.

7. Write an equivalent fraction with the given denominator.

$$\frac{5}{12} = \frac{}{60}$$

8. Add: $\frac{3}{8} + \frac{5}{12} + \frac{9}{16}$

9. What is $5\frac{7}{12}$ increased by $3\frac{7}{18}$?

10. Subtract: $9\frac{5}{9} - 3\frac{11}{12}$

11. Multiply: $\frac{9}{16} \times \frac{4}{27}$

12. Find the product of $2\frac{1}{8}$ and $4\frac{5}{17}$.

13. Divide: $\frac{11}{12} \div \frac{3}{4}$

14. What is $2\frac{3}{8}$ divided by $2\frac{1}{2}$?

15. Simplify: $\left(\frac{2}{3}\right)^2 \cdot \left(\frac{3}{4}\right)^3$

16. Simplify: $\left(\frac{2}{3}\right)^2 - \left(\frac{2}{3} - \frac{1}{2}\right) + 2$

17. Write 65.0309 in words.

18. Add:
$$\begin{array}{r} 379.006 \\ 27.523 \\ 9.8707 \\ + \ 88.2994 \end{array}$$

19. What is 29.005 decreased by 7.9286?

20. Multiply: 9.074
   $\times$ 6.09

21. Divide: $8.09\overline{)17.42963}$
Round to the nearest thousandth

22. Convert $\frac{11}{15}$ to a decimal. Round to the nearest thousandth.

23. Convert $0.16\frac{2}{3}$ to a fraction.

24. Place the correct symbol, $<$ or $>$, between the two numbers.
$\frac{8}{9}$ 0.98

25. Vacation The graph at the right shows the number of vacation days per year that are legally mandated in several countries. How many more vacation days does Sweden mandate than Germany?

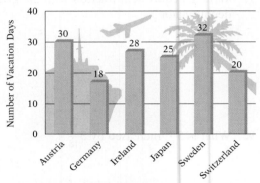

Number of Legally Mandated Vacation Days
Sources: Economic Policy Institute; *World Almanac*

26. Health A patient is put on a diet to lose 24 pounds in 3 months. The patient loses $9\frac{1}{2}$ pounds the first month and $6\frac{3}{4}$ pounds the second month. How much weight must this patient lose the third month to achieve the goal?

27. Banking You have a checking account balance of $814.35. You then write checks for $42.98, $16.43, and $137.56. Find your checking account balance after you write the checks.

28. Mechanics A machine lathe takes 0.017 inch from a brass bushing that is 1.412 inches thick. Find the resulting thickness of the bushing.

Dana White/PhotoEdit, Inc.

29. Taxes The state income tax on your business is $820 plus 0.08 times your profit. You made a profit of $64,860 last year. Find the amount of income tax you paid last year.

30. Finances You bought a camera costing $410.96. The down payment was $40, and the balance is to be paid in 8 equal monthly payments. Find the monthly payment.

Ratio and Proportion

OBJECTIVES

SECTION 4.1
A To write the ratio of two quantities in simplest form
B To solve application problems

SECTION 4.2
A To write rates
B To write unit rates
C To solve application problems

SECTION 4.3
A To determine whether a proportion is true
B To solve proportions
C To solve application problems

ARE YOU READY?

Take the Chapter 4 Prep Test to find out if you are ready to learn to:

- Write ratios, rates, and unit rates
- Solve proportions

PREP TEST

Do these exercises to prepare for Chapter 4.

1. Simplify: $\dfrac{8}{10}$

2. Simplify: $\dfrac{450}{650 + 250}$

3. Write as a decimal: $\dfrac{372}{15}$

4. Which is greater, 4×33 or 62×2?

5. Complete: $? \times 5 = 20$

SECTION

4.1   Ratio

OBJECTIVE A   **To write the ratio of two quantities in simplest form**

Point of Interest

In the 1990s, the major-league pitchers with the best strikeout-to-walk ratios (having pitched a minimum of 100 innings) were

Dennis Eckersley   6.46:1
Shane Reynolds   4.13:1
Greg Maddux   4:1
Bret Saberhagen   3.92:1
Rod Beck   3.81:1

The best single-season strikeout-to-walk ratio for starting pitchers in the same period was that of Bret Saberhagen, 11:1. (*Source:* Elias Sports Bureau)

Quantities such as 3 feet, 12 cents, and 9 cars are number quantities written with units.

$$\begin{array}{l} 3 \ \text{feet} \\ 12 \ \text{cents} \\ 9 \ \text{cars} \\ \uparrow \\ \text{units} \end{array}$$

These are some examples of units. Shirts, dollars, trees, miles, and gallons are further examples.

A **ratio** is a comparison of two quantities that have the *same* units. This comparison can be written three different ways:

1. As a fraction
2. As two numbers separated by a colon (:)
3. As two numbers separated by the word *to*

The ratio of the lengths of two boards, one 8 feet long and the other 10 feet long, can be written as

1. $\dfrac{8 \ \text{feet}}{10 \ \text{feet}} = \dfrac{8}{10} = \dfrac{4}{5}$

2. 8 feet:10 feet = 8:10 = 4:5

3. 8 feet to 10 feet = 8 to 10 = 4 to 5

Writing the **simplest form of a ratio** means writing it so that the two numbers have no common factor other than 1.

This ratio means that the smaller board is $\dfrac{4}{5}$ the length of the longer board.

EXAMPLE • 1

Write the comparison $6 to $8 as a ratio in simplest form using a fraction, a colon, and the word *to*.

Solution $\dfrac{\$6}{\$8} = \dfrac{6}{8} = \dfrac{3}{4}$

$6:$8 = 6:8 = 3:4
$6 to $8 = 6 to 8 = 3 to 4

YOU TRY IT • 1

Write the comparison 20 pounds to 24 pounds as a ratio in simplest form using a fraction, a colon, and the word *to*.

Your solution

EXAMPLE • 2

Write the comparison 18 quarts to 6 quarts as a ratio in simplest form using a fraction, a colon, and the word *to*.

Solution $\dfrac{18 \ \text{quarts}}{6 \ \text{quarts}} = \dfrac{18}{6} = \dfrac{3}{1}$

18 quarts:6 quarts = 18:6 = 3:1
18 quarts to 6 quarts = 18 to 6 = 3 to 1

YOU TRY IT • 2

Write the comparison 64 miles to 8 miles as a ratio in simplest form using a fraction, a colon, and the word *to*.

Your solution

Solutions on p. S10

OBJECTIVE B To solve application problems

© Charles O'Rear/Corbis

Use the table below for Example 3 and You Try It 3.

Board Feet of Wood at a Lumber Store			
Pine	Ash	Oak	Cedar
20,000	18,000	10,000	12,000

EXAMPLE · 3

Find, as a fraction in simplest form, the ratio of the number of board feet of pine to the number of board feet of oak.

Strategy

To find the ratio, write the ratio of board feet of pine (20,000) to board feet of oak (10,000) in simplest form.

Solution

$$\frac{20,000}{10,000} = \frac{2}{1}$$

The ratio is $\frac{2}{1}$.

YOU TRY IT · 3

Find, as a fraction in simplest form, the ratio of the number of board feet of cedar to the number of board feet of ash.

Your strategy

Your solution

EXAMPLE · 4

The cost of building a patio cover was $500 for labor and $700 for materials. What, as a fraction in simplest form, is the ratio of the cost of materials to the total cost for labor and materials?

Strategy

To find the ratio, write the ratio of the cost of materials ($700) to the total cost ($500 + $700) in simplest form.

Solution

$$\frac{\$700}{\$500 + \$700} = \frac{700}{1200} = \frac{7}{12}$$

The ratio is $\frac{7}{12}$.

YOU TRY IT · 4

A company spends $600,000 a month for television advertising and $450,000 a month for radio advertising. What, as a fraction in simplest form, is the ratio of the cost of radio advertising to the total cost of radio and television advertising?

Your strategy

Your solution

Solutions on p. S10

4.1 EXERCISES

OBJECTIVE A   To write the ratio of two quantities in simplest form

For Exercises 1 to 18, write the comparison as a ratio in simplest form using a fraction, a colon (:), and the word *to*.

1. 3 pints to 15 pints

2. 6 pounds to 8 pounds

3. $40 to $20

4. 10 feet to 2 feet

5. 3 miles to 8 miles

6. 2 hours to 3 hours

7. 6 minutes to 6 minutes

8. 8 days to 12 days

9. 35 cents to 50 cents

10. 28 inches to 36 inches

11. 30 minutes to 60 minutes

12. 25 cents to 100 cents

13. 32 ounces to 16 ounces

14. 12 quarts to 4 quarts

15. 30 yards to 12 yards

16. 12 quarts to 18 quarts

17. 20 gallons to 28 gallons

18. 14 days to 7 days

 19. To write a ratio that compares 3 days to 3 weeks, change 3 weeks into an equivalent number of _____.

 20. Is the ratio 3 : 4 the same as the ratio 4 : 3?

OBJECTIVE B   To solve application problems

For Exercises 21 to 23, write ratios in simplest form using a fraction.

Family Budget						
Housing	Food	Transportation	Taxes	Utilities	Miscellaneous	Total
$1600	$800	$600	$700	$300	$800	$4800

21. **Budgets** Use the table to find the ratio of housing costs to total expenses.

22. **Budgets** Use the table to find the ratio of food costs to total expenses.

23. **Budgets** Use the table to find the ratio of utilities costs to food costs.

 24. Refer to the table above. Write a verbal description of the ratio represented by 1 : 2. (*Hint:* There is more than one answer.)

25. **Facial Hair** Using the data in the news clipping at the right and the figure 50 million for the number of adult males in the United States, write the ratio of the number of men who participated in Movember to the number of adult males in the U.S. (*Source: Time,* February 18, 2008) Write the ratio as a fraction in simplest form.

Grow a Mustache, Save a Life

Last fall, in an effort to raise money for the Prostate Cancer Foundation, approximately 2000 men participated in a month-long mustache-growing competition. The event was dubbed Movember.

Source: Time, February 18, 2008

26. **Real Estate** A house with an original value of $180,000 increased in value to $220,000 in 5 years. What is the ratio of the increase in value to the original value of the house?

27. **Energy Prices** The price of gasoline jumped from $2.70 per gallon to $3.24 per gallon in 1 year. What is the ratio of the increase in price to the original price?

28. **Sports** National Collegiate Athletic Association (NCAA) statistics show that for every 154,000 high school seniors playing basketball, only 4000 will play college basketball as first-year students. Write the ratio of the number of first-year students playing college basketball to the number of high school seniors playing basketball.

29. **Sports** NCAA statistics show that for every 2800 college seniors playing college basketball, only 50 will play as rookies in the National Basketball Association. Write the ratio of the number of National Basketball Association rookies to the number of college seniors playing basketball.

Mike Powell/Allsport Concepts/Getty Images

Female Vocalists The table at the right shows the concert earnings for Madonna, Barbra Streisand, and Celine Dion for performances between June 2006 and June 2007.

30. Find the ratio of the amount earned by Celine Dion to the amount earned by Barbra Stresand. Write the ratio in simplest form using the word *to.*

31. Find the ratio of the amount earned by Madonna to the total amount earned by the three women. Write the ratio in simplest form using the word *to.*

32. **Consumerism** In a recent year, women spent $2 million on swimwear and purchased 92,000 swimsuits. During the same year, men spent $500,000 on swimwear and purchased 37,000 swimsuits. (*Source:* NPD Group) **a.** Find the ratio of the amount men spent on swimwear to the amount women spent on swimwear. **b.** Find the ratio of the amout men spent on swimwear to the total amount men and women spent on swimwear. Write the ratios as fractions in simplest form.

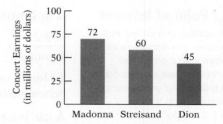

Earnings from Concerts, June 2006 to June 2007

Source: Time, Feburary 18, 2008

Applying the Concepts

33. Is the value of a ratio always less than 1? Explain.

SECTION

4.2 Rates

OBJECTIVE A   To write rates

 Point of Interest

Listed below are rates at which various crimes are committed in our nation.

Crime	Every
Larceny	4 seconds
Burglary	14 seconds
Robbery	60 seconds
Rape	6 minutes
Murder	31 minutes

A **rate** is a comparison of two quantities that have *different* units. A rate is written as a fraction.

A distance runner ran 26 miles in 4 hours. The distance-to-time rate is written

$$\frac{26 \text{ miles}}{4 \text{ hours}} = \frac{13 \text{ miles}}{2 \text{ hours}}$$

Writing the **simplest form of a rate** means writing it so that the two numbers that form the rate have no common factor other than 1.

EXAMPLE · 1

Write "6 roof supports for every 9 feet" as a rate in simplest form.

Solution

$$\frac{6 \text{ supports}}{9 \text{ feet}} = \frac{2 \text{ supports}}{3 \text{ feet}}$$

YOU TRY IT · 1

Write "15 pounds of fertilizer for 12 trees" as a rate in simplest form.

Your solution

Solution on p. S11

OBJECTIVE B   To write unit rates

 Point of Interest

According to a Gallup Poll, women see doctors more often than men do. On average, men visit the doctor 3.8 times per year, whereas women go to the doctor 5.8 times per year.

A **unit rate** is a rate in which the number in the denominator is 1.

$$\frac{\$3.25}{1 \text{ pound}}$$ or $3.25/pound is read "$3.25 per pound."

To find a unit rate, divide the number in the numerator of the rate by the number in the denominator of the rate.

A car traveled 344 miles on 16 gallons of gasoline. To find the miles per gallon (unit rate), divide the numerator of the rate by the denominator of the rate.

$$\frac{344 \text{ miles}}{16 \text{ gallons}}$$ is the rate. $16\overline{)344.0}$ 21.5 21.5 miles/gallon is the unit rate.

EXAMPLE · 2

Write "300 feet in 8 seconds" as a unit rate.

Solution

$$\frac{300 \text{ feet}}{8 \text{ seconds}}$$ $8\overline{)300.0}$ 37.5

37.5 feet/second

YOU TRY IT · 2

Write "260 miles in 8 hours" as a unit rate.

Your solution

Solution on p. S11

OBJECTIVE C To solve application problems

HOW TO 1 The table at the right shows air fares for some routes in the continental United States. Find the cost per mile for the four routes in order to determine the most expensive route and the least expensive route on the basis of mileage flown.

Long Routes	Miles	Fare
New York–Los Angeles	2475	$683
San Francisco–Dallas	1464	$536
Denver–Pittsburgh	1302	$525
Minneapolis–Hartford	1050	$483

PPT

Denver Airport

Integrating Technology

To calculate the costs per mile using a calculator, perform four divisions:

683 ÷ 2475 =

536 ÷ 1464 =

525 ÷ 1302 =

483 ÷ 1050 =

In each case, round the number in the display to the nearest hundredth.

Strategy
To find the cost per mile, divide the fare by the miles flown for each route. Compare the costs per mile to determine the most expensive and least expensive routes per mile.

Solution

New York–Los Angeles $\dfrac{683}{2475} \approx 0.28$

San Francisco–Dallas $\dfrac{536}{1464} \approx 0.37$

Denver–Pittsburgh $\dfrac{525}{1302} \approx 0.40$

Minneapolis–Hartford $\dfrac{483}{1050} = 0.46$

$0.28 < 0.37 < 0.40 < 0.46$

The Minneapolis–Hartford route is the most expensive per mile, and the New York–Los Angeles route is the least expensive per mile.

EXAMPLE 3

As an investor, Jung Ho purchased 100 shares of stock for $1500. One year later, Jung sold the 100 shares for $1800. What was his profit per share?

Strategy
To find Jung's profit per share:

- Find the total profit by subtracting the original cost ($1500) from the selling price ($1800).
- Find the profit per share (unit rate) by dividing the total profit by the number of shares of stock (100).

Solution
$1800 - 1500 = 300$

$300 \div 100 = 3$

Jung Ho's profit was $3/share.

YOU TRY IT 3

Erik Peltier, a jeweler, purchased 5 ounces of a gold alloy for $1625. Later, he sold the 5 ounces for $1720. What was Erik's profit per ounce?

Your strategy

Your solution

Solution on p. S11

4.2 EXERCISES

OBJECTIVE A   To write rates

For Exercises 1 to 8, write each phrase as a rate in simplest form.

1. 3 pounds of meat for 4 people

2. 30 ounces in 24 glasses

3. $80 for 12 boards

4. 84 cents for 3 candy bars

5. 300 miles on 15 gallons

6. 88 feet in 8 seconds

7. 16 gallons in 2 hours

8. 25 ounces in 5 minutes

 9. For television advertising rates, what units are **a.** in the numerator and **b.** in the denominator?

OBJECTIVE B   To write unit rates

 For Exercises 10 to 12, complete the unit rate.

10. 5 miles in ___ hour

11. 15 feet in ___ second

12. 5 grams of fat in ___ serving

For Exercises 13 to 22, write each phrase as a unit rate.

13. 10 feet in 4 seconds

14. 816 miles in 6 days

15. $3900 earned in 4 weeks

16. $51,000 earned in 12 months

17. 1100 trees planted on 10 acres

18. 3750 words on 15 pages

19. $131.88 earned in 7 hours

20. 628.8 miles in 12 hours

21. 409.4 miles on 11.5 gallons of gasoline

22. $11.05 for 3.4 pounds

OBJECTIVE C   To solve application problems

Miles per Dollar One measure of how expensive it is to drive your car is calculated as miles per dollar, which is the number of miles you drive on 1 dollar's worth of gasoline.

23. Suppose you get 26 miles per gallon of gasoline and gasoline costs $3.49 per gallon. Calculate your miles per dollar. Round to the nearest tenth.

24. Suppose you get 23 miles per gallon of gasoline and gasoline costs $3.15 per gallon. It costs you $44.10 to fill the tank. Calculate your miles per dollar. Round to the nearest tenth.

25. Corn Production See the news clipping at the right. Find the average number of bushels harvested from each acre of corn grown in Iowa. Round to the nearest hundredth.

26. Consumerism The Pierre family purchased a 250-pound side of beef for $365.75 and had it packaged. During the packaging, 75 pounds of beef were discarded as waste. What was the cost per pound for the packaged beef?

27. Manufacturing Regency Computer produced 5000 thumb drives for $13,268.16. Of the disks made, 122 did not meet company standards. What was the cost per disk for those disks that met company standards?

28. Advertising The advertising fee for a 30-second spot on the TV show *Deal or No Deal* is $165,000. The show averages 16.1 million viewers. (*Source: USA Today, December 18, 2006*) What is the advertiser's cost per viewer for a 30-second ad? Round to the nearest cent.

29. Demography The table at the right shows the population and area of three countries. The population density of a country is the number of people per square mile.

a. Which country has the least population density?

b. How many more people per square mile are there in India than in the United States? Round to the nearest whole number.

AP Images

Country	Population	Area (in square miles)
Australia	20,264,000	2,968,000
India	1,129,866,000	1,269,000
United States	301,140,000	3,619,000

Exchange Rates Another application of rates is in the area of international trade. Suppose a company in Canada purchases a shipment of sneakers from an American company. The Canadian company must exchange Canadian dollars for U.S. dollars in order to pay for the order. The number of Canadian dollars that are equivalent to 1 U.S. dollar is called the **exchange rate.**

30. The table at the right shows the exchange rates per U.S. dollar for three foreign countries and for the euro at the time of this writing.
a. How many euros would be paid for an order of American computer hardware costing $120,000?
b. Calculate the cost, in Japanese yen, of an American car costing $34,000.

Exchange Rates per U.S. Dollar	
Australian Dollar	1.0694
Canadian Dollar	1.0179
Japanese Yen	105.3300
The Euro	0.6483

31. Use the table in Exercise 30. What does the quantity 1.0179×2500 represent?

Applying the Concepts

32. Compensation You have a choice of receiving a wage of $34,000 per year, $2840 per month, $650 per week, or $18 per hour. Which pay choice would you take? Assume a 40-hour work week with 52 weeks per year.

33. The price–earnings ratio of a company's stock is one measure used by stock market analysts to assess the financial well-being of the company. Explain the meaning of the price–earnings ratio.

SECTION

4.3   Proportions

OBJECTIVE A   **To determine whether a proportion is true**

Point of Interest

Proportions were studied by the earliest mathematicians. Clay tablets uncovered by archaeologists show evidence of proportions in Egyptian and Babylonian cultures dating from 1800 B.C.

A **proportion** is an expression of the equality of two ratios or rates.

$$\frac{50 \text{ miles}}{4 \text{ gallons}} = \frac{25 \text{ miles}}{2 \text{ gallons}}$$

Note that the units of the numerators are the same and the units of the denominators are the same.

$$\frac{3}{6} = \frac{1}{2}$$

This is the equality of two ratios.

A proportion is **true** if the fractions are equal when written in lowest terms.

In any true proportion, the **cross products** are equal.

> **HOW TO · 1**   Is $\frac{2}{3} = \frac{8}{12}$ a true proportion?
>
> $$\frac{2}{3} \diagup\hspace{-0.6em}\diagdown \frac{8}{12} \quad \begin{array}{l} 3 \times 8 = 24 \\ 2 \times 12 = 24 \end{array}$$
>
> The cross products *are* equal.
> $\frac{2}{3} = \frac{8}{12}$ is a true proportion.

A proportion is **not true** if the fractions are not equal when reduced to lowest terms.

If the cross products are not equal, then the proportion is not true.

> **HOW TO · 2**   Is $\frac{4}{5} = \frac{8}{9}$ a true proportion?
>
> $$\frac{4}{5} \diagup\hspace{-0.6em}\diagdown \frac{8}{9} \quad \begin{array}{l} 5 \times 8 = 40 \\ 4 \times 9 = 36 \end{array}$$
>
> The cross products *are not* equal.
> $\frac{4}{5} = \frac{8}{9}$ is not a true proportion.

EXAMPLE · 1

Is $\frac{5}{8} = \frac{10}{16}$ a true proportion?

Solution

$$\frac{5}{8} \diagup\hspace{-0.6em}\diagdown \frac{10}{16} \quad \begin{array}{l} 8 \times 10 = 80 \\ 5 \times 16 = 80 \end{array}$$

The cross products are equal.
The proportion is true.

YOU TRY IT · 1

Is $\frac{6}{10} = \frac{9}{15}$ a true proportion?

Your solution

EXAMPLE · 2

Is $\frac{62 \text{ miles}}{4 \text{ gallons}} = \frac{33 \text{ miles}}{2 \text{ gallons}}$ a true proportion?

Solution

$$\frac{62}{4} \diagup\hspace{-0.6em}\diagdown \frac{33}{2} \quad \begin{array}{l} 4 \times 33 = 132 \\ 62 \times 2 = 124 \end{array}$$

The cross products are not equal.
The proportion is not true.

YOU TRY IT · 2

Is $\frac{\$32}{6 \text{ hours}} = \frac{\$90}{8 \text{ hours}}$ a true proportion?

Your solution

Solutions on p. S11

OBJECTIVE B   To solve proportions

Tips for Success

An important element of success is practice. We cannot do anything well if we do not practice it repeatedly. Practice is crucial to success in mathematics. In this objective you are learning a new skill: how to solve a proportion. You will need to practice this skill over and over again in order to be successful at it.

Sometimes one of the numbers in a proportion is unknown. In this case, it is necessary to *solve* the proportion.

To **solve a proportion,** find a number to replace the unknown so that the proportion is true.

HOW TO 3   Solve: $\frac{9}{6} = \frac{3}{n}$

$$\frac{9}{6} = \frac{3}{n}$$
$9 \times n = 6 \times 3$ • Find the cross products.
$9 \times n = 18$
$n = 18 \div 9$ • Think of $9 \times n = 18$ as
$n = 2$

Check:

$\frac{9}{6} \quad \frac{3}{2}$ $6 \times 3 = 18$
$9 \times 2 = 18$

EXAMPLE • 3

Solve $\frac{n}{12} = \frac{25}{60}$ and check.

Solution

$n \times 60 = 12 \times 25$ • Find the cross
$n \times 60 = 300$ products. Then
$n = 300 \div 60$ solve for n.
$n = 5$

Check:

$\frac{5}{12} \quad \frac{25}{60}$ $12 \times 25 = 300$
$5 \times 60 = 300$

YOU TRY IT • 3

Solve $\frac{n}{14} = \frac{3}{7}$ and check.

Your solution

EXAMPLE • 4

Solve $\frac{4}{9} = \frac{n}{16}$. Round to the nearest tenth.

Solution

$4 \times 16 = 9 \times n$ • Find the cross
$64 = 9 \times n$ products. Then
$64 \div 9 = n$ solve for n.
$7.1 \approx n$

Note: A rounded answer is an approximation. Therefore, the answer to a check will not be exact.

YOU TRY IT • 4

Solve $\frac{5}{7} = \frac{n}{20}$. Round to the nearest tenth.

Your solution

Solutions on p. S11

EXAMPLE · 5

Solve $\frac{28}{52} = \frac{7}{n}$ and check.

Solution

$28 \times n = 52 \times 7$ • Find the cross
$28 \times n = 364$ products. Then
$\quad\quad n = 364 \div 28$ solve for n.
$\quad\quad n = 13$

Check:

$$\frac{28}{52} \diagdown\!\!\!\!\diagup \begin{matrix} 7 \\ \hline 13 \end{matrix} \rightarrow \begin{matrix} 52 \times 7 = 364 \\ 28 \times 13 = 364 \end{matrix}$$

YOU TRY IT · 5

Solve $\frac{15}{20} = \frac{12}{n}$ and check.

Your solution

EXAMPLE · 6

Solve $\frac{15}{n} = \frac{8}{3}$. Round to the nearest hundredth.

Solution

$15 \times 3 = n \times 8$
$\quad\quad 45 = n \times 8$
$45 \div 8 = n$
$\quad 5.63 \approx n$

YOU TRY IT · 6

Solve $\frac{12}{n} = \frac{7}{4}$. Round to the nearest hundredth.

Your solution

EXAMPLE · 7

Solve $\frac{n}{9} = \frac{3}{1}$ and check.

Solution

$n \times 1 = 9 \times 3$
$n \times 1 = 27$
$\quad\quad n = 27 \div 1$
$\quad\quad n = 27$

Check:

$$\frac{27}{9} \diagdown\!\!\!\!\diagup \begin{matrix} 3 \\ \hline 1 \end{matrix} \rightarrow \begin{matrix} 9 \times 3 = 27 \\ 27 \times 1 = 27 \end{matrix}$$

YOU TRY IT · 7

Solve $\frac{n}{12} = \frac{4}{1}$ and check.

Your solution

Solutions on p. S11

OBJECTIVE C   To solve application problems

The application problems in this objective require you to write and solve a proportion. When setting up a proportion, remember to keep the same units in the numerator and the same units in the denominator.

EXAMPLE · 8

The dosage of a certain medication is 2 ounces for every 50 pounds of body weight. How many ounces of this medication are required for a person who weighs 175 pounds?

Strategy

To find the number of ounces of medication for a person weighing 175 pounds, write and solve a proportion using n to represent the number of ounces of medication for a 175-pound person.

Solution

$$\frac{2 \text{ ounces}}{50 \text{ pounds}} = \frac{n \text{ ounces}}{175 \text{ pounds}}$$

$2 \times 175 = 50 \times n$

$350 = 50 \times n$

$350 \div 50 = n$

$7 = n$

• The unit "ounces" is in the numerator. The unit "pounds" is in the denominator.

A 175-pound person requires 7 ounces of medication.

YOU TRY IT · 8

Three tablespoons of a liquid plant fertilizer are to be added to every 4 gallons of water. How many tablespoons of fertilizer are required for 10 gallons of water?

Your strategy

Your solution

EXAMPLE · 9

A mason determines that 9 cement blocks are required for a retaining wall 2 feet long. At this rate, how many cement blocks are required for a retaining wall that is 24 feet long?

Strategy

To find the number of cement blocks for a retaining wall 24 feet long, write and solve a proportion using n to represent the number of blocks required.

Solution

$$\frac{9 \text{ cement blocks}}{2 \text{ feet}} = \frac{n \text{ cement blocks}}{24 \text{ feet}}$$

$9 \times 24 = 2 \times n$

$216 = 2 \times n$

$216 \div 2 = n$

$108 = n$

A 24-foot retaining wall requires 108 cement blocks.

YOU TRY IT · 9

Twenty-four jars can be packed in 6 identical boxes. At this rate, how many jars can be packed in 15 boxes?

Your strategy

Your solution

Solutions on p. S11

4.3 EXERCISES

OBJECTIVE A   **To determine whether a proportion is true**

For Exercises 1 to 18, determine whether the proportion is true or not true.

1. $\dfrac{4}{8} = \dfrac{10}{20}$

2. $\dfrac{39}{48} = \dfrac{13}{16}$

3. $\dfrac{7}{8} = \dfrac{11}{12}$

4. $\dfrac{15}{7} = \dfrac{17}{8}$

5. $\dfrac{27}{8} = \dfrac{9}{4}$

6. $\dfrac{3}{18} = \dfrac{4}{19}$

7. $\dfrac{45}{135} = \dfrac{3}{9}$

8. $\dfrac{3}{4} = \dfrac{54}{72}$

9. $\dfrac{50 \text{ miles}}{2 \text{ gallons}} = \dfrac{25 \text{ miles}}{1 \text{ gallon}}$

10. $\dfrac{16 \text{ feet}}{10 \text{ seconds}} = \dfrac{24 \text{ feet}}{15 \text{ seconds}}$

11. $\dfrac{6 \text{ minutes}}{5 \text{ cents}} = \dfrac{30 \text{ minutes}}{25 \text{ cents}}$

12. $\dfrac{16 \text{ pounds}}{12 \text{ days}} = \dfrac{20 \text{ pounds}}{14 \text{ days}}$

13. $\dfrac{\$15}{4 \text{ pounds}} = \dfrac{\$45}{12 \text{ pounds}}$

14. $\dfrac{270 \text{ trees}}{6 \text{ acres}} = \dfrac{90 \text{ trees}}{2 \text{ acres}}$

15. $\dfrac{300 \text{ feet}}{4 \text{ rolls}} = \dfrac{450 \text{ feet}}{7 \text{ rolls}}$

16. $\dfrac{1 \text{ gallon}}{4 \text{ quarts}} = \dfrac{7 \text{ gallons}}{28 \text{ quarts}}$

17. $\dfrac{\$65}{5 \text{ days}} = \dfrac{\$26}{2 \text{ days}}$

18. $\dfrac{80 \text{ miles}}{2 \text{ hours}} = \dfrac{110 \text{ miles}}{3 \text{ hours}}$

 19. Suppose that in a true proportion you switch the numerator of the first fraction with the denominator of the second fraction. Must the result be another true proportion?

 20. Write a true proportion in which the cross products are equal to 36.

OBJECTIVE B   To solve proportions

21. Consider the proportion $\frac{n}{7} = \frac{9}{21}$ in Exercise 23. In lowest terms, $\frac{9}{21} = \frac{3}{7}$. Will solving the proportion $\frac{n}{7} = \frac{3}{7}$ give the same result for n as found in Exercise 23?

For Exercises 22 to 41, solve. Round to the nearest hundredth, if necessary.

22. $\frac{n}{4} = \frac{6}{8}$

23. $\frac{n}{7} = \frac{9}{21}$

24. $\frac{12}{18} = \frac{n}{9}$

25. $\frac{7}{21} = \frac{35}{n}$

26. $\frac{6}{n} = \frac{24}{36}$

27. $\frac{3}{n} = \frac{15}{10}$

28. $\frac{n}{6} = \frac{2}{3}$

29. $\frac{5}{12} = \frac{n}{144}$

30. $\frac{n}{5} = \frac{7}{8}$

31. $\frac{4}{n} = \frac{9}{5}$

32. $\frac{5}{12} = \frac{n}{8}$

33. $\frac{36}{20} = \frac{12}{n}$

34. $\frac{n}{15} = \frac{21}{12}$

35. $\frac{40}{n} = \frac{15}{8}$

36. $\frac{28}{8} = \frac{12}{n}$

37. $\frac{n}{30} = \frac{65}{120}$

38. $\frac{0.3}{5.6} = \frac{n}{25}$

39. $\frac{1.3}{16} = \frac{n}{30}$

40. $\frac{0.7}{9.8} = \frac{3.6}{n}$

41. $\frac{1.9}{7} = \frac{13}{n}$

OBJECTIVE C   To solve application problems

42. Jesse walked 3 miles in 40 minutes. Let n be the number of miles Jesse can walk in 60 minutes at the same rate. To determine how many miles Jesse can walk in 60 minutes, a student used the proportion $\frac{40}{3} = \frac{60}{n}$. Is this a valid proportion to use in solving this problem?

For Exercises 43 to 61, solve. Round to the nearest hundredth.

43. **Nutrition** A 6-ounce package of Puffed Wheat contains 600 calories. How many calories are in a 0.5-ounce serving of the cereal?

44. **Health** Using the data at the right and a figure of 300 million for the number of Americans, determine the number of morbidly obese Americans.

45. **Fuel Efficiency** A car travels 70.5 miles on 3 gallons of gas. Find the distance the car can travel on 14 gallons of gas.

46. **Landscaping** Ron Stokes uses 2 pounds of fertilizer for every 100 square feet of lawn for landscape maintenance. At this rate, how many pounds of fertilizer did he use on a lawn that measures 3500 square feet?

47. **Gardening** A nursery prepares a liquid plant food by adding 1 gallon of water for each 2 ounces of plant food. At this rate, how many gallons of water are required for 25 ounces of plant food?

48. **Masonry** A brick wall 20 feet in length contains 1040 bricks. At the same rate, how many bricks would it take to build a wall 48 feet in length?

49. **Cartography** The scale on the map at the right is "1.25 inches equals 10 miles." Find the distance between Carlsbad and Del Mar, which are 2 inches apart on the map.

50. **Architecture** The scale on the plans for a new house is "1 inch equals 3 feet." Find the width and the length of a room that measures 5 inches by 8 inches on the drawing.

51. **Medicine** The dosage for a medication is $\frac{1}{3}$ ounce for every 40 pounds of body weight. At this rate, how many ounces of medication should a physician prescribe for a patient who weighs 150 pounds? Write the answer as a decimal.

52. **Banking** A bank requires a monthly payment of $33.45 on a $2500 loan. At the same rate, find the monthly payment on a $10,000 loan.

53. **Elections** A pre-election survey showed that 2 out of every 3 eligible voters would cast ballots in the county election. At this rate, how many people in a county of 240,000 eligible voters would vote in the election?

54. **Interior Design** A paint manufacturer suggests using 1 gallon of paint for every 400 square feet of wall. At this rate, how many gallons of paint would be required for a room that has 1400 square feet of wall?

55. **Insurance** A 60-year-old male can obtain $10,000 of life insurance for $35.35 per month. At this rate, what is the monthly cost for $50,000 of life insurance?

In the News

Number of Obese Americans Increasing

In the past 20 years, the number of obese Americans (those at least 30 pounds overweight) has doubled. The number of morbidly obese (those at least 100 pounds overweight) has quadrupled to 1 in 50.

Source: Time, July 9, 2006

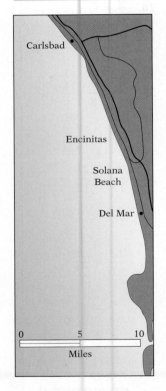

Carlsbad

Encinitas

Solana Beach

Del Mar

0   5   10

Miles

56. Food Waste At the rate given in the news clipping, find the cost of food wasted yearly by **a.** the average family of three and **b.** the average family of five.

In the News

How Much Food Do You Waste?

In the United States, the estimated cost of food wasted each year by the average family of four is $590.

Source: University of Arizona

57. Manufacturing Suppose a computer chip manufacturer knows from experience that in an average production run of 2000 circuit boards, 60 will be defective. How many defective circuit boards can be expected in a run of 25,000 circuit boards?

58. Investments You own 240 shares of stock in a computer company. The company declares a stock split of 5 shares for every 3 owned. How many shares of stock will you own after the stock split?

59. Physics The ratio of weight on the moon to weight on Earth is 1:6. If a bowling ball weighs 16 pounds on Earth, what would it weigh on the moon?

60. Automobiles When engineers designed a new car, they first built a model of the car. The ratio of the size of a part on the model to the actual size of the part is 2:5. If a door is 1.3 feet long on the model, what is the length of the door on the car?

61. Investments Carlos Capasso owns 50 shares of Texas Utilities that pay dividends of $153. At this rate, what dividend would Carlos receive after buying 300 additional shares of Texas Utilities?

Applying the Concepts

62. Publishing In the first quarter of 2008, *USA Today* reported that Eckhart Tolle's *A New Earth* outsold John Grisham's *The Appeal* by 3.7 copies to 1. Explain how a proportion can be used to determine the number of copies of *A New Earth* sold given the number of copies of *The Appeal* sold.

63. Social Security According to the Social Security Administration, the numbers of workers per retiree in the future are expected to be as given in the table below.

Year	2020	2030	2040
Number of workers per retiree	2.5	2.1	2.0

Why is the shrinking number of workers per retiree of importance to the Social Security Administration?

64. Elections A survey of voters in a city claimed that 2 people of every 5 who voted cast a ballot in favor of city amendment A and that 3 people of every 4 who voted cast a ballot against amendment A. Is this possible? Explain your answer.

65. Write a word problem that requires solving a proportion to find the answer.

FOCUS ON PROBLEM SOLVING

Looking for a Pattern

A very useful problem-solving strategy is looking for a pattern.

Problem A legend says that a peasant invented the game of chess and gave it to a very rich king as a present. The king so enjoyed the game that he gave the peasant the choice of anything in the kingdom. The peasant's request was simple: "Place one grain of wheat on the first square, 2 grains on the second square, 4 grains on the third square, 8 on the fourth square, and continue doubling the number of grains until the last square of the chessboard is reached." How many grains of wheat must the king give the peasant?

Solution A chessboard consists of 64 squares. To find the total number of grains of wheat on the 64 squares, we begin by looking at the amount of wheat on the first few squares.

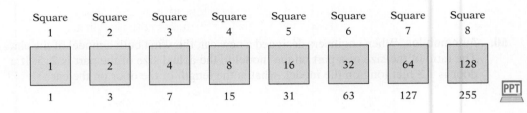

Square 1	Square 2	Square 3	Square 4	Square 5	Square 6	Square 7	Square 8
1	2	4	8	16	32	64	128
1	3	7	15	31	63	127	255

The bottom row of numbers represents the sum of the number of grains of wheat up to and including that square. For instance, the number of grains of wheat on the first 7 squares is $1 + 2 + 4 + 8 + 16 + 32 + 64 = 127$.

Notice that the number of grains of wheat on a square can be expressed as a power of 2.

The number of grains on square $n = 2^{n-1}$.

For example, the number of grains on square $7 = 2^{7-1} = 2^6 = 64$.

A second pattern of interest is that **the number *below* a square** (the total number of grains up to and including that square) **is 1 less than the number of grains of wheat *on the next square*.** For example, the number *below* square 7 is 1 less than the number on square 8 ($128 - 1 = 127$). From this observation, the number of grains of wheat on the first 8 squares is the number on square 8 (128) plus 1 less than the number on square 8 (127): The total number of grains of wheat on the first 8 squares is $128 + 127 = 255$.

From this observation,

$$\begin{array}{ccc} \text{Number of grains of} & \text{number of grains} & \text{1 less than the number} \\ \text{wheat on the chessboard} = & \text{on square 64} + & \text{of grains on square 64} \end{array}$$

$$= 2^{64-1} + (2^{64-1} - 1)$$
$$= 2^{63} + 2^{63} - 1 \approx 18{,}000{,}000{,}000{,}000{,}000{,}000{,}000$$

To give you an idea of the magnitude of this number, this is more wheat than has been produced in the world since chess was invented.

The same king decided to have a banquet in the long banquet room of the palace to celebrate the invention of chess. The king had 50 square tables, and each table could seat only one person on each side. The king pushed the tables together to form one long banquet table. How many people could sit at this table? *Hint:* Try constructing a pattern by using 2 tables, 3 tables, and 4 tables.

PROJECTS AND GROUP ACTIVITIES

The Golden Ratio There are certain designs that have been repeated over and over in both art and architecture. One of these involves the **golden rectangle.**

A golden rectangle is drawn at the right. Begin with a square that measures, say, 2 inches on a side. Let *A* be the midpoint of a side (halfway between two corners). Now measure the distance from *A* to *B*. Place this length along the bottom of the square, starting at *A*. The resulting rectangle is a golden rectangle.

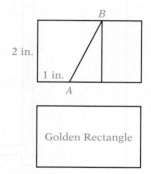

Golden Rectangle

The **golden ratio** is the ratio of the length of the golden rectangle to its width. If you have drawn the rectangle following the procedure above, you will find that the golden ratio is approximately 1.6 to 1.

The golden ratio appears in many different situations. Some historians claim that some of the great pyramids of Egypt are based on the golden ratio. The drawing at the right shows the Pyramid of Giza, which dates from approximately 2600 B.C. The ratio of the height to a side of the base is approximately 1.6 to 1.

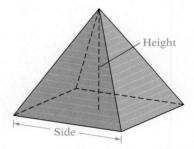

1. There are instances of the golden rectangle in the Mona Lisa painted by Leonardo da Vinci. Do some research on this painting and write a few paragraphs summarizing your findings.

2. What do 3 × 5 and 5 × 8 index cards have to do with the golden rectangle?

3. What does the United Nations Building in New York City have to do with the golden rectangle?

4. When was the Parthenon in Athens, Greece, built? What does the front of that building have to do with the golden rectangle?

Drawing the Floor Plans for a Building

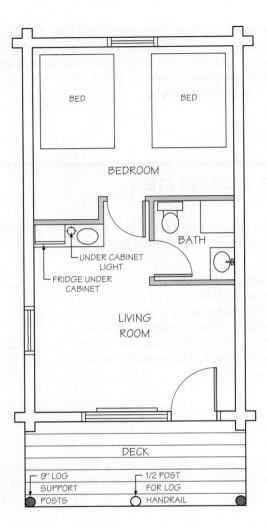

BED

BED

BEDROOM

BATH

UNDER CABINET LIGHT

FRIDGE UNDER CABINET

LIVING ROOM

DECK

9" LOG SUPPORT POSTS

1/2 POST FOR LOG HANDRAIL

The drawing at the left is a sketch of the floor plan for a cabin at a resort in the mountains of Utah. The measurements are missing. Assume that you are the architect and will finish the drawing. You will have to decide the size of the rooms and put in the measurements to scale.

Design a cabin that you would like to own. Select a scale and draw all the rooms to scale.

If you are interested in architecture, visit an architect who is using CAD (computer-aided design) software to create a floor plan. Computer technology has revolutionized the field of architectural design.

The U.S. House of Representatives

The framers of the Constitution decided to use a ratio to determine the number of representatives from each state. It was determined that each state would have one representative for every 30,000 citizens, with a minimum of one representative. Congress has changed this ratio over the years, and we now have 435 representatives.

Find the number of representatives from your state. Determine the ratio of citizens to representatives. Also do this for the most populous state and for the least populous state.

You might consider getting information on the number of representatives for each state and the populations of different states via the Internet.

CHAPTER 4

SUMMARY

KEY WORDS	EXAMPLES
A *ratio* is the comparison of two quantities with the same units. A ratio can be written in three ways: as a fraction, as two numbers separated by a colon (:), or as two numbers separated by the word *to*. A ratio is in *simplest form* when the two numbers do not have a common factor. [4.1A, p. 174]	The comparison 16 to 24 ounces can be written as a ratio in simplest form as $\frac{2}{3}$, 2:3, or 2 to 3.
A *rate* is the comparison of two quantities with different units. A rate is written as a fraction. A rate is in *simplest form* when the numbers that form the rate do not have a common factor. [4.2A, p. 178]	You earned $63 for working 6 hours. The rate is written in simplest form as $\frac{\$21}{2 \text{ hours}}$.
A *unit rate* is a rate in which the number in the denominator is 1. [4.2B, p. 178]	You traveled 144 miles in 3 hours. The unit rate is 48 miles per hour.
A *proportion* is an expression of the equality of two ratios or rates. A proportion is true if the fractions are equal when written in lowest terms; in any true proportion, the *cross products* are equal. A proportion is not true if the fractions are not equal when written in lowest terms; if the cross products are not equal, the proportion is not true. [4.3A, p. 182]	The proportion $\frac{3}{5} = \frac{12}{20}$ is true because the cross products are equal: $3 \times 20 = 5 \times 12$. The proportion $\frac{3}{4} = \frac{12}{20}$ is not true because the cross products are not equal: $3 \times 20 \neq 4 \times 12$.

ESSENTIAL RULES AND PROCEDURES	EXAMPLES
To find a unit rate, divide the number in the numerator of the rate by the number in the denominator of the rate. [4.2B, p. 178]	You earned $41 for working 4 hours. $$41 \div 4 = 10.25$$ The unit rate is $10.25/hour.
To solve a proportion, find a number to replace the unknown so that the proportion is true. [4.3B, p. 183]	$$\frac{6}{24} = \frac{9}{n}$$ $6 \times n = 24 \times 9$ • Find the cross products. $6 \times n = 216$ $n = 216 \div 6$ $n = 36$
To set up a proportion, keep the same units in the numerator and the same units in the denominator. [4.3C, p. 184]	Three machines fill 5 cereal boxes per minute. How many boxes can 8 machines fill per minute? $$\frac{3 \text{ machines}}{5 \text{ cereal boxes}} = \frac{8 \text{ machines}}{n \text{ cereal boxes}}$$

CHAPTER 4

CONCEPT REVIEW

Test your knowledge of the concepts presented in this chapter. Answer each question.
Then check your answers against the ones provided in the Answer Section.

1. If the units in a comparison are different, is it a ratio or a rate?

2. How do you find a unit rate?

3. How do you write the ratio $\frac{6}{7}$ using a colon?

4. How do you write the ratio 12 : 15 in simplest form?

5. How do you write the rate $\frac{342 \text{ miles}}{9.5 \text{ gallons}}$ as a unit rate?

6. When is a proportion true?

7. How do you solve a proportion?

8. How do the units help you to set up a proportion?

9. How do you check the solution of a proportion?

10. How do you write the ratio 19 : 6 as a fraction?

CHAPTER 4

REVIEW EXERCISES

1. Determine whether the proportion is true or not true.

$$\frac{2}{9} = \frac{10}{45}$$

2. Write the comparison 32 dollars to 80 dollars as a ratio in simplest form using a fraction, a colon (:), and the word *to*.

3. Write "250 miles in 4 hours" as a unit rate.

4. Determine whether the proportion is true or not true.

$$\frac{8}{15} = \frac{32}{60}$$

5. Solve the proportion.

$$\frac{16}{n} = \frac{4}{17}$$

6. Write "$500 earned in 40 hours" as a unit rate.

7. Write "$8.75 for 5 pounds" as a unit rate.

8. Write the comparison 8 feet to 28 feet as a ratio in simplest form using a fraction, a colon (:), and the word *to*.

9. Solve the proportion.

$$\frac{n}{8} = \frac{9}{2}$$

10. Solve the proportion. Round to the nearest hundredth.

$$\frac{18}{35} = \frac{10}{n}$$

11. Write the comparison 6 inches to 15 inches as a ratio in simplest form using a fraction, a colon (:), and the word *to*.

12. Determine whether the proportion is true or not true.

$$\frac{3}{8} = \frac{10}{24}$$

13. Write "$35 in 4 hours" as a rate in simplest form.

14. Write "326.4 miles on 12 gallons" as a unit rate.

15. Write the comparison 12 days to 12 days as a ratio in simplest form using a fraction, a colon (:), and the word *to*.

16. Determine whether the proportion is true or not true.

$$\frac{5}{7} = \frac{25}{35}$$

17. Solve the proportion. Round to the nearest hundredth.

$$\frac{24}{11} = \frac{n}{30}$$

18. Write "100 miles in 3 hours" as a rate in simplest form.

19. **Business** In 5 years, the price of a calculator went from $80 to $48. What is the ratio, as a fraction in simplest form, of the decrease in price to the original price?

20. **Taxes** The property tax on a $245,000 home is $4900. At the same rate, what is the property tax on a home valued at $320,000?

21. **Consumerism** Rita Sterling bought a computer system for $2400. Five years later, she sold the computer for $900. Find the ratio of the amount she received for the computer to the cost of the computer.

22. **Manufacturing** The total cost of manufacturing 1000 camera phones was $36,600. Of the phones made, 24 did not pass inspection. What is the cost per phone of the phones that *did* pass inspection?

23. **Masonry** A brick wall 40 feet in length contains 448 concrete blocks. At the same rate, how many blocks would it take to build a wall that is 120 feet in length?

24. **Advertising** A retail computer store spends $30,000 a year on radio advertising and $12,000 on newspaper advertising. Find the ratio, as a fraction in simplest form, of radio advertising to newspaper advertising.

25. **Consumerism** A 15-pound turkey costs $13.95. What is the cost per pound?

26. **Travel** Mahesh drove 198.8 miles in 3.5 hours. Find the average number of miles he drove per hour.

27. **Insurance** An insurance policy costs $9.87 for every $1000 of insurance. At this rate, what is the cost of $50,000 of insurance?

28. **Investments** Pascal Hollis purchased 80 shares of stock for $3580. What was the cost per share?

29. **Landscaping** Monique uses 1.5 pounds of fertilizer for every 200 square feet of lawn. How many pounds of fertilizer will she have to use on a lawn that measures 3000 square feet?

30. **Real Estate** A house had an original value of $160,000, but its value increased to $240,000 in 2 years. Find the ratio, as a fraction in simplest form, of the increase to the original value.

CHAPTER 4

TEST

1. Write "$46,036.80 earned in 12 months" as a unit rate.

2. Write the comparison 40 miles to 240 miles as a ratio in simplest form using a fraction, a colon (:), and the word *to*.

3. Write "18 supports for every 8 feet" as a rate in simplest form.

4. Determine whether the proportion is true or not true.
$$\frac{40}{125} = \frac{5}{25}$$

5. Write the comparison 12 days to 8 days as a ratio in simplest form using a fraction, a colon (:), and the word *to*.

6. Solve the proportion.
$$\frac{5}{12} = \frac{60}{n}$$

7. Write "256.2 miles on 8.4 gallons of gas" as a unit rate.

8. Write the comparison 27 dollars to 81 dollars as a ratio in simplest form using a fraction, a colon (:), and the word *to*.

9. Determine whether the proportion is true or not true.
$$\frac{5}{14} = \frac{25}{70}$$

10. Solve the proportion.
$$\frac{n}{18} = \frac{9}{4}$$

11. Write "$81 for 6 boards" as a rate in simplest form.

12. Write the comparison 18 feet to 30 feet as a ratio in simplest form using a fraction, a colon (:), and the word *to*.

13. **Investments** Fifty shares of a utility stock pay a dividend of $62.50. At the same rate, what is the dividend paid on 500 shares of the utility stock?

14. **Electricity** A transformer has 40 turns in the primary coil and 480 turns in the secondary coil. State the ratio of the number of turns in the primary coil to the number of turns in the secondary coil.

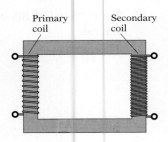

Primary coil   Secondary coil

15. **Travel** A plane travels 2421 miles in 4.5 hours. Find the plane's speed in miles per hour.

16. **Physiology** A research scientist estimates that the human body contains 88 pounds of water for every 100 pounds of body weight. At this rate, estimate the number of pounds of water in a college student who weighs 150 pounds.

17. **Business** If 40 feet of lumber costs $69.20, what is the per-foot cost of the lumber?

18. **Medicine** The dosage of a certain medication is $\frac{1}{4}$ ounce for every 50 pounds of body weight. How many ounces of this medication are required for a person who weighs 175 pounds? Write the answer as a decimal.

19. **Sports** A basketball team won 20 games and lost 5 games during the season. Write, as a fraction in simplest form, the ratio of the number of games won to the total number of games played.

20. **Manufacturing** A computer manufacturer discovers through experience that an average of 3 defective hard drives are found in every 100 hard drives manufactured. How many defective hard drives are expected to be found in the production of 1200 hard drives?

CUMULATIVE REVIEW EXERCISES

1. Subtract: 20,095
                 − 10,937

2. Write $2 \cdot 2 \cdot 2 \cdot 2 \cdot 3 \cdot 3 \cdot 3$ in exponential notation.

3. Simplify: $4 - (5 - 2)^2 \div 3 + 2$

4. Find the prime factorization of 160.

5. Find the LCM of 9, 12, and 18.

6. Find the GCF of 28 and 42.

7. Write $\frac{40}{64}$ in simplest form.

8. Find $4\frac{7}{15}$ more than $3\frac{5}{6}$.

9. What is $4\frac{5}{9}$ less than $10\frac{1}{6}$?

10. Multiply: $\frac{11}{12} \times 3\frac{1}{11}$

11. Find the quotient of $3\frac{1}{3}$ and $\frac{5}{7}$.

12. Simplify: $\left(\frac{2}{5} + \frac{3}{4}\right) \div \frac{3}{2}$

13. Write 4.0709 in words.

14. Round 2.09762 to the nearest hundredth.

15. Divide: $8.09\overline{)16.0976}$
    Round to the nearest thousandth.

16. Convert $0.06\frac{2}{3}$ to a fraction.

17. Write the comparison 25 miles to 200 miles as a ratio in simplest form using a fraction.

18. Write "87 cents for 6 pencils" as a rate in simplest form.

19. Write "250.5 miles on 7.5 gallons of gas" as a unit rate.

20. Solve $\frac{40}{n} = \frac{160}{17}$.

21. Travel A car traveled 457.6 miles in 8 hours. Find the car's speed in miles per hour.

22. Solve the proportion.

$$\frac{12}{5} = \frac{n}{15}$$

23. Banking You had $1024 in your checking account. You then wrote checks for $192 and $88. What is your new checking account balance?

24. Finance Malek Khatri buys a tractor for $32,360. A down payment of $5000 is required. The balance remaining is paid in 48 equal monthly installments. What is the monthly payment?

25. Homework Assignments Yuko is assigned to read a book containing 175 pages. She reads $\frac{2}{5}$ of the book during Thanksgiving vacation. How many pages of the assignment remain to be read?

26. Real Estate A building contractor bought $2\frac{1}{3}$ acres of land for $84,000. What was the cost of each acre?

27. Consumerism Benjamin Eli bought a shirt for $45.58 and a tie for $19.18. He used a $100 bill to pay for the purchases. Find the amount of change.

28. Compensation If you earn an annual salary of $41,619, what is your monthly salary?

29. Erosion A soil conservationist estimates that a river bank is eroding at the rate of 3 inches every 6 months. At this rate, how many inches will be eroded in 50 months?

30. Medicine The dosage of a certain medication is $\frac{1}{2}$ ounce for every 50 pounds of body weight. How many ounces of this medication are required for a person who weighs 160 pounds? Write the answer as a decimal.

Percents

OBJECTIVES

SECTION 5.1
A To write a percent as a fraction or a decimal
B To write a fraction or a decimal as a percent

SECTION 5.2
A To find the amount when the percent and the base are given
B To solve application problems

SECTION 5.3
A To find the percent when the base and amount are given
B To solve application problems

SECTION 5.4
A To find the base when the percent and amount are given
B To solve application problems

SECTION 5.5
A To solve percent problems using proportions
B To solve application problems

ARE YOU READY?

Take the Chapter 5 Prep Test to find out if you are ready to learn to:

- Convert fractions, decimals, and percents
- Solve percent problems using the basic percent equation
- Solve percent problems using proportions

PREP TEST

Do these exercises to prepare for Chapter 5.

For Exercises 1 to 6, multiply or divide.

1. $19 \times \dfrac{1}{100}$

2. 23×0.01

3. 0.47×100

4. $0.06 \times 47,500$

5. $60 \div 0.015$

6. $8 \div \dfrac{1}{4}$

7. Multiply $\dfrac{5}{8} \times 100$. Write the answer as a decimal.

8. Write $\dfrac{200}{3}$ as a mixed number.

9. Divide $28 \div 16$. Write the answer as a decimal.

SECTION

5.1 Introduction to Percents

OBJECTIVE A   **To write a percent as a fraction or a decimal**

Percent means "parts of 100." In the figure at the right, there are 100 parts. Because 13 of the 100 parts are shaded, 13% of the figure is shaded. The symbol % is the **percent sign.**

In most applied problems involving percents, it is necessary either to rewrite a percent as a fraction or a decimal or to rewrite a fraction or a decimal as a percent.

To write a percent as a fraction, remove the percent sign and multiply by $\frac{1}{100}$.

$$13\% = 13 \times \frac{1}{100} = \frac{13}{100}$$

To write a percent as a decimal, remove the percent sign and multiply by 0.01.

✓ **Take Note**

Recall that division is defined as multiplication by the reciprocal. Therefore, multiplying by $\frac{1}{100}$ is equivalent to dividing by 100.

$$13\% \quad = \quad 13 \times 0.01 \quad = \quad 0.13$$

Move the decimal point two places to the left. Then remove the percent sign.

EXAMPLE • 1

a. Write 120% as a fraction.
b. Write 120% as a decimal.

Solution

a. $120\% = 120 \times \frac{1}{100} = \frac{120}{100}$

$= 1\frac{1}{5}$

b. $120\% = 120 \times 0.01 = 1.2$
Note that percents larger than 100 are greater than 1.

YOU TRY IT • 1

a. Write 125% as a fraction.
b. Write 125% as a decimal.

Your solution

EXAMPLE • 2

Write $16\frac{2}{3}\%$ as a fraction.

Solution $16\frac{2}{3}\% = 16\frac{2}{3} \times \frac{1}{100}$

$= \frac{50}{3} \times \frac{1}{100} = \frac{50}{300} = \frac{1}{6}$

YOU TRY IT • 2

Write $33\frac{1}{3}\%$ as a fraction.

Your solution

EXAMPLE • 3

Write 0.5% as a decimal.

Solution $0.5\% = 0.5 \times 0.01 = 0.005$

YOU TRY IT • 3

Write 0.25% as a decimal.

Your solution

Solutions on pp. S11–S12

OBJECTIVE B To write a fraction or a decimal as a percent

A fraction or a decimal can be written as a percent by multiplying by 100%.

HOW TO · 1 Write $\frac{3}{8}$ as a percent.

$$\frac{3}{8} = \frac{3}{8} \times 100\% = \frac{3}{8} \times \frac{100}{1}\% = \frac{300}{8}\% = 37\frac{1}{2}\% \text{ or } 37.5\%$$

HOW TO · 2 Write 0.37 as a percent.

$$0.37 \quad = \quad 0.37 \times 100\% \quad = \quad 37\%$$

Move the decimal point two places to the right. Then write the percent sign.

EXAMPLE · 4

Write 0.015, 2.15, and $0.33\frac{1}{3}$ as percents.

Solution

$$0.015 = 0.015 \times 100\%$$
$$= 1.5\%$$

$$2.15 = 2.15 \times 100\% = 215\%$$

$$0.33\frac{1}{3} = 0.33\frac{1}{3} \times 100\%$$

$$= 33\frac{1}{3}\%$$

YOU TRY IT · 4

Write 0.048, 3.67, and $0.62\frac{1}{2}$ as percents.

Your solution

EXAMPLE · 5

Write $\frac{2}{3}$ as a percent.
Write the remainder in fractional form.

Solution $\quad \frac{2}{3} = \frac{2}{3} \times 100\% = \frac{200}{3}\%$

$$= 66\frac{2}{3}\%$$

YOU TRY IT · 5

Write $\frac{5}{6}$ as a percent.
Write the remainder in fractional form.

Your solution

EXAMPLE · 6

Write $2\frac{2}{7}$ as a percent.
Round to the nearest tenth.

Solution $\quad 2\frac{2}{7} = \frac{16}{7} = \frac{16}{7} \times 100\%$

$$= \frac{1600}{7}\% \approx 228.6\%$$

YOU TRY IT · 6

Write $1\frac{4}{9}$ as a percent.
Round to the nearest tenth.

Your solution

Solutions on p. S12

5.1 EXERCISES

For Exercises 1 to 16, write as a fraction and as a decimal.

1. 25%   **2.** 40%   **3.** 130%   **4.** 150%

5. 100%   **6.** 87%   **7.** 73%   **8.** 45%

9. 383%   **10.** 425%   **11.** 70%   **12.** 55%

13. 88%   **14.** 64%   **15.** 32%   **16.** 18%

For Exercises 17 to 28, write as a fraction.

17. $66\frac{2}{3}\%$ **18.** $12\frac{1}{2}\%$ **19.** $83\frac{1}{3}\%$ **20.** $3\frac{1}{8}\%$ **21.** $11\frac{1}{9}\%$ **22.** $\frac{3}{8}\%$

23. $45\frac{5}{11}\%$ **24.** $15\frac{3}{8}\%$ **25.** $4\frac{2}{7}\%$ **26.** $5\frac{3}{4}\%$ **27.** $6\frac{2}{3}\%$ **28.** $8\frac{2}{3}\%$

For Exercises 29 to 40, write as a decimal.

29. 6.5%   **30.** 9.4%   **31.** 12.3%   **32.** 16.7%   **33.** 0.55%   **34.** 0.45%

35. 8.25%   **36.** 6.75%   **37.** 5.05%   **38.** 3.08%   **39.** 2%   **40.** 7%

41. When a certain percent is written as a fraction, the result is an improper fraction. Is the percent less than, equal to, or greater than 100%?

For Exercises 42 to 53, write as a percent.

42. 0.16   **43.** 0.73   **44.** 0.05   **45.** 0.01   **46.** 1.07   **47.** 2.94

48. 0.004   **49.** 0.006   **50.** 1.012   **51.** 3.106   **52.** 0.8   **53.** 0.7

For Exercises 54 to 65, write as a percent. If necessary, round to the nearest tenth of a percent.

54. $\dfrac{27}{50}$ **55.** $\dfrac{37}{100}$ **56.** $\dfrac{1}{3}$ **57.** $\dfrac{2}{5}$ **58.** $\dfrac{5}{8}$ **59.** $\dfrac{1}{8}$

60. $\dfrac{1}{6}$ **61.** $1\dfrac{1}{2}$ **62.** $\dfrac{7}{40}$ **63.** $1\dfrac{2}{3}$ **64.** $1\dfrac{7}{9}$ **65.** $\dfrac{7}{8}$

For Exercises 66 to 73, write as a percent. Write the remainder in fractional form.

66. $\dfrac{15}{50}$ **67.** $\dfrac{12}{25}$ **68.** $\dfrac{7}{30}$ **69.** $\dfrac{1}{3}$

70. $2\dfrac{3}{8}$ **71.** $1\dfrac{2}{3}$ **72.** $2\dfrac{1}{6}$ **73.** $\dfrac{7}{8}$

 74. Does a mixed number represent a percent greater than 100% or less than 100%?

 75. A decimal number less than 0 has zeros in the tenths and hundredths places. Does the decimal represent a percent greater than 1% or less than 1%?

76. Write the part of the square that is shaded as a fraction, as a decimal, and as a percent. Write the part of the square that is not shaded as a fraction, as a decimal, and as a percent.

Applying the Concepts

 77. **The Food Industry** In a survey conducted by Opinion Research Corp. for Lloyd's Barbeque Co., people were asked to name their favorite barbeque side dishes. 38% named corn on the cob, 35% named cole slaw, 11% named corn bread, and 10% named fries. What percent of those surveyed named something other than corn on the cob, cole slaw, corn bread, or fries?

78. **Consumerism** A sale on computers advertised $\dfrac{1}{3}$ off the regular price. What percent of the regular price does this represent?

79. **Consumerism** A suit was priced at 50% off the regular price. What fraction of the regular price does this represent?

80. **Elections** If $\dfrac{2}{5}$ of the population voted in an election, what percent of the population did not vote?

5.2   Percent Equations: Part 1

OBJECTIVE A   **To find the amount when the percent and the base are given**

A real estate broker receives a payment that is 4% of a $285,000 sale. To find the amount the broker receives requires answering the question "4% of $285,000 is what?"

This sentence can be written using mathematical symbols and then solved for the unknown number.

4%	of	$285,000	is	what?
↓	↓	↓	↓	↓

of   is written as × (times)
is   is written as = (equals)
what is written as n (the unknown number)

$$\boxed{\begin{array}{c}\text{Percent}\\4\%\end{array}} \times \boxed{\begin{array}{c}\text{base}\\285{,}000\end{array}} = \boxed{\begin{array}{c}\text{amount}\\n\end{array}}$$

$$0.04 \times 285{,}000 = n$$
$$11{,}400 = n$$

Note that the percent is written as a decimal.

The broker receives a payment of $11,400.

The solution was found by solving the **basic percent equation** for amount.

The Basic Percent Equation

$$\boxed{\text{Percent}} \times \boxed{\text{base}} = \boxed{\text{amount}}$$

In most cases, the percent is written as a decimal before the basic percent equation is solved. However, some percents are more easily written as a fraction than as a decimal. For example,

$$33\frac{1}{3}\% = \frac{1}{3} \qquad 66\frac{2}{3}\% = \frac{2}{3} \qquad 16\frac{2}{3}\% = \frac{1}{6} \qquad 83\frac{1}{3}\% = \frac{5}{6}$$

EXAMPLE · 1

Find 5.7% of 160.

Solution
Percent × base = amount
$$0.057 \times 160 = n$$
$$9.12 = n$$

• The word *Find* is used instead of the words *what is*.

YOU TRY IT · 1

Find 6.3% of 150.

Your solution

EXAMPLE · 2

What is $33\frac{1}{3}\%$ of 90?

Solution
Percent × base = amount
$$\frac{1}{3} \times 90 = n$$
$$30 = n$$

• $33\frac{1}{3}\% = \frac{1}{3}$

YOU TRY IT · 2

What is $16\frac{2}{3}\%$ of 66?

Your solution

Solutions on p. S12

OBJECTIVE B   To solve application problems

Solving percent problems requires identifying the three elements of the basic percent equation. Recall that these three parts are the *percent,* the *base,* and the *amount.* Usually the base follows the phrase "percent of."

 During a recent year, Americans gave $212 billion to charities. The circle graph at the right shows where that money came from. Use these data for Example 3 and You Try It 3.

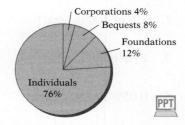

Corporations 4%
Bequests 8%
Foundations 12%
Individuals 76%

Charitable Giving
Sources: American Association of Fundraising Counsel; AP

EXAMPLE · 3

How much of the amount given to charities came from individuals?

Strategy
To determine the amount that came from individuals, write and solve the basic percent equation using n to represent the amount. The percent is 76%. The base is $212 billion.

Solution
Percent × base = amount
$$76\% \times 212 = n$$
$$0.76 \times 212 = n$$
$$161.12 = n$$

Individuals gave $161.12 billion to charities.

YOU TRY IT · 3

How much of the amount given to charities was given by corporations?

Your strategy

Your solution

EXAMPLE · 4

A quality control inspector found that 1.2% of 2500 camera phones inspected were defective. How many camera phones inspected were not defective?

Strategy
To find the number of nondefective phones:
• Find the number of defective phones. Write and solve the basic percent equation using n to represent the number of defective phones (amount). The percent is 1.2% and the base is 2500.
• Subtract the number of defective phones from the number of phones inspected (2500).

Solution
$$1.2\% \times 2500 = n$$
$$0.012 \times 2500 = n$$
$$30 = n \text{ defective phones}$$

$$2500 - 30 = 2470$$

2470 camera phones were not defective.

YOU TRY IT · 4

An electrician's hourly wage was $33.50 before an 8% raise. What is the new hourly wage?

Your strategy

Your solution

Solutions on p. S12

5.2 EXERCISES

OBJECTIVE A   To find the amount when the percent and the base are given

1. 8% of 100 is what?

2. 16% of 50 is what?

3. 27% of 40 is what?

4. 52% of 95 is what?

5. 0.05% of 150 is what?

6. 0.075% of 625 is what?

7. 125% of 64 is what?

8. 210% of 12 is what?

9. Find 10.7% of 485.

10. Find 12.8% of 625.

11. What is 0.25% of 3000?

12. What is 0.06% of 250?

13. 80% of 16.25 is what?

14. 26% of 19.5 is what?

15. What is $1\frac{1}{2}$% of 250?

16. What is $5\frac{3}{4}$% of 65?

17. $16\frac{2}{3}$% of 120 is what?

18. What is $66\frac{2}{3}$% of 891?

19. Which is larger: 5% of 95, or 75% of 6?

20. Which is larger: 112% of 5, or 0.45% of 800?

21. Which is smaller: 79% of 16, or 20% of 65?

22. Which is smaller: 15% of 80, or 95% of 15?

 23. Is 15% of a number greater than or less than the number?

 24. Is 150% of a number greater than or less than the number?

OBJECTIVE B   To solve application problems

 25. Read Exercise 26. Without doing any calculations, determine whether the number of people in the United States aged 18 to 24 who do not have health insurance is *less than, equal to,* or *greater than* 44 million.

 26. **Health Insurance** Approximately 30% of the 44 million people in the United States who do not have health insurance are between the ages of 18 and 24. (*Source:* U.S. Census Bureau) About how many people in the United States aged 18 to 24 do not have health insurance?

 27. **Aviation** The Federal Aviation Administration reported that 55,422 new student pilots were flying single-engine planes last year. The number of new student pilots flying single-engine planes this year is 106% of the number flying single-engine planes last year. How many new student pilots are flying single-engine planes this year?

28. **Jewelry** An 18-carat yellow-gold necklace contains 75% gold, 16% silver, and 9% copper. If the necklace weighs 25 grams, how many grams of copper are in the necklace?

29. **Jewelry** Fourteen-carat yellow gold contains 58.5% gold, 17.5% silver, and 24% copper. If a jeweler has a 50-gram piece of 14-carat yellow gold, how many grams of gold, silver, and copper are in the piece?

30. **Lifestyles** There are 114 million households in the United States. Opposite-sex cohabitating couples comprise 4.4% of these households. (*Source:* Families and Living Arrangements) Find the number of opposite-sex cohabitating couples who maintain households in the United States. Round to the nearest million.

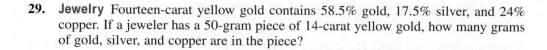

31. **e-Filed Tax Returns** See the news clipping at the right. How many of the 128 million returns were filed electronically? Round to the nearest million.

> **In the News**
>
> **More Taxpayers Filing Electronically**
>
> The IRS reported that, as of May 4, it has received 128 million returns. Sixty percent of the returns were filed electronically.
>
> *Source:* IRS

32. **Taxes** A sales tax of 6% of the cost of a car is added to the purchase price of $29,500. What is the total cost of the car, including sales tax?

33. **Email** The number of email messages sent each day has risen to 171 billion, of which 71% are spam. (*Source:* FeedsFarm.com) How many email messages sent per day are not spam?

34. **Prison Population** The prison population in the United States is 1,596,127 people. Male prisoners comprise 91% of this population. (*Source: Time,* March 17, 2008) How many inmates are male? How many are female?

35. **Entertainment** A USA TODAY.com online poll asked 8878 Internet users, "Would you use software to cut out objectionable parts of movies?" 29.8% of the respondents answered yes. How many respondents did not answer yes to the question? Round to the nearest whole number.

Applying the Concepts

36. **Jewelry** Eighteen-carat white gold contains 75% gold, 15% silver, and 10% platinum. A jeweler wants to make a 2-ounce, 18-carat, white gold ring. If gold costs $900 per ounce, silver costs $17.20 per ounce, and platinum costs $1900 per ounce, what is the cost of the metal used to make the ring?

SECTION

5.3 Percent Equations: Part II

OBJECTIVE A **To find the percent when the base and amount are given**

A recent promotional game at a grocery store listed the probability of winning a prize as "1 chance in 2." A percent can be used to describe the chance of winning. This requires answering the question "What percent of 2 is 1?"

The chance of winning can be found by solving the basic percent equation for *percent*.

Integrating Technology

The percent key **%** on a scientific calculator moves the decimal point to the right two places when pressed after a multiplication or division computation. For the example at the right, enter

1 ÷ 2 % =

The display reads 50.

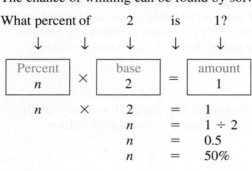

$$n \times 2 = 1$$
$$n = 1 \div 2$$
$$n = 0.5$$
$$n = 50\%$$

• The solution must be written as a percent in order to answer the question.

There is a 50% chance of winning a prize.

EXAMPLE • 1

What percent of 40 is 30?

Solution
$$\text{Percent} \times \text{base} = \text{amount}$$
$$n \times 40 = 30$$
$$n = 30 \div 40$$
$$n = 0.75$$
$$n = 75\%$$

YOU TRY IT • 1

What percent of 32 is 16?

Your solution

EXAMPLE • 2

What percent of 12 is 27?

Solution
$$\text{Percent} \times \text{base} = \text{amount}$$
$$n \times 12 = 27$$
$$n = 27 \div 12$$
$$n = 2.25$$
$$n = 225\%$$

YOU TRY IT • 2

What percent of 15 is 48?

Your solution

EXAMPLE • 3

25 is what percent of 75?

Solution
$$\text{Percent} \times \text{base} = \text{amount}$$
$$n \times 75 = 25$$
$$n = 25 \div 75$$
$$n = \frac{1}{3} = 33\frac{1}{3}\%$$

YOU TRY IT • 3

30 is what percent of 45?

Your solution

Solutions on p. S12

27. Agriculture According to the U.S. Department of Agriculture, of the 63 billion pounds of vegetables produced in the United States in 1 year, 16 billion pounds were wasted. What percent of the vegetables produced were wasted? Round to the nearest tenth of a percent.

28. Wind Energy In a recent year, wind machines in the United States generated 17.8 billion kilowatt-hours of electricity, enough to serve over 1.6 million households. The nation's total electricity production that year was 4,450 billion kilowatt-hours. (*Source:* Energy Information Administration) What percent of the total energy production was generated by wind machines?

29. Diabetes Approximately 7% of the American population has diabetes. Within this group, 14.6 million are diagnosed, while 6.2 million are undiagnosed. (*Source:* The National Diabetes Education Program) What percent of Americans with diabetes have not been diagnosed with the disease? Round to the nearest tenth of a percent.

30. Internal Revenue Service See the news clipping at the right. Given that the number of millionaires in the United States is 9.3 million, what percent of U.S. millionaires were audited by the IRS? Round to the nearest hundredth of a percent.

> **In the News**
>
> **More Millionaires Audited**
>
> The Internal Revenue Service reported that 17,015 millionaires were audited this year. This figure is 33% more than last year.
>
> *Source:* The Internal Revenue Service; TSN Financial Services

31. Construction In a test of the breaking strength of concrete slabs for freeway construction, 3 of the 200 slabs tested did not meet safety requirements. What percent of the slabs did meet safety requirements?

Applying the Concepts

Pets The graph at the right shows several categories of average lifetime costs of dog ownership. Use this graph for Exercises 32 to 34. Round answers to the nearest tenth of a percent.

32. What percent of the total amount is spent on food?

33. What percent of the total is spent on veterinary care?

34. What percent of the total is spent on all categories except training?

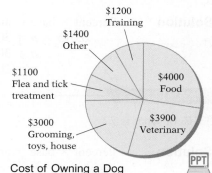

Cost of Owning a Dog
Source: Based on data from the American Kennel Club, *USA Today* research

35. Sports The Fun in the Sun organization claims to have taken a survey of 350 people, asking them to give their favorite outdoor temperature for hiking. The results are given in the table at the right. Explain why these results are not possible.

Favorite Temperature	Percent
Greater than 90	5%
80–89	28%
70–79	35%
60–69	32%
Below 60	13%

SECTION

5.4   Percent Equations: Part III

OBJECTIVE A   To find the base when the percent and amount are given

Tips for Success

After completing this objective, you will have learned to solve the basic percent equation for each of the three elements: percent, base, and amount. You will need to be able to recognize these three different types of problems. To test yourself, try the Chapter 5 Review Exercises.

In 1780, the population of Virginia was 538,000. This was 19% of the total population of the United States at that time. To find the total population at that time, you must answer the question "19% of what number is 538,000?"

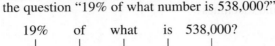

19%	of	what	is	538,000?
↓	↓	↓	↓	↓

| Percent 19% | | base n | | amount 538,000 |

$$0.19 \times n = 538{,}000$$
$$n = 538{,}000 \div 0.19$$
$$n \approx 2{,}832{,}000$$

• The population of the United States in 1780 can be found by solving the basic percent equation for the base.

The population of the United States in 1780 was approximately 2,832,000.

EXAMPLE · 1

18% of what is 900?

Solution
$$\text{Percent} \times \text{base} = \text{amount}$$
$$0.18 \times n = 900$$
$$n = 900 \div 0.18$$
$$n = 5000$$

YOU TRY IT · 1

86% of what is 215?

Your solution

EXAMPLE · 2

30 is 1.5% of what?

Solution
$$\text{Percent} \times \text{base} = \text{amount}$$
$$0.015 \times n = 30$$
$$n = 30 \div 0.015$$
$$n = 2000$$

YOU TRY IT · 2

15 is 2.5% of what?

Your solution

EXAMPLE · 3

$33\frac{1}{3}\%$ of what is 7?

Solution
$$\text{Percent} \times \text{base} = \text{amount}$$
$$\frac{1}{3} \times n = 7$$
$$n = 7 \div \frac{1}{3}$$
$$n = 21$$

• Note that the percent is written as a fraction.

YOU TRY IT · 3

$16\frac{2}{3}\%$ of what is 5?

Your solution

Solutions on p. S13

OBJECTIVE B   To solve application problems

To solve percent problems, it is necessary to identify the percent, the base, and the amount. Usually the base follows the phrase "percent of."

EXAMPLE · 4

A business office bought a used copy machine for $900, which was 75% of the original cost. What was the original cost of the copier?

Strategy

To find the original cost of the copier, write and solve the basic percent equation using n to represent the original cost (base). The percent is 75% and the amount is $900.

Solution

$75\% \times n = 900$
$0.75 \times n = 900$
$n = 900 \div 0.75$
$n = 1200$

The original cost of the copier was $1200.

YOU TRY IT · 4

A used car has a value of $10,458, which is 42% of the car's original value. What was the car's original value?

Your strategy

Your solution

EXAMPLE · 5

A carpenter's wage this year is $26.40 per hour, which is 110% of last year's wage. What was the increase in the hourly wage over last year?

Strategy

To find the increase in the hourly wage over last year:

• Find last year's wage. Write and solve the basic percent equation using n to represent last year's wage (base). The percent is 110% and the amount is $26.40.
• Subtract last year's wage from this year's wage (26.40).

Solution

$110\% \times n = 26.40$
$1.10 \times n = 26.40$
$n = 26.40 \div 1.10$
$n = 24.00$ • **Last year's wage**

$26.40 - 24.00 = 2.40$

The increase in the hourly wage was $2.40.

YOU TRY IT · 5

Chang's Sporting Goods has a tennis racket on sale for $89.60, which is 80% of the original price. What is the difference between the original price and the sale price?

Your strategy

Your solution

Solutions on p. S13

5.4 EXERCISES

OBJECTIVE A   To find the base when the percent and amount are given

1. 12% of what is 9?

2. 38% of what is 171?

3. 8 is 16% of what?

4. 54 is 90% of what?

5. 10 is 10% of what?

6. 37 is 37% of what?

7. 30% of what is 25.5?

8. 25% of what is 21.5?

9. 2.5% of what is 30?

10. 10.4% of what is 52?

11. 125% of what is 24?

12. 180% of what is 21.6?

13. 18 is 240% of what?

14. 24 is 320% of what?

15. 4.8 is 15% of what?

16. 87.5 is 50% of what?

17. 25.6 is 12.8% of what?

18. 45.014 is 63.4% of what?

19. 30% of what is 2.7?

20. 78% of what is 3.9?

21. 84 is $16\frac{2}{3}$% of what?

22. 120 is $33\frac{1}{3}$% of what?

 23. Consider the question "*P*% of what number is 50?" If the percent *P* is greater than 100%, is the unknown number greater than 50 or less than 50?

OBJECTIVE B   To solve application problems

 24. Read Exercise 25. Without doing any calculations, determine whether the number of travelers who allowed their children to miss school to go on a trip is *less than, equal to,* or *greater than* 1.738 million.

 25. **Travel** Of the travelers who, during a recent year, allowed their children to miss school to go along on a trip, approximately 1.738 million allowed their children to miss school for more than a week. This represented 11% of the travelers who allowed their children to miss school. (*Source:* Travel Industry Association) About how many travelers allowed their children to miss school to go along on a trip?

26. e-Commerce Using the information in the news clipping at the right, calculate the total retail sales during the fourth quarter of last year. Round to the nearest billion.

27. Marathons In 2008, 98.2% of the runners who started the Boston Marathon, or 21,963 people, crossed the finish line. (*Source:* www.bostonmarathon.org) How many runners started the Boston Marathon in 2008?

28. Education In the United States today, 23.1% of women and 27.5% of men have earned a bachelor's or graduate degree. (*Source:* Census Bureau) How many women in the United States have earned a bachelor's or graduate degree?

29. Wind-Powered Ships Using the information in the news clipping at the right, calculate the cargo ships' daily fuel bill.

Courtesy SkySails

30. Taxes A TurboTax online survey asked people how they planned to use their tax refunds. Seven hundred forty people, or 22% of the respondents, said they would save the money. How many people responded to the survey?

31. Manufacturing During a quality control test, Micronics found that 24 computer boards were defective. This amount was 0.8% of the computer boards tested.
a. How many computer boards were tested?
b. How many of the computer boards tested were not defective?

32. Directory Assistance Of the calls a directory assistance operator received, 441 were requests for telephone numbers listed in the current directory. This accounted for 98% of the calls for assistance that the operator received.
a. How many calls did the operator receive?
b. How many telephone numbers requested were not listed in the current directory?

Applying the Concepts

33. Nutrition The table at the right contains nutrition information about a breakfast cereal. The amount of thiamin in one serving of this cereal with skim milk is 0.45 milligram. Find the recommended daily allowance of thiamin for an adult.

34. Increase a number by 10%. Now decrease the number by 10%. Is the result the original number? Explain.

In the News

eCommerce on the Rise

Retail e-commerce sales for the fourth quarter of last year exceeded e-commerce sales for the first three quarters of the year. E-commerce sales during October, November, and December totaled $35.3 billion, or 3.4% of total retail sales during the quarter.

Source: Service Sector Statistics

In the News

Kite-Powered Cargo Ships

In January 2008, the first cargo ship partially powered by a giant kite set sail from Germany bound for Venezuela. The 1722-square-foot kite helped to propel the ship, which consequently used 20% less fuel, cutting approximately $1600 from the ship's daily fuel bill.

Source: The Internal Revenue Service; TSN Financial Services

NUTRITION INFORMATION

SERVING SIZE: 1.4 OZ WHEAT FLAKES WITH 0.4 OZ. RAISINS: 39.4 g. ABOUT 1/2 CUP
SERVINGS PER PACKAGE: 14

	CEREAL & RAISINS	WITH 1/2 CUP VITAMINS A & D SKIM MILK

PERCENTAGE OF U.S. RECOMMENDED DAILY ALLOWANCES (U.S. RDA)

	CEREAL & RAISINS	WITH 1/2 CUP SKIM MILK
PROTEIN	4	15
VITAMIN A	15	20
VITAMIN C	**	2
THIAMIN	25	30
RIBOFLAVIN	25	35
NIACIN	25	35
CALCIUM	**	15
IRON	100	100
VITAMIN D	10	25
VITAMIN B$_6$	25	25
FOLIC ACID	25	25
VITAMIN B$_{12}$	25	30
PHOSPHOROUS	10	15
MAGNESIUM	10	20
ZINC	25	30
COPPER	2	4

* 2% MILK SUPPLIES AN ADDITIONAL 20 CALORIES, 2 g FAT, AND 10 mg CHOLESTEROL.
** CONTAINS LESS THAN 2% OF THE U.S. RDA OF THIS NUTRIENT

SECTION

5.5 Percent Problems: Proportion Method

OBJECTIVE A To solve percent problems using proportions

Problems that can be solved using the basic percent equation can also be solved using proportions.

The proportion method is based on writing two ratios. One ratio is the percent ratio, written as $\frac{\text{percent}}{100}$. The second ratio is the amount-to-base ratio, written as $\frac{\text{amount}}{\text{base}}$. These two ratios form the proportion

$$\frac{\text{percent}}{100} = \frac{\text{amount}}{\text{base}}$$

To use the proportion method, first identify the percent, the amount, and the base (the base usually follows the phrase "percent of").

Integrating Technology

To use a calculator to solve the proportions at the right for n, enter

23	**x**	45	**÷**	100	**=**
100	**x**	4	**÷**	25	**=**
100	**x**	12	**÷**	60	**=**

What is 23% of 45?

$$\frac{23}{100} = \frac{n}{45}$$
$$23 \times 45 = 100 \times n$$
$$1035 = 100 \times n$$
$$1035 \div 100 = n$$
$$10.35 = n$$

What percent of 25 is 4?

$$\frac{n}{100} = \frac{4}{25}$$
$$n \times 25 = 100 \times 4$$
$$n \times 25 = 400$$
$$n = 400 \div 25$$
$$n = 16$$

12 is 60% of what number?

$$\frac{60}{100} = \frac{12}{n}$$
$$60 \times n = 100 \times 12$$
$$60 \times n = 1200$$
$$n = 1200 \div 60$$
$$n = 20$$

EXAMPLE 1

15% of what is 7? Round to the nearest hundredth.

Solution
$$\frac{15}{100} = \frac{7}{n}$$
$$15 \times n = 100 \times 7$$
$$15 \times n = 700$$
$$n = 700 \div 15$$
$$n \approx 46.67$$

YOU TRY IT 1

26% of what is 22? Round to the nearest hundredth.

Your solution

EXAMPLE 2

30% of 63 is what?

Solution
$$\frac{30}{100} = \frac{n}{63}$$
$$30 \times 63 = 100 \times n$$
$$1890 = 100 \times n$$
$$1890 \div 100 = n$$
$$18.90 = n$$

YOU TRY IT 2

16% of 132 is what?

Your solution

Solutions on p. S13

OBJECTIVE B   To solve application problems

EXAMPLE • 3

An antiques dealer found that 86% of the 250 items that were sold during one month sold for under $1000. How many items sold for under $1000?

Strategy

To find the number of items that sold for under $1000, write and solve a proportion using n to represent the number of items sold for less than $1000 (amount). The percent is 86%, and the base is 250.

Solution

$$\frac{86}{100} = \frac{n}{250}$$
$$86 \times 250 = 100 \times n$$
$$21{,}500 = 100 \times n$$
$$21{,}500 \div 100 = n$$
$$215 = n$$

215 items sold for under $1000.

YOU TRY IT • 3

Last year it snowed 64% of the 150 days of the ski season at a resort. How many days did it snow?

Your strategy

Your solution

EXAMPLE • 4

In a test of the strength of nylon rope, 5 pieces of the 25 pieces tested did not meet the test standards. What percent of the nylon ropes tested did meet the standards?

Strategy

To find the percent of ropes tested that met the standards:

• Find the number of ropes that met the test standards (25 − 5).

• Write and solve a proportion using n to represent the percent of ropes that met the test standards. The base is 25, and the amount is the number of ropes that met the standards.

Solution

$25 - 5 = 20$ ropes met test standards

$$\frac{n}{100} = \frac{20}{25}$$
$$n \times 25 = 100 \times 20$$
$$n \times 25 = 2000$$
$$n = 2000 \div 25$$
$$n = 80$$

80% of the ropes tested did meet the test standards.

YOU TRY IT • 4

The Rincon Fire Department received 24 false alarms out of a total of 200 alarms received. What percent of the alarms received were not false alarms?

Your strategy

Your solution

5.5 EXERCISES

OBJECTIVE A   To solve percent problems using proportions

1. 26% of 250 is what?

2. What is 18% of 150?

3. 37 is what percent of 148?

4. What percent of 150 is 33?

5. 68% of what is 51?

6. 126 is 84% of what?

7. What percent of 344 is 43?

8. 750 is what percent of 50?

9. 82 is 20.5% of what?

10. 2.4% of what is 21?

11. What is 6.5% of 300?

12. 96% of 75 is what?

13. 7.4 is what percent of 50?

14. What percent of 1500 is 693?

15. 50.5% of 124 is what?

16. What is 87.4% of 255?

17. 33 is 220% of what?

18. 160% of what is 40?

19. **a.** Which equation(s) below can be used to answer the question "What is 12% of 75?"
    b. Which equation(s) below can be used to answer the question "75 is 12% of what?"

    (i) $\dfrac{12}{100} = \dfrac{75}{n}$ (ii) $0.12 \times 75 = n$ (iii) $\dfrac{12}{100} = \dfrac{n}{75}$ (iv) $0.12 \times n = 75$

OBJECTIVE B   To solve application problems

20. Read Exercise 21. Without doing any calculations, determine whether the length of time the drug will be effective is *less than* or *greater than* 6 hours.

21. **Medicine** A manufacturer of an anti-inflammatory drug claims that the drug will be effective for 6 hours. An independent testing service determined that the drug was effective for only 80% of the length of time claimed by the manufacturer. Find the length of time the drug will be effective as determined by the testing service.

22. **Geography** The land area of North America is approximately 9,400,000 square miles. This represents approximately 16% of the total land area of the world. What is the approximate total land area of the world?

23. **Girl Scout Cookies** Using the information in the news clipping at the right, calculate the cash generated annually **a.** from sales of Thin Mints and **b.** from sales of Trefoil shortbread cookies.

Jeff Greenberg/age fotostock

24. **Charities** The American Red Cross spent $185,048,179 for administrative expenses. This amount was 3.16% of its total revenue. Find the American Red Cross's total revenue. Round to the nearest hundred million.

25. **Poultry** In a recent year, North Carolina produced 1,300,000,000 pounds of turkey. This was 18.6% of the U.S. total in that year. Calculate the U.S. total turkey production for that year. Round to the nearest billion.

26. **Mining** During 1 year, approximately 2,240,000 ounces of gold went into the manufacturing of electronic equipment in the United States. This is 16% of all the gold mined in the United States that year. How many ounces of gold were mined in the United States that year?

27. **Education** See the news clipping at the right. What percent of the baby boomers living in the United States have some college experience but have not earned a college degree? Round to the nearest tenth of a percent.

28. **Demography** According to a 25-city survey of the status of hunger and homelessness by the U.S. Conference of Mayors, 41% of the homeless in the United States are single men, 41% are families with children, 13% are single women, and 5% are unaccompanied minors. How many homeless people in the United States are single men?

29. **Police Officers** The graph at the right shows the causes of death for all police officers killed in the line of duty during a recent year. What percent of the deaths were due to traffic accidents? Round to the nearest tenth of a percent.

Applying the Concepts

30. **The Federal Government** In the 110th Senate, there were 49 Republicans, 49 Democrats, and 2 Independents. In the 110th House of Representatives, there were 202 Republicans, 233 Democrats, and 0 Independents. Which had the larger percentage of Republicans, the 110th Senate or the 110th House of Representatives?

In the News

Thin Mints Biggest Seller

Every year, sales from all the Girl Scout cookies sold by about 2.7 million girls total $700 million. The most popular cookie is Thin Mints, which earn 25% of total sales, while sales of the Trefoil shortbread cookies represent only 9% of total sales.

Source: Southwest Airlines Spirit Magazine 2007

In the News

Over Half of Baby Boomers Have College Experience

Of the 78 million baby boomers living in the United States, 45 million have some college experience but no college degree. Twenty million baby boomers have one or more college degrees.

Sources: The National Center for Education Statistics; U.S. Census Bureau; *McCook Daily Gazette*

Causes of Death for Police Officers Killed in the Line of Duty

Source: International Union of Police Associations

FOCUS ON PROBLEM SOLVING

Using a Calculator as a Problem-Solving Tool

A calculator is an important tool for problem solving. Here are a few problems to solve with a calculator. You may need to research some of the questions to find information you do not know.

1. Choose any single-digit positive number. Multiply the number by 1507 and 7373. What is the answer? Choose another positive single-digit number and again multiply by 1507 and 7373. What is the answer? What pattern do you see? Why does this work?

2. The gross domestic product in 2007 was $13,841,300,000. Is this more or less than the amount of money that would be placed on the last square of a standard checkerboard if 1 cent were placed on the first square, 2 cents were placed on the second square, 4 cents were placed on the third square, 8 cents were placed on the fourth square, and so on, until the 64th square was reached?

3. Which of the reciprocals of the first 16 natural numbers have a terminating-decimal representation and which have a repeating-decimal representation?

4. What is the largest natural number n for which 4^n . $1 \cdot 2 \cdot 3 \cdot 4 \cdot 5 \cdot \cdots \cdot n$?

5. If $1000 bills are stacked one on top of another, is the height of $1 billion less than or greater than the height of the Washington Monument?

6. What is the value of $1 + \cfrac{1}{1 + \cfrac{1}{1 + \cfrac{1}{1 + \cfrac{1}{1 + 1}}}}$?

7. Calculate 15^2, 35^2, 65^2, and 85^2. Study the results. Make a conjecture about a relationship between a number ending in 5 and its square. Use your conjecture to find 75^2 and 95^2. Does your conjecture work for 125^2?

8. Find the sum of the first 1000 natural numbers. (*Hint:* You could just start adding $1 + 2 + 3 + \cdots$, but even if you performed one operation every 3 seconds, it would take you an hour to find the sum. Instead, try pairing the numbers and then adding the pairs. Pair 1 and 1000, 2 and 999, 3 and 998, and so on. What is the sum of each pair? How many pairs are there? Use this information to answer the original question.)

9. For a borrower to qualify for a home loan, a bank requires that the monthly mortgage payment be less than 25% of the borrower's monthly take-home income. A laboratory technician has deductions for taxes, insurance, and retirement that amount to 25% of the technician's monthly gross income. What minimum monthly income must this technician earn to receive a bank loan that has a mortgage payment of $1200 per month?

Using Estimation as a Problem-Solving Tool

You can use your knowledge of rounding, your understanding of percent, and your experience with the basic percent equation to quickly estimate the answer to a percent problem. Here is an example.

> **HOW TO · 1**   What is 11.2% of 978?
>
> Round the given numbers. $11.2\% \approx 10\%$
> $978 \approx 1000$
>
> Mentally calculate with the rounded numbers. $10\% \text{ of } 1000 = \frac{1}{10} \text{ of } 1000 = 100$
>
> 11.2% of 978 is approximately 100.

✓ **Take Note**

The exact answer is $0.112 \times 978 = 109.536$. The exact answer 109.536 is close to the approximation of 100.

For Exercises 1 to 8, state which quantity is greater.

1. 49% of 51, or 201% of 15
2. 99% of 19, or 22% of 55
3. 8% of 31, or 78% of 10
4. 24% of 402, or 76% of 205
5. 10.2% of 51, or 20.9% of 41
6. 51.8% of 804, or 25.3% of 1223
7. 26% of 39.217, or 9% of 85.601
8. 66% of 31.807, or 33% of 58.203

For Exercises 9 to 12, use estimation to provide an approximate number.

9. A company found that 24% of its 2096 employees favored a new dental plan. How many employees favored the new dental plan?

10. A local newspaper reported that 52.3% of the 29,875 eligible voters in the town voted in the last election. How many people voted in the last election?

11. 19.8% of the 2135 first-year students at a community college have part-time jobs. How many of the first-year students at the college have part-time jobs?

12. A couple made a down payment of 33% of the $310,000 cost of a home. Find the down payment.

© Ariel Skelly/Corbis

PROJECTS AND GROUP ACTIVITIES

Health

The American College of Sports Medicine (ACSM) recommends that you know how to determine your target heart rate in order to get the full benefit of exercise. Your **target heart rate** is the rate at which your heart should beat during any aerobic exercise such as running, cycling, fast walking, or participating in an aerobics class. According to the ACSM, you should reach your target rate and then maintain it for 20 minutes or more to achieve cardiovascular fitness. The intensity level varies for different individuals. A sedentary person might begin at the 60% level and gradually work up to 70%, whereas athletes and very fit individuals might work at the 85% level. The ACSM suggests that you calculate both 50% and 85% of your maximum heart rate. This will give you the low and high ends of the range within which your heart rate should stay.

To calculate your target heart rate:

	Example
Subtract your age from 220. This is your maximum heart rate.	$220 - 20 = 200$
Multiply your maximum heart rate by 50%. This is the low end of your range.	$200(0.50) = 100$
Divide the low end by 6. This is your low 10-second heart rate.	$100 \div 6 \approx 17$
Multiply your maximum heart rate by 85%. This is the high end of your range.	$200(0.85) = 170$
Divide the high end by 6. This is your high 10-second heart rate.	$170 \div 6 \approx 28$

1. Why are the low end and high end divided by 6 in order to determine the low and high 10-second heart rates?

2. Calculate your target heart rate, both the low and high end of your range.

Consumer Price Index

The consumer price index (CPI) is a percent that is written without the percent sign. For instance, a CPI of 160.1 means 160.1%. This number means that an item that cost $100 between 1982 and 1984 (the base years) would cost $160.10 today. Determining the cost is an application of the basic percent equation.

$$\text{Percent} \times \text{base} = \text{amount}$$
$$\text{CPI} \times \text{cost in base year} = \text{cost today}$$
$$1.601 \times 100 = 160.1 \qquad \bullet\ 160.1\% = 1.601$$

The table below gives the CPI for various products in March of 2008. If you have Internet access, you can obtain current data for the items below, as well as other items not on this list, by visiting the website of the Bureau of Labor Statistics.

Product	*CPI*
All items	213.5
Food and beverages	209.7
Housing	214.4
Clothes	120.9
Transportation	195.2
Medical care	363.0
Entertainment[1]	112.7
Education[1]	121.8

[1]Indexes on December 1997 = 100

1. Of the items listed, are there any items that in 2008 cost more than twice as much as they cost during the base year? If so, which items?

2. Of the items listed, are there any items that in 2008 cost more than one-and-one-half times as much as they cost during the base years but less than twice as much as they cost during the base years? If so, which items?

3. If the cost for textbooks for one semester was $120 in the base years, how much did similar textbooks cost in 2008? Use the "Education" category.

4. If a new car cost $40,000 in 2008, what would a comparable new car have cost during the base years? Use the "Transportation" category.

5. If a movie ticket cost $10 in 2008, what would a comparable movie ticket have cost during the base years? Use the "Entertainment" category.

6. The base year for the CPI was 1967 before the change to 1982–1984. If 1967 were still used as the base year, the CPI for all items in 2008 (not just those listed above) would be 639.6.

   a. Using the base year of 1967, explain the meaning of a CPI of 639.6.

   b. Using the base year of 1967 and a CPI of 639.6, if textbooks cost $75 for one semester in 1967, how much did similar textbooks cost in 2008?

   c. Using the base year of 1967 and a CPI of 639.6, if a family's food budget in 2008 is $1000 per month, what would a comparable family budget have been in 1967?

CHAPTER 5

SUMMARY

KEY WORDS	**EXAMPLES**
Percent means "parts of 100." [5.1A, p. 202]	23% means 23 of 100 equal parts.

ESSENTIAL RULES AND PROCEDURES	**EXAMPLES**
To write a percent as a fraction, drop the percent sign and multiply by $\frac{1}{100}$. [5.1A, p. 202]	$56\% = 56\left(\frac{1}{100}\right) = \frac{56}{100} = \frac{14}{25}$
To write a percent as a decimal, drop the percent sign and multiply by 0.01. [5.1A, p. 202]	$87\% = 87(0.01) = 0.87$
To write a fraction as a percent, multiply by 100%. [5.1B, p. 203]	$\frac{7}{20} = \frac{7}{20}(100\%) = \frac{700}{20}\% = 35\%$
To write a decimal as a percent, multiply by 100%. [5.1B, p. 203]	$0.325 = 0.325(100\%) = 32.5\%$
The Basic Percent Equation [5.2A, p. 206] The basic percent equation is $\qquad$ Percent $\times$ base = amount Solving percent problems requires identifying the three elements of this equation. Usually the base follows the phrase "percent of."	8% of 250 is what number? Percent $\times$ base = amount $0.08 \times 250 = n$ $20 = n$
Proportion Method of Solving a Percent Problem [5.5A, p. 218] The following proportion can be used to solve percent problems. $\qquad \dfrac{\text{percent}}{100} = \dfrac{\text{amount}}{\text{base}}$ To use the proportion method, first identify the percent, the amount, and the base. The base usually follows the phrase "percent of."	8% of 250 is what number? $\dfrac{\text{percent}}{100} = \dfrac{\text{amount}}{\text{base}}$ $\dfrac{8}{100} = \dfrac{n}{250}$ $8 \times 250 = 100 \times n$ $2000 = 100 \times n$ $2000 \div 100 = n$ $20 = n$

CHAPTER 5

CONCEPT REVIEW

Test your knowledge of the concepts presented in this chapter. Answer each question.
Then check your answers against the ones provided in the Answer Section.

1. How do you write 197% as a fraction?

2. How do you write 6.7% as a decimal?

3. How do you write $\frac{9}{5}$ as a percent?

4. How do you write 56.3 as a percent?

5. What is the basic percent equation?

6. What percent of 40 is 30? Did you multiply or divide?

7. Find 11.7% of 532. Did you multiply or divide?

8. 36 is 240% of what number? Did you multiply or divide?

9. How do you use the proportion method to solve a percent problem?

10. What percent of 1400 is 763? Use the proportion method to solve.

CHAPTER 5

REVIEW EXERCISES

1. What is 30% of 200?

2. 16 is what percent of 80?

3. Write $1\frac{3}{4}$ as a percent.

4. 20% of what is 15?

5. Write 12% as a fraction.

6. Find 22% of 88.

7. What percent of 20 is 30?

8. $16\frac{2}{3}\%$ of what is 84?

9. Write 42% as a decimal.

10. What is 7.5% of 72?

11. $66\frac{2}{3}\%$ of what is 105?

12. Write 7.6% as a decimal.

13. Find 125% of 62.

14. Write $16\frac{2}{3}\%$ as a fraction.

15. Use the proportion method to find what percent of 25 is 40.

16. 20% of what number is 15? Use the proportion method.

17. Write 0.38 as a percent.

18. 78% of what is 8.5? Round to the nearest tenth.

19. What percent of 30 is 2.2? Round to the nearest tenth of a percent.

20. What percent of 15 is 92? Round to the nearest tenth of a percent.

21. Education Trent missed 9 out of 60 questions on a history exam. What percent of the questions did he answer correctly? Use the proportion method.

22. Advertising A company used 7.5% of its $60,000 advertising budget for newspaper advertising. How much of the advertising budget was spent for newspaper advertising?

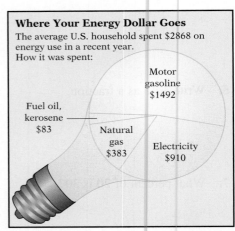

Where Your Energy Dollar Goes
The average U.S. household spent $2868 on energy use in a recent year. How it was spent:

Motor gasoline $1492

Fuel oil, kerosene $83

Natural gas $383

Electricity $910

23. Energy The graph at the right shows the amounts that the average U.S. household spends for energy use. What percent of these costs is for electricity? Round to the nearest tenth of a percent.

24. Consumerism Joshua purchased a camcorder for $980 and paid a sales tax of 6.25% of the cost. What was the total cost of the camcorder?

Source: Energy Information Administration

25. Health In a survey of 350 women and 420 men, 275 of the women and 300 of the men reported that they wore sunscreen often. To the nearest tenth of a percent, what percent of the women wore sunscreen often?

26. Demography It is estimated that the world's population will be 9,100,000,000 by the year 2050. This is 149% of the population in 2000. (*Source:* U.S. Census Bureau). What was the world's population in 2000? Round to the nearest hundred million.

27. Computers A computer system can be purchased for $1800. This is 60% of what the computer cost 4 years ago. What was the cost of the computer 4 years ago? Use the proportion method.

28. Agriculture In a recent year, Wisconsin growers produced 281.72 million pounds of cranberries. This represented 49.25% of the total cranberry crop in the United States that year. Find the total cranberry crop in the United States that year. Round to the nearest million.

CHAPTER 5

TEST

1. Write 97.3% as a decimal.

2. Write $83\frac{1}{3}\%$ as a fraction.

3. Write 0.3 as a percent.

4. Write 1.63 as a percent.

5. Write $\frac{3}{2}$ as a percent.

6. Write $\frac{2}{3}$ as a percent.

7. What is 77% of 65?

8. 47.2% of 130 is what?

9. Which is larger:
7% of 120, or 76% of 13?

10. Which is smaller:
13% of 200, or 212% of 12?

11. **Advertising** A travel agency uses 6% of its $750,000 budget for advertising. What amount of the budget is spent on advertising?

12. **Agriculture** During the packaging process for vegetables, spoiled vegetables are discarded by an inspector. In one day an inspector found that 6.4% of the 1250 pounds of vegetables were spoiled. How many pounds of vegetables were not spoiled?

Nutrition The table at the right contains nutrition information about a breakfast cereal. Solve Exercises 13 and 14 with information taken from this table.

13. The recommended amount of potassium per day for an adult is 3000 milligrams (mg). What percent, to the nearest tenth of a percent, of the daily recommended amount of potassium is provided by one serving of this cereal with skim milk?

NUTRITION INFORMATION

SERVING SIZE: 1.4 OZ WHEAT FLAKES WITH 0.4 OZ. RAISINS: 39.4 g. ABOUT 1/2 CUP
SERVINGS PER PACKAGE: 14

	CEREAL & RAISINS	WITH 1/2 CUP VITAMINS A & D SKIM MILK
CALORIES	120	180
PROTEIN, g	3	7
CARBOHYDRATE, g	28	34
FAT, TOTAL, g	1	1*
UNSATURATED, g 1		
SATURATED, g 0		
CHOLESTEROL, mg	0	0*
SODIUM, mg	125	190
POTASSIUM, mg	240	440

* 2% MILK SUPPLIES AN ADDITIONAL 20 CALORIES. 2 g FAT, AND 10 mg CHOLESTEROL. ** CONTAINS LESS THAN 2% OF THE U.S. RDA OF THIS NUTRIENT

14. The daily recommended number of calories for a 190-pound man is 2200 calories. What percent, to the nearest tenth of a percent, of the daily recommended number of calories is provided by one serving of this cereal with 2% milk?

15. Employment The Urban Center Department Store has 125 permanent employees and must hire an additional 20 temporary employees for the holiday season. What percent of the number of permanent employees is the number hired as temporary employees for the holiday season?

16. Education Conchita missed 7 out of 80 questions on a math exam. What percent of the questions did she answer correctly? Round to the nearest tenth of a percent.

17. 12 is 15% of what?

18. 42.5 is 150% of what? Round to the nearest tenth.

19. Manufacturing A manufacturer of PDAs found 384 defective PDAs during a quality control study. This amount was 1.2% of the PDAs tested. Find the number of PDAs tested.

20. Real Estate A new house was bought for $285,000. Five years later the house sold for $456,000. The increase was what percent of the original price?

21. 123 is 86% of what number? Use the proportion method. Round to the nearest tenth.

22. What percent of 12 is 120? Use the proportion method.

23. Wages An administrative assistant receives a wage of $16.24 per hour. This amount is 112% of last year's wage. What is the dollar increase in the hourly wage over last year? Use the proportion method.

24. Demography A city has a population of 71,500. Ten years ago the population was 32,500. The population now is what percent of the population 10 years ago? Use the proportion method.

25. Fees The annual license fee on a car is 1.4% of the value of the car. If the license fee during a year is $350, what is the value of the car? Use the proportion method.

CUMULATIVE REVIEW EXERCISES

1. Simplify: $18 \div (7 - 4)^2 + 2$

2. Find the LCM of 16, 24, and 30.

3. Find the sum of $2\frac{1}{3}$, $3\frac{1}{2}$, and $4\frac{5}{8}$.

4. Subtract: $27\frac{5}{12} - 14\frac{9}{16}$

5. Multiply: $7\frac{1}{3} \times 1\frac{5}{7}$

6. What is $\frac{14}{27}$ divided by $1\frac{7}{9}$?

7. Simplify: $\left(\frac{3}{4}\right)^3 \cdot \left(\frac{8}{9}\right)^2$

8. Simplify: $\left(\frac{2}{3}\right)^2 - \left(\frac{3}{8} - \frac{1}{3}\right) \div \frac{1}{2}$

9. Round 3.07973 to the nearest hundredth.

10. Subtract:
$$\begin{array}{r} 3.0902 \\ - 1.9706 \end{array}$$

11. Divide: $0.032\overline{)1.097}$
Round to the nearest ten-thousandth.

12. Convert $3\frac{5}{8}$ to a decimal.

13. Convert 1.75 to a fraction.

14. Place the correct symbol, $<$ or $>$, between the two numbers.
$\frac{3}{8}$ 0.87

15. Solve the proportion $\frac{3}{8} = \frac{20}{n}$.
Round to the nearest tenth.

16. Write "$153.60 earned in 8 hours" as a unit rate.

17. Write $18\frac{1}{3}\%$ as a fraction.

18. Write $\frac{5}{6}$ as a percent.

19. 16.3% of 120 is what?

20. 24 is what percent of 18?

21. 12.4 is 125% of what?

22. What percent of 35 is 120? Round to the nearest tenth.

23. **Taxes** Sergio has an income of $740 per week. One-fifth of his income is deducted for income tax payments. Find his take-home pay.

24. **Finance** Eunice bought a used car for $12,530, with a down payment of $2000. The balance was paid in 36 equal monthly payments. Find the monthly payment.

25. **Taxes** The gasoline tax is $.41 a gallon. Find the number of gallons of gasoline used during a month in which $172.20 was paid in gasoline taxes.

26. **Taxes** The real estate tax on a $344,000 home is $6880. At the same rate, find the real estate tax on a home valued at $500,000.

27. **Lodging** The graph at the right shows the breakdown of the locations of the 53,500 hotels throughout the United States. How many hotels in the United States are located along highways?

28. **Elections** A survey of 300 people showed that 165 people favored a certain candidate for mayor. What percent of the people surveyed did not favor this candidate?

29. **Television** According to the Cabletelevision Advertising Bureau, cable households watch television 36.5% of the time. On average, how many hours per week do cable households spend watching TV? Round to the nearest tenth.

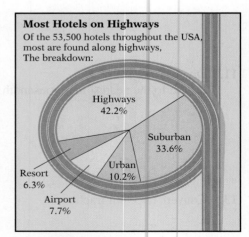

Most Hotels on Highways
Of the 53,500 hotels throughout the USA, most are found along highways, The breakdown:

Highways 42.2%
Suburban 33.6%
Urban 10.2%
Resort 6.3%
Airport 7.7%

Source: American Hotel and Lodging Association

30. **Health** The Environmental Protection Agency found that 990 out of 5500 children tested had levels of lead in their blood that exceeded federal guidelines. What percent of the children tested had levels of lead in the blood that exceeded federal standards?

Applications for Business and Consumers

OBJECTIVES

ARE YOU READY?

Take the Chapter 6 Prep Test to find out if you are ready to learn to:

- Find unit cost, total cost, and the most economical purchase
- Find percent increase and percent decrease and apply them to markup and discount
- Calculate simple interest and compound interest
- Calculate expenses associated with buying and owning a home or a car
- Calculate commissions, wages, and salaries
- Calculate checkbook balances and balance a checkbook

PREP TEST

Do these exercises to prepare for Chapter 6.

For Exercises 1 to 6, add, subtract, multiply, or divide.

1. Divide: $3.75 \div 5$

2. Multiply: 3.47×15

3. Subtract: $874.50 - 369.99$

4. Multiply: $0.065 \times 150,000$

5. Multiply: $1500 \times 0.06 \times 0.5$

6. Add: $1372.47 + 36.91 + 5.00 + 2.86$

7. Divide $10 \div 3$. Round to the nearest hundredth.

8. Divide $345 \div 570$. Round to the nearest thousandth.

9. Place the correct symbol, $<$ or $>$, between the two numbers.
$0.379 \quad 0.397$

SECTION

6.1 Applications to Purchasing

OBJECTIVE A   To find unit cost

Frequently, stores promote items for purchase by advertising, say, 2 Red Baron Bake to Rise Pizzas for $10.50 or 5 cans of StarKist tuna for $4.25. The **unit cost** is the cost of *one* Red Baron Pizza or of *one* can of StarKist tuna. To find the unit cost, divide the total cost by the number of units.

2 pizzas for $10.50	5 cans for $4.25
$10.50 \div 2 = 5.25$	$4.25 \div 5 = 0.85$
$5.25 is the cost of one pizza.	$.85 is the cost of one can.
Unit cost: $5.25 per pizza	Unit cost: $.85 per can

EXAMPLE · 1

Find the unit cost. Round to the nearest tenth of a cent.
a. 3 gallons of mint chip ice cream for $17
b. 4 ounces of Crest toothpaste for $2.29

Strategy
To find the unit cost, divide the total cost by the number of units.

Solution
a. $17 \div 3 \approx 5.667$
  $5.667 per gallon
b. $2.29 \div 4 = 0.5725$
  $.573 per ounce

YOU TRY IT · 1

Find the unit cost. Round to the nearest tenth of a cent.
a. 8 size-AA Energizer batteries for $7.67
b. 15 ounces of Suave shampoo for $2.29

Your strategy

Your solution

Solution on p. S13

OBJECTIVE B   To find the most economical purchase

Comparison shoppers often find the most economical buy by comparing unit costs.

One store is selling 6 twelve-ounce cans of ginger ale for $2.99, and a second store is selling 24 twelve-ounce cans of ginger ale for $11.79. To find the better buy, compare the unit costs.

$2.99 \div 6 \approx 0.498$	$11.79 \div 24 \approx 0.491$
Unit cost: $.498 per can	Unit cost: $.491 per can

Because $.491 < $.498, the better buy is 24 cans for $11.79.

EXAMPLE • 2

Find the more economical purchase: 5 pounds of nails for $4.80, or 4 pounds of nails for $3.78.

Strategy

To find the more economical purchase, compare the unit costs.

Solution

4.80 ÷ 5 = 0.96
3.78 ÷ 4 = 0.945
$.945 < $.96

The more economical purchase is 4 pounds for $3.78.

YOU TRY IT • 2

Find the more economical purchase: 6 cans of fruit for $8.70, or 4 cans of fruit for $6.96.

Your strategy

Your solution

Solution on p. S14

OBJECTIVE C   To find total cost

An installer of floor tile found the unit cost of identical floor tiles at three stores.

Store 1	Store 2	Store 3
$1.22 per tile	$1.18 per tile	$1.28 per tile

By comparing the unit costs, the installer determined that store 2 would provide the most economical purchase.

The installer also uses the unit cost to find the total cost of purchasing 300 floor tiles at store 2. The **total cost** is found by multiplying the unit cost by the number of units purchased.

Unit cost	×	number of units	=	total cost
1.18	×	300	=	354

The total cost is $354.

EXAMPLE • 3

Clear redwood lumber costs $5.43 per foot. How much would 25 feet of clear redwood cost?

Strategy

To find the total cost, multiply the unit cost (5.43) by the number of units (25).

Solution

Unit cost	×	number of units	=	total cost
5.43	×	25	=	135.75

The total cost is $135.75.

YOU TRY IT • 3

Pine saplings cost $9.96 each. How much would 7 pine saplings cost?

Your strategy

Your solution

Solution on p. S14

6.1 EXERCISES

OBJECTIVE A　　To find unit cost

For Exercises 1 to 10, find the unit cost. Round to the nearest tenth of a cent.

1. Heinz B·B·Q sauce, 18 ounces for $.99

2. Birds-eye maple, 6 feet for $18.75

3. Diamond walnuts, $2.99 for 8 ounces

4. A&W root beer, 6 cans for $2.99

5. Ibuprofen, 50 tablets for $3.99

6. Visine eye drops, 0.5 ounce for $3.89

7. Adjustable wood clamps, 2 for $13.95

8. Corn, 6 ears for $2.85

9. Cheerios cereal, 15 ounces for $2.99

10. Doritos Cool Ranch chips, 14.5 ounces for $2.99

11. A store advertises a "buy one, get one free" sale on pint containers of ice cream. How would you find the unit cost of one pint of ice cream?

OBJECTIVE B　　To find the most economical purchase

For Exercises 12 to 21, suppose your local supermarket offers the following products at the given prices. Find the more economical purchase.

12. Sutter Home pasta sauce, 25.5 ounces for $3.29, or Muir Glen Organic pasta sauce, 26 ounces for $3.79

13. Kraft mayonnaise, 40 ounces for $3.98, or Springfield mayonnaise, 32 ounces for $3.39

14. Ortega salsa, 20 ounces for $3.29 or 12 ounces for $1.99

15. L'Oréal shampoo, 13 ounces for $4.69, or Cortexx shampoo, 12 ounces for $3.99

16. Golden Sun vitamin E, 200 tablets for $12.99 or 400 tablets for $18.69

17. Ultra Mr. Clean, 20 ounces for $2.67, or Ultra Spic and Span, 14 ounces for $2.19

18. 16 ounces of Kraft cheddar cheese for $4.37, or 9 ounces of Land O'Lakes cheddar cheese for $2.29

19. Bertolli olive oil, 34 ounces for $9.49, or Pompeian olive oil, 8 ounces for $2.39

20. Maxwell House coffee, 4 ounces for $3.99, or Sanka coffee, 2 ounces for $2.39

21. Wagner's vanilla extract, $3.95 for 1.5 ounces, or Durkee vanilla extract, 1 ounce for $2.84

 For Exercises 22 and 23, suppose a box of Tea A contains twice as many tea bags as a box of Tea B. Decide which box of tea is the more economical purchase.

22. The price of a box of Tea A is less than twice the price of a box of Tea B.

23. The price of a box of Tea B is greater than half the price of a box of Tea A.

OBJECTIVE C   To find total cost

24. If sliced bacon costs $4.59 per pound, find the total cost of 3 pounds.

25. Used red brick costs $.98 per brick. Find the total cost of 75 bricks.

26. Kiwi fruit cost $.43 each. Find the total cost of 8 kiwi.

27. Boneless chicken filets cost $4.69 per pound. Find the cost of 3.6 pounds. Round to the nearest cent.

28. Herbal tea costs $.98 per ounce. Find the total cost of 6.5 ounces.

29. If Stella Swiss Lorraine cheese costs $5.99 per pound, find the total cost of 0.65 pound. Round to the nearest cent.

30. Red Delicious apples cost $1.29 per pound. Find the total cost of 2.1 pounds. Round to the nearest cent.

31. Choice rib eye steak costs $9.49 per pound. Find the total cost of 2.8 pounds. Round to the nearest cent.

 32. Suppose a store flyer advertises cantaloupes as "buy one, get one free." True or false? The total cost of 6 cantaloupes at the sale price is the same as the total cost of 3 cantaloupes at the regular price.

Applying the Concepts

 33. Explain in your own words the meaning of unit pricing.

 34. What is the UPC (Universal Product Code) and how is it used?

ISBN 0-395-75524-7

SECTION

6.2　Percent Increase and Percent Decrease

OBJECTIVE A　　To find percent increase

Percent increase is used to show how much a quantity has increased over its original value. The statements "Food prices increased by 2.3% last year" and "City council members received a 4% pay increase" are examples of percent increase.

Point of Interest

According to the U.S. Census Bureau, the number of persons aged 65 and over in the United States will increase to about 82.0 million by 2050, a 136% increase from 2000.

HOW TO 1　　According to the Energy Information Administration, the number of alternative-fuel vehicles increased from approximately 277,000 to 352,000 in four years. Find the percent increase in alternative-fuel vehicles. Round to the nearest percent.

$$\boxed{\begin{array}{c}\text{New}\\\text{value}\end{array}}-\boxed{\begin{array}{c}\text{original}\\\text{value}\end{array}}=\boxed{\begin{array}{c}\text{amount of}\\\text{increase}\end{array}}$$

$$352{,}000\ -\ 277{,}000\ =\ 75{,}000$$

Amount of increase (75,000)

Original value (277,000)

New value (352,000)

Now solve the basic percent equation for percent.

$$\text{Percent}\ \times\ \text{base}\ =\ \text{amount}$$

$$\boxed{\begin{array}{c}\text{Percent}\\\text{increase}\end{array}}\times\boxed{\begin{array}{c}\text{original}\\\text{value}\end{array}}=\boxed{\begin{array}{c}\text{amount of}\\\text{increase}\end{array}}$$

$$n\quad\times\quad277{,}000\quad=75{,}000$$
$$n=75{,}000\div277{,}000$$
$$n\approx0.27$$

The number of alternative-fuel vehicles increased by approximately 27%.

EXAMPLE 1

The average wholesale price of coffee increased from $2 per pound to $3 per pound in one year. What was the percent increase in the price of 1 pound of coffee?

Strategy

To find the percent increase:

• Find the amount of the increase.
• Solve the basic percent equation for *percent*.

Solution

$$\boxed{\begin{array}{c}\text{New}\\\text{value}\end{array}}-\boxed{\begin{array}{c}\text{original}\\\text{value}\end{array}}=\boxed{\begin{array}{c}\text{amount of}\\\text{increase}\end{array}}$$

$$3\quad-\quad2\quad=\quad1$$

Percent × base = amount
$$n\quad\times\ 2\ =\ \ 1$$
$$n=1\div2$$
$$n=0.5=50\%$$

The percent increase was 50%.

YOU TRY IT 1

The average price of gasoline rose from $3.46 to $3.83 in 5 months. What was the percent increase in the price of gasoline? Round to the nearest percent.

Your strategy

Your solution

Solution on p. S14

EXAMPLE • 2

Chris Carley was earning $13.50 an hour as a nursing assistant before receiving a 10% increase in pay. What is Chris's new hourly pay?

Strategy

To find the new hourly wage:

• Solve the basic percent equation for *amount*.
• Add the amount of the increase to the original wage.

Solution

Percent × base = amount
0.10 ×13.50 =   *n*
         1.35 =   *n*

The amount of the increase was $1.35.
13.50 + 1.35 = 14.85

The new hourly wage is $14.85.

YOU TRY IT • 2

Yolanda Liyama was making a wage of $12.50 an hour as a baker before receiving a 14% increase in hourly pay. What is Yolanda's new hourly wage?

Your strategy

Your solution

Solution on p. S14

OBJECTIVE B   **To apply percent increase to business—markup**

Some of the expenses involved in operating a business are salaries, rent, equipment, and utilities. To pay these expenses and earn a profit, a business must sell a product at a higher price than it paid for the product.

Cost is the price a business pays for a product, and **selling price** is the price at which a business sells a product to a customer. The difference between selling price and cost is called **markup.**

| Selling price | − | cost | = | markup |

or

| Cost | + | markup | = | selling price |

Markup is frequently expressed as a percent of a product's cost. This percent is called the **markup rate.**

| Markup rate | × | cost | = | markup |

Point of Interest

According to *Managing a Small Business,* from Liraz Publishing Company, goods in a store are often marked up 50% to 100% of the cost. This allows a business to make a profit of 5% to 10%.

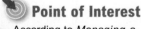

David Madison/STONE/Getty Images

HOW TO • 2   Suppose Bicycles Galore purchases an AMP Research B-5 bicycle for $2119.20 and sells it for $2649. What markup rate does Bicycles Galore use?

| Selling price | − | cost | = | markup |
| 2649.00 | − | 2119.20 | = | 529.80 |

Percent	×	base	=	amount
Markup rate	×	cost	=	markup
n	×	2119.20	=	529.80

n = 529.80 ÷ 2119.20 = 0.25

• First find the markup.

• Then solve the basic percent equation for *percent.*

The markup rate is 25%.

EXAMPLE · 3

The manager of a sporting goods store determines that a markup rate of 36% is necessary to make a profit. What is the markup on a pair of skis that costs the store $225?

Strategy

To find the markup, solve the basic percent equation for *amount*.

Solution

Percent	×	base	=	amount
Markup rate	×	cost	=	markup

$$0.36 \quad × \quad 225 \quad = \quad n$$
$$81 = n$$

The markup is $81.

YOU TRY IT · 3

A bookstore manager determines that a markup rate of 20% is necessary to make a profit. What is the markup on a book that costs the bookstore $32?

Your strategy

Your solution

EXAMPLE · 4

A plant nursery bought a yellow twig dogwood for $9.50 and used a markup rate of 46%. What is the selling price?

Strategy

To find the selling price:
• Find the markup by solving the basic percent equation for *amount*.
• Add the markup to the cost.

Solution

Percent	×	base	=	amount
Markup rate	×	cost	=	markup

$$0.46 \quad × \quad 9.50 \quad = \quad n$$
$$4.37 = n$$

Cost	+	markup	=	selling price

$$9.50 \quad + \quad 4.37 \quad = \quad 13.87$$

The selling price is $13.87.

YOU TRY IT · 4

A clothing store bought a leather jacket for $72 and used a markup rate of 55%. What is the selling price?

Your strategy

Your solution

Solutions on p. S14

OBJECTIVE C   **To find percent decrease**

Percent decrease is used to show how much a quantity has decreased from its original value. The statements "The number of family farms decreased by 2% last year" and "There has been a 50% decrease in the cost of a Pentium chip" are examples of percent decrease.

HOW TO · 3 During a 2-year period, the value of U.S. agricultural products exported decreased from approximately $60.6 billion to $52.0 billion. Find the percent decrease in the value of U.S. agricultural exports. Round to the nearest tenth of a percent.

$$
\boxed{\begin{array}{c}\text{Original}\\ \text{value}\end{array}} - \boxed{\begin{array}{c}\text{new}\\ \text{value}\end{array}} = \boxed{\begin{array}{c}\text{amount of}\\ \text{decrease}\end{array}}
$$

$$60.6 \quad - \quad 52.0 \quad = \quad 8.6$$

Now solve the basic percent equation for percent.

$$\text{Percent} \quad \times \quad \text{base} \quad = \quad \text{amount}$$

$$
\boxed{\begin{array}{c}\text{Percent}\\ \text{decrease}\end{array}} \times \boxed{\begin{array}{c}\text{original}\\ \text{value}\end{array}} = \boxed{\begin{array}{c}\text{amount of}\\ \text{decrease}\end{array}}
$$

$$n \quad \times \quad 60.6 \quad = \quad 8.6$$
$$n = 8.6 \div 60.6$$
$$n \approx 0.142$$

The value of agricultural exports decreased approximately 14.2%.

Amount of decrease (8.6)

New value (52.0)

Original value (60.6)

 Tips for Success
Note in the example below that solving a word problem involves stating a strategy and using the strategy to find a solution. If you have difficulty with a word problem, write down the known information. Be very specific. Write out a phrase or sentence that states what you are trying to find. See *AIM for Success* at the front of the book.

EXAMPLE · 5

 During an 8-year period, the population of Baltimore, Maryland, decreased from approximately 736,000 to 646,000. Find the percent decrease in Baltimore's population. Round to the nearest tenth of a percent.

Strategy
To find the percent decrease:
- Find the amount of the decrease.
- Solve the basic percent equation for *percent*.

Solution

$$
\boxed{\begin{array}{c}\text{Original}\\ \text{value}\end{array}} - \boxed{\begin{array}{c}\text{new}\\ \text{value}\end{array}} = \boxed{\begin{array}{c}\text{amount of}\\ \text{decrease}\end{array}}
$$

$$736{,}000 \quad - \quad 646{,}000 \quad = \quad 90{,}000$$

$$\text{Percent} \times \text{base} = \text{amount}$$
$$n \times 736{,}000 = 90{,}000$$
$$n = 90{,}000 \div 736{,}000$$
$$n \approx 0.122$$

Baltimore's population decreased approximately 12.2%.

YOU TRY IT · 5

During an 8-year period, the population of Norfolk, Virginia, decreased from approximately 261,000 to 215,000. Find the percent decrease in Norfolk's population. Round to the nearest tenth of a percent.

Your strategy

Your solution

EXAMPLE • 6

The total sales for December for a stationery store were $96,000. For January, total sales showed an 8% decrease from December's sales. What were the total sales for January?

Strategy

To find the total sales for January:
- Find the amount of decrease by solving the basic percent equation for *amount*.
- Subtract the amount of decrease from the December sales.

Solution

Percent × base = amount
0.08 × 96,000 = n
            7680 = n

The decrease in sales was $7680.

96,000 − 7680 = 88,320

The total sales for January were $88,320.

YOU TRY IT • 6

Fog decreased the normal 5-mile visibility at an airport by 40%. What was the visibility in the fog?

Your strategy

Your solution

Solution on p. S14

OBJECTIVE D   To apply percent decrease to business—discount

To promote sales, a store may reduce the regular price of some of its products temporarily. The reduced price is called the **sale price.** The difference between the regular price and the sale price is called the **discount.**

| Regular price | − | sale price | = | discount |

or

| Regular price | − | discount | = | sale price |

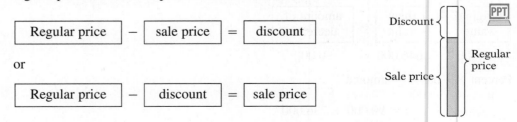

Discount is frequently stated as a percent of a product's regular price. This percent is called the **discount rate.**

| Discount rate | × | regular price | = | discount |

EXAMPLE • 7

A GE 25-inch stereo television that regularly sells for $299 is on sale for $250. Find the discount rate. Round to the nearest tenth of a percent.

Strategy

To find the discount rate:

- Find the discount.
- Solve the basic percent equation for *percent*.

Solution

Regular price	−	sale price	=	discount
299	−	250	=	49

Percent	×	base	=	amount
Discount rate	×	regular price	=	discount
n	×	299	=	49

$$n = 49 \div 299$$
$$n \approx 0.164$$

The discount rate is 16.4%.

YOU TRY IT • 7

A white azalea that regularly sells for $12.50 is on sale for $10.99. Find the discount rate. Round to the nearest tenth of a percent.

Your strategy

Your solution

EXAMPLE • 8

A 20-horsepower lawn mower is on sale for 25% off the regular price of $1525. Find the sale price.

Strategy

To find the sale price:

- Find the discount by solving the basic percent equation for *amount*.
- Subtract to find the sale price.

Solution

Percent	×	base	=	amount
Discount rate	×	regular price	=	discount
0.25	×	1525	=	n

$$381.25 = n$$

Regular price	−	discount	=	sale price
1525	−	381.25	=	1143.75

The sale price is $1143.75.

YOU TRY IT • 8

A hardware store is selling a Newport security door for 15% off the regular price of $225. Find the sale price.

Your strategy

Your solution

Solutions on p. S15

6.2 EXERCISES

OBJECTIVE A   **To find percent increase**

Solve. If necessary, round percents to the nearest tenth of a percent.

 1. **Bison** See the news clipping at the right. Find the percent increase in human consumption of bison from 2005 to the date of this news article.

2. **Fuel Efficiency** An automobile manufacturer increased the average mileage on a car from 17.5 miles per gallon to 18.2 miles per gallon. Find the percent increase in mileage.

3. **Business** In the 1990s, the number of Target stores increased from 420 stores to 914 stores. (*Source:* Target) What was the percent increase in the number of Target stores in the 1990s?

4. **Demography** The graph at the right shows the number of unmarried American couples living together. (*Source:* U.S. Census Bureau) Find the percent increase in the number of unmarried couples living together from 1980 to 2000.

5. **Sports** In 1924, the number of events in the Winter Olympics was 14. The 2006 Winter Olympics in Salt Lake City included 84 medal events. (*Source:* David Wallenchinsky's *The Complete Book of the Winter Olympics*) Find the percent increase in the number of events in the Winter Olympics from 1924 to 2006.

6. **Television** During 1 year, the number of people subscribing to direct broadcasting satellite systems increased 87%. If the number of subscribers at the beginning of the year was 2.3 million, how many subscribers were there at the end of the year?

7. **Pets** In a recent year, Americans spent $35.9 billion on their pets. This was up from $17 billion a decade earlier. (*Source: Time,* February 4, 2008) Find the percent increase in the amount Americans spent on their pets during the 10-year period.

8. **Demography** From 1970 to 2000, the average age of American mothers giving birth to their first child rose 16.4%. (*Source:* Centers for Disease Control and Prevention) If the average age in 1970 was 21.4 years, what was the average age in 2000? Round to the nearest tenth.

 9. **Compensation** A welder earning $12 per hour is given a 10% raise. To find the new wage, we can multiply $12 by 0.10 and add the product to $12. Can the new wage be found by multiplying $12 by 1.10?

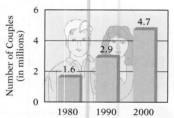

Unmarried U.S. Couples Living Together

OBJECTIVE B   To apply percent increase to business—markup

The three important markup equations are:

   (1) Selling price − cost = markup
   (2) Cost + markup = selling price
   (3) Markup rate × cost = markup

For Exercises 10 and 11, list, in the order in which they will be used, the equations needed to solve each problem.

10. A book that cost the seller $17 is sold for $23. Find the markup rate.

11. A DVD that cost the seller $12 has a markup rate of 55%. Find the selling price.

12. A window air conditioner cost AirRite Air Conditioning Systems $285. Find the markup on the air conditioner if the markup rate is 25% of the cost.

13. The manager of Brass Antiques has determined that a markup rate of 38% is necessary for a profit to be made. What is the markup on a brass doorknob that costs $45?

14. Computer Inc. uses a markup of $975 on a computer system that costs $3250. What is the markup rate on this system?

15. Saizon Pen & Office Supply uses a markup of $12 on a calculator that costs $20. What markup rate does this amount represent?

16. Giant Photo Service uses a markup rate of 48% on its Model ZA cameras, which cost the shop $162. What is the selling price?

17. The Circle R golf pro shop uses a markup rate of 45% on a set of Tour Pro golf clubs that costs the shop $210. What is the selling price?

18. Resner Builders' Hardware uses a markup rate of 42% for a table saw that costs $225. What is the selling price of the table saw?

19. Brad Burt's Magic Shop uses a markup rate of 48%. What is the selling price of a telescoping sword that costs $50?

OBJECTIVE C   To find percent decrease

Solve. If necessary, round to the nearest tenth of a percent.

20. **Law School** Use the news clipping at the right to find the percent decrease in the number of people who took the LSATs in the last three years.

21. **Travel** A new bridge reduced the normal 45-minute travel time between two cities by 18 minutes. What percent decrease does this represent?

> **In the News**
>
> **Fewer Students Take LSATs**
>
> This year 137,444 people took the Law School Admission Test (LSATs). Three years ago, the LSATs were administered to 148,014 people.
>
> *Source:* Law School Admission Council

22. Energy By installing energy-saving equipment, the Pala Rey Youth Camp reduced its normal $800-per-month utility bill by $320. What percent decrease does this amount represent?

	1990 Census	2000 Census	2005 Population Estimate
Chicago	1,783,726	2,896,016	2,842,518
Detroit	1,027,974	951,270	886,671
Phildelphia	1,585,577	1,517,550	1,463,281

Source: Census Bureau

23. Urban Populations The table at the right above shows the populations of three cities in the United States.
   a. Find the percent decrease in the population of Detroit from 1990 to 2005.
   b. Find the percent decrease in the population of Philadelphia from 1990 to 2005.
   c. Find the percent decrease in the population of Chicago from 2000 to 2005.

24. Missing Persons See the news clipping at the right. Find the percent decrease over the last 10 years in the number of people entered into the National Crime Information Center's Missing Person File.

> **In the News**
>
> **Missing-Person Cases Decrease**
>
> This year, 834,536 missing-person cases were entered into the National Crime Information Center's Missing Person File. Ten years ago, the number was 969,264.
>
> *Source:* National Crime Information Center

25. Depreciation It is estimated that the value of a new car is reduced 30% after 1 year of ownership. Using this estimate, find how much value a $28,200 new car loses after 1 year.

26. Employment A department store employs 1200 people during the holiday. At the end of the holiday season, the store reduces the number of employees by 45%. What is the decrease in the number of employees?

27. Finance Juanita's average monthly expense for gasoline was $176. After joining a car pool, she was able to reduce the expense by 20%.
   a. What was the amount of the decrease?
   b. What is the average monthly gasoline bill now?

28. Investments An oil company paid a dividend of $1.60 per share. After a reorganization, the company reduced the dividend by 37.5%.
   a. What was the amount of the decrease?
   b. What is the new dividend?

29. The Military In 2000, the Pentagon revised its account of the number of Americans killed in the Korean War from 54,246 to 36,940. (*Source: Time,* June 12, 2000) What is the percent decrease in the reported number of military personnel killed in the Korean War? Round to nearest tenth of a percent.

30. In a math class, the average grade on the second test was 5% lower than the average grade on the first test. What should you multiply the first test average by to find the difference between the average grades on the two tests?

OBJECTIVE D   **To apply percent decrease to business—discount**

The three important discount equations are:
  (1) Regular price − sale price = discount
  (2) Regular price − discount = sale price
  (3) Discount rate × regular price = discount

For Exercises 31 and 32, list, in the order in which they will be used, the equations needed to solve each problem.

31. Shoes that regularly sell for $65 are on sale for 15% off the regular price. Find the sale price.

32. A radio with a regular price of $89 is on sale for $59. Find the discount rate.

33. The Austin College Bookstore is giving a discount of $8 on calculators that normally sell for $24. What is the discount rate?

34. A discount clothing store is selling a $72 sport jacket for $24 off the regular price. What is the discount rate?

35. A disc player that regularly sells for $400 is selling for 20% off the regular price. What is the discount?

36. Dacor Appliances is selling its $450 washing machine for 15% off the regular price. What is the discount?

37. An electric grill that regularly sells for $140 is selling for $42 off the regular price. What is the discount rate?

38. Quick Service Gas Station has its regularly priced $125 tune-up on sale for 16% off the regular price.
  a. What is the discount?
  b. What is the sale price?

39. Tomatoes that regularly sell for $1.25 per pound are on sale for 20% off the regular price.
  a. What is the discount?
  b. What is the sale price?

40. An outdoor supply store has its regularly priced $160 sleeping bags on sale for $120. What is the discount rate?

41. Standard Brands ceiling paint that regularly sells for $20 per gallon is on sale for $16 per gallon. What is the discount rate?

Applying the Concepts

42. Business A promotional sale at a department store offers 25% off the sale price. The sale price itself is 25% off the regular price. Is this the same as a sale that offers 50% off the regular price? If not, which sale gives the better price? Explain your answer.

SECTION

6.3 Interest

OBJECTIVE A

To calculate simple interest

When you deposit money in a bank—for example, in a savings account—you are permitting the bank to use your money. The bank may use the deposited money to lend customers the money to buy cars or make renovations on their homes. The bank pays you for the privilege of using your money. The amount paid to you is called **interest.** If you are the one borrowing money from the bank, the amount you pay for the privilege of using that money is also called interest.

The original amount deposited or borrowed is called the **principal.** The amount of interest paid is usually given as a percent of the principal. The percent used to determine the amount of interest is the **interest rate.**

> ✓ **Take Note**
>
> If you deposit $1000 in a savings account paying 5% interest, the $1000 is the principal and 5% is the interest rate.

Interest paid on the original principal is called **simple interest**. To calculate simple interest, multiply the principal by the interest rate per period by the number of time periods. In this objective, we are working with annual interest rates, so the time periods are years. The simple interest formula for an annual interest rate is given below.

> **Simple Interest Formula for Annual Interest Rates**
>
> Principal × annual interest rate × time (in years) = interest

---

Interest rates are generally given as percents. Before performing calculations involving an interest rate, write the interest rate as a decimal.

> **HOW TO 1** Calculate the simple interest due on a 2-year loan of $1500 that has an annual interest rate of 7.5%.
>
> | Principal | × | annual interest rate | × | time (in years) | = | interest |
> |:---:|:---:|:---:|:---:|:---:|:---:|:---:|
> | 1500 | × | 0.075 | × | 2 | = | 225 |
>
> The simple interest due is $225.

When we borrow money, the total amount to be repaid to the lender is the sum of the principal and the interest. This amount is called the **maturity value of a loan.**

> **Maturity Value Formula for Simple Interest Loans**
>
> Principal + interest = maturity value

In the example above, the simple interest due on the loan of $1500 was $225. The maturity value of the loan is therefore $1500 + $225 = $1725.

Take Note

The time of the loan must be in years. Eight months is $\frac{8}{12}$ of a year.

See Example 1. The time of the loan must be in years. 180 days is $\frac{180}{365}$ of a year.

HOW TO 2 Calculate the maturity value of a simple interest, 8-month loan of $8000 if the annual interest rate is 9.75%.

First find the interest due on the loan.

Principal	×	annual interest rate	×	time (in years)	=	interest
8000	×	0.0975	×	$\frac{8}{12}$	=	520

Find the maturity value.

Principal	+	interest	=	maturity value
8000	+	520	=	8520

The maturity value of the loan is $8520.

The monthly payment on a loan can be calculated by dividing the maturity value by the length of the loan in months.

Monthly Payment on a Simple Interest Loan

Maturity value ÷ length of the loan in months = monthly payment

---

In the example above, the maturity value of the loan is $8520. To find the monthly payment on the 8-month loan, divide 8520 by 8.

Maturity value	÷	length of the loan in months	=	monthly payment
8520	÷	8	=	1065

The monthly payment on the loan is $1065.

EXAMPLE · 1

Kamal borrowed $500 from a savings and loan association for 180 days at an annual interest rate of 7%. What is the simple interest due on the loan?

Strategy

To find the simple interest due, multiply the principal (500) times the annual interest rate (7% = 0.07) times the time in years (180 days = $\frac{180}{365}$ year).

Solution

Principal	×	annual interest rate	×	time (in years)	=	interest
500	×	0.07	×	$\frac{180}{365}$	≈	17.26

The simple interest due is $17.26.

YOU TRY IT · 1

A company borrowed $15,000 from a bank for 18 months at an annual interest rate of 8%. What is the simple interest due on the loan?

Your strategy

Your solution

Solution on p. S15

EXAMPLE · 2

Calculate the maturity value of a simple interest, 9-month loan of $4000 if the annual interest rate is 8.75%.

Strategy
To find the maturity value:
- Use the simple interest formula to find the simple interest due.
- Find the maturity value by adding the principal and the interest.

Solution

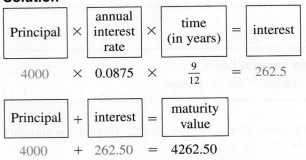

Principal	×	annual interest rate	×	time (in years)	=	interest
4000	×	0.0875	×	$\frac{9}{12}$	=	262.5

Principal	+	interest	=	maturity value
4000	+	262.50	=	4262.50

The maturity value is $4262.50.

YOU TRY IT · 2

Calculate the maturity value of a simple interest, 90-day loan of $3800. The annual interest rate is 6%.

Your strategy

Your solution

EXAMPLE · 3

The simple interest due on a 3-month loan of $1400 is $26.25. Find the monthly payment on the loan.

Strategy
To find the monthly payment:
- Find the maturity value by adding the principal and the interest.
- Divide the maturity value by the length of the loan in months (3).

Solution
Principal + interest = maturity value
1400 + 26.25 = 1426.25

Maturity value ÷ length of the loan = payment
1426.25 ÷ 3 ≈ 475.42

The monthly payment is $475.42.

YOU TRY IT · 3

The simple interest due on a 1-year loan of $1900 is $152. Find the monthly payment on the loan.

Your strategy

Your solution

Solutions on p. S15

OBJECTIVE B **To calculate finance charges on a credit card bill**

When a customer uses a credit card to make a purchase, the customer is actually receiving a loan. Therefore, there is frequently an added cost to the consumer who purchases on credit. This may be in the form of an annual fee and interest charges on purchases. The interest charges on purchases are called **finance charges**.

The finance charge on a credit card bill is calculated using the simple interest formula. In the last objective, the interest rates were annual interest rates. However, credit card companies generally issue *monthly* bills and express interest rates on credit card purchases as *monthly* interest rates. Therefore, when using the simple interest formula to calculate finance charges on credit card purchases, use a monthly interest rate and express the time in months.

Note: In the simple interest formula, the time must be expressed in the same period as the rate. For an *annual* interest rate, the time must be expressed in years. For a *monthly* interest rate, the time must be expressed in months.

EXAMPLE • 4	**YOU TRY IT • 4**
A credit card company charges a customer 1.5% per month on the unpaid balance of charges on the credit card. What is the finance charge in a month in which the customer has an unpaid balance of $254?	The credit card that Francesca uses charges her 1.6% per month on her unpaid balance. Find the finance charge when her unpaid balance for the month is $1250.
Strategy	**Your strategy**
To find the finance charge, multiply the principal, or unpaid balance (254), times the monthly interest rate (1.5%) times the number of months (1).	
Solution	**Your solution**

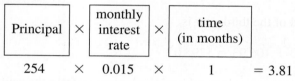

Principal	×	monthly interest rate	×	time (in months)	
254	×	0.015	×	1	= 3.81

The finance charge is $3.81.

Solution on p. S15

OBJECTIVE C   To calculate compound interest

Usually, the interest paid on money deposited or borrowed is compound interest. **Compound interest** is computed not only on the original principal but also on interest already earned. Here is an illustration.

Suppose $1000 is invested for 3 years at an annual interest rate of 9% compounded annually. Because this is an *annual* interest rate, we will calculate the interest earned each year.

During the first year, the interest earned is calculated as follows:

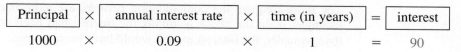

Principal	×	annual interest rate	×	time (in years)	=	interest
1000	×	0.09	×	1	=	90

At the end of the first year, the total amount in the account is

$$1000 + 90 = 1090$$

During the second year, the interest earned is calculated on the amount in the account at the end of the first year.

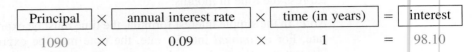

Principal	×	annual interest rate	×	time (in years)	=	interest
1090	×	0.09	×	1	=	98.10

Note that the interest earned during the second year ($98.10) is greater than the interest earned during the first year ($90). This is because the interest earned during the first year was added to the original principal, and the interest for the second year was calculated using this sum. If the account earned simple interest, the interest earned would be the same every year ($90).

At the end of the second year, the total amount in the account is the sum of the amount in the account at the end of the first year and the interest earned during the second year.

$$1090 + 98.10 = 1188.10$$

The interest earned during the third year is calculated using the amount in the account at the end of the second year ($1188.10).

Principal	×	annual interest rate	×	time (in years)	=	interest
1188.10	×	0.09	×	1	≈	106.93

The amount in the account at the end of the third year is

$$1188.10 + 106.93 = 1295.03$$

> **✓ Take Note**
>
> The interest earned each year keeps increasing. This is the effect of compound interest.

To find the interest earned for the three years, subtract the original principal from the new principal.

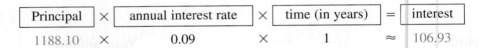

New principal	−	original principal	=	interest earned
1295.03	−	1000	=	295.03

Note that the compound interest earned is $295.03. The simple interest earned on the investment would have been only $1000 × 0.09 × 3 = $270.

In this example, the interest was compounded annually. However, interest can be compounded

Compounding periods:	annually (once a year)
	semiannually (twice a year)
	quarterly (four times a year)
	monthly (12 times a year)
	daily (365 times a year)

The more frequent the compounding periods, the more interest the account earns. For example, if, in the above example, the interest had been compounded quarterly rather than annually, the interest earned would have been greater.

Calculating compound interest can be very tedious, so there are tables that can be used to simplify these calculations. A portion of a Compound Interest Table is given in the Appendix.

HOW TO 3 What is the value after 5 years of $1000 invested at 7% annual interest, compounded quarterly?

To find the interest earned, multiply the original principal (1000) by the factor found in the Compound Interest Table. To find the factor, first find the table headed "Compounded Quarterly" in the Compound Interest Table in the Appendix. Then look at the number where the 7% column and the 5-year row meet.

	Compounded Quarterly						
	4%	*5%*	*6%*	*7%*	*8%*	*9%*	*10%*
1 year	1.04060	1.05094	1.06136	1.07186	1.08243	1.09308	1.10381
5 years	1.22019	1.28204	1.34686	**1.41478**	1.48595	1.56051	1.63862
10 years	1.48886	1.64362	1.81402	2.00160	2.20804	2.43519	2.68506
15 years	1.81670	2.10718	2.44322	2.83182	3.28103	3.80013	4.39979
20 years	2.21672	2.70148	3.29066	4.00639	4.87544	5.93015	7.20957

The factor is 1.41478.

$$1000 \times 1.41478 = 1414.78$$

The value of the investment after 5 years is $1414.78.

EXAMPLE 5

An investment of $650 pays 8% annual interest, compounded semiannually. What is the interest earned in 5 years?

Strategy

To find the interest earned:

- Find the new principal by multiplying the original principal (650) by the factor found in the Compound Interest Table (1.48024).
- Subtract the original principal from the new principal.

Solution

$650 \times 1.48024 \approx 962.16$

The new principal is $962.16.

$962.16 - 650 = 312.16$

The interest earned is $312.16.

YOU TRY IT 5

An investment of $1000 pays 6% annual interest, compounded quarterly. What is the interest earned in 20 years?

Your strategy

Your solution

Solution on pp. S15–S16

6.3 EXERCISES

OBJECTIVE A   To calculate simple interest

1. A 2-year student loan of $10,000 is made at an annual simple interest rate of 4.25%. The simple interest on the loan is $850. Identify **a.** the principal, **b.** the interest, **c.** the interest rate, and **d.** the time period of the loan.

2. A contractor obtained a 9-month loan for $80,000 at an annual simple interest rate of 9.75%. The simple interest on the loan is $5850. Identify **a.** the principal, **b.** the interest, **c.** the interest rate, and **d.** the time period of the loan.

3. Find the simple interest Jacob Zucker owes on a 2-year student loan of $8000 at an annual interest rate of 6%.

4. Find the simple interest Kara Tanamachi owes on a $1\frac{1}{2}$-year loan of $1500 at an annual interest rate of 7.5%.

© Richard Cummins/Corbis

5. To finance the purchase of 15 new cars, the Tropical Car Rental Agency borrowed $100,000 for 9 months at an annual interest rate of 4.5%. What is the simple interest due on the loan?

6. A home builder obtained a preconstruction loan of $50,000 for 8 months at an annual interest rate of 9.5%. What is the simple interest due on the loan?

7. A bank lent Gloria Masters $20,000 at an annual interest rate of 8.8%. The period of the loan was 9 months. Find the simple interest due on the loan.

8. Eugene Madison obtained an 8-month loan of $4500 at an annual interest rate of 6.2%. Find the simple interest Eugene owes on the loan.

9. Jorge Elizondo took out a 75-day loan of $7500 at an annual interest rate of 5.5%. Find the simple interest due on the loan.

10. Kristi Yang borrowed $15,000. The term of the loan was 90 days, and the annual simple interest rate was 7.4%. Find the simple interest due on the loan.

11. The simple interest due on a 4-month loan of $4800 is $320. What is the maturity value of the loan?

12. The simple interest due on a 60-day loan of $6500 is $80.14. Find the maturity value of the loan.

13. William Carey borrowed $12,500 for 8 months at an annual simple interest rate of 4.5%. Find the total amount due on the loan.

14. You arrange for a 9-month bank loan of $9000 at an annual simple interest rate of 8.5%. Find the total amount you must repay to the bank.

15. Capital City Bank approves a home-improvement loan application for $14,000 at an annual simple interest rate of 5.25% for 270 days. What is the maturity value of the loan?

16. A credit union lends a member $5000 for college tuition. The loan is made for 18 months at an annual simple interest rate of 6.9%. What is the maturity value of this loan?

17. Action Machining Company purchased a robot-controlled lathe for $225,000 and financed the full amount at 8% annual simple interest for 4 years. The simple interest on the loan is $72,000. Find the monthly payment.

18. For the purchase of an entertainment center, a $1900 loan is obtained for 2 years at an annual simple interest rate of 9.4%. The simple interest due on the loan is $357.20. What is the monthly payment on the loan?

19. To attract new customers, Heller Ford is offering car loans at an annual simple interest rate of 4.5%.
   a. Find the interest charged to a customer who finances a car loan of $12,000 for 2 years.
   b. Find the monthly payment.

20. Cimarron Homes Inc. purchased a snow plow for $57,000 and financed the full amount for 5 years at an annual simple interest rate of 9%.
   a. Find the interest due on the loan.
   b. Find the monthly payment.

21. Dennis Pappas decided to build onto his present home instead of buying a new, larger house. He borrowed $142,000 for $5\frac{1}{2}$ years at an annual simple interest rate of 7.5%. Find the monthly payment.

22. Rosalinda Johnson took out a 6-month, $12,000 loan. The annual simple interest rate on the loan was 8.5%. Find the monthly payment.

23. Student A and Student B borrow the same amount of money at the same annual interest rate. Student A has a 2-year loan and Student B has a 1-year loan. In each case, state whether the first quantity is *less than, equal to,* or *greater than* the second quantity.
   a. Student A's principal; Student B's principal
   b. Student A's maturity value; Student B's maturity value
   c. Student A's monthly payment; Student B's monthly payment

Knut Platon/STONE/Getty Images

OBJECTIVE B   To calculate finance charges on a credit card bill

24. A credit card company charges a customer 1.25% per month on the unpaid balance of charges on the credit card. What is the finance charge in a month in which the customer has an unpaid balance of $118.72?

25. The credit card that Dee Brown uses charges her 1.75% per month on her unpaid balance. Find the finance charge when her unpaid balance for the month is $391.64.

26. What is the finance charge on an unpaid balance of $12,368.92 on a credit card that charges 1.5% per month on any unpaid balance?

27. Suppose you have an unpaid balance of $995.04 on a credit card that charges 1.2% per month on any unpaid balance. What finance charge do you owe the company?

28. A credit card customer has an unpaid balance of $1438.20. What is the difference between monthly finance charges of 1.15% per month on the unpaid balance and monthly finance charges of 1.85% per month?

29. One credit card company charges 1.25% per month on any unpaid balance, and a second company charges 1.75%. What is the difference between the finance charges that these two companies assess on an unpaid balance of $687.45?

 Your credit card company requires a minimum monthly payment of $10. You plan to pay off the balance on your credit card by paying the minimum amount each month and making no further purchases using this credit card. For Exercises 30 and 31, state whether the finance charge for the second month will be *less than, equal to,* or *greater than* the finance charge for the first month, and state whether you will eventually be able to pay off the balance.

30. The finance charge for the first month was less than $10.

31. The finance charge for the first month was exactly $10.

OBJECTIVE C   To calculate compound interest

32. North Island Federal Credit Union pays 4% annual interest, compounded daily, on time savings deposits. Find the value after 1 year of $750 deposited in this account.

33. Tanya invested $2500 in a tax-sheltered annuity that pays 8% annual interest, compounded daily. Find the value of her investment after 20 years.

34. Sal Travato invested $3000 in a corporate retirement account that pays 6% annual interest, compounded semiannually. Find the value of his investment after 15 years.

35. To replace equipment, a farmer invested $20,000 in an account that pays 7% annual interest, compounded monthly. What is the value of the investment after 5 years?

36. Green River Lodge invests $75,000 in a trust account that pays 8% interest, compounded quarterly.
 a. What will the value of the investment be in 5 years?
 b. How much interest will be earned in the 5 years?

37. To save for retirement, a couple deposited $3000 in an account that pays 7% annual interest, compounded daily.
 a. What will the value of the investment be in 10 years?
 b. How much interest will be earned in the 10 years?

38. To save for a child's education, the Petersens deposited $2500 into an account that pays 6% annual interest, compounded daily. Find the amount of interest earned on this account over a 20-year period.

39. How much interest is earned in 2 years on $4000 deposited in an account that pays 6% interest, compounded quarterly?

40. The compound interest factor for a 5-year investment at an annual interest rate of 6%, compounded semiannually, is 1.34392. What does the expression $3500 - (3500 \times 1.34392)$ represent?

Applying the Concepts

41. Banking At 4 P.M. on July 31, you open a savings account that pays 5% annual interest and you deposit $500 in the account. Your deposit is credited as of August 1. At the beginning of September, you receive a statement from the bank that shows that during the month of August, you received $2.12 in interest. The interest has been added to your account, bringing the total on deposit to $502.12. At the beginning of October, you receive a statement from the bank that shows that during the month of September, you received $2.06 in interest on the $502.12 on deposit. Explain why you received less interest during the second month when there was more money on deposit.

42. Banking Suppose you have a savings account that earns interest at the rate of 6% per year, compounded monthly. On January 1, you open this account with a deposit of $100.
 a. On February 1, you deposit an additional $100 into the account. What is the value of the account after the deposit?
 b. On March 1, you deposit an additional $100 into the account. What is the value of the account after the deposit?
 Note: This type of savings plan, wherein equal amounts ($100) are saved at equal time intervals (every month), is called an annuity.

SECTION

6.4 Real Estate Expenses

OBJECTIVE A   **To calculate the initial expenses of buying a home**

One of the largest investments most people ever make is the purchase of a home. The major initial expense in the purchase is the **down payment,** which is normally a percent of the purchase price. This percent varies among banks, but it usually ranges from 5% to 25%.

The **mortgage** is the amount that is borrowed to buy real estate. The mortgage amount is the difference between the purchase price and the down payment.

HOW TO · 1   A home is purchased for $140,000, and a down payment of $21,000 is made. Find the mortgage.

Purchase price	−	down payment	=	mortgage
140,000	−	21,000	=	119,000

The mortgage is $119,000.

✓ **Take Note**

Because *points* means percent, a loan origination fee of $2\frac{1}{2}$ points $= 2\frac{1}{2}\% =$ 2.5% = 0.025.

Another initial expense in buying a home is the **loan origination fee,** which is a fee that the bank charges for processing the mortgage papers. The loan origination fee is usually a percent of the mortgage and is expressed in **points,** which is the term banks use to mean percent. For example, "5 points" means "5 percent."

Points	×	mortgage	=	loan origination fee

EXAMPLE · 1

A house is purchased for $250,000, and a down payment, which is 20% of the purchase price, is made. Find the mortgage.

Strategy

To find the mortgage:

• Find the down payment by solving the basic percent equation for *amount.*
• Subtract the down payment from the purchase price.

Solution

Percent	×	base	=	amount
Percent	×	purchase price	=	down payment
0.20	×	250,000	=	n
		50,000	=	n

Purchase price	−	down payment	=	mortgage
250,000	−	50,000	=	200,000

The mortgage is $200,000.

YOU TRY IT · 1

An office building is purchased for $1,500,000, and a down payment, which is 25% of the purchase price, is made. Find the mortgage.

Your strategy

Your solution

Solution on p. S16

| EXAMPLE • 2 | | YOU TRY IT • 2 |

A home is purchased with a mortgage of $165,000. The buyer pays a loan origination fee of $3\frac{1}{2}$ points. How much is the loan origination fee?

Strategy

To find the loan origination fee, solve the basic percent equation for *amount*.

Solution

Percent × base = amount

| Points | × | mortgage | = | fee |

0.035 × 165,000 = *n*

5775 = *n*

The loan origination fee is $5775.

The mortgage on a real estate investment is $180,000. The buyer paid a loan origination fee of $4\frac{1}{2}$ points. How much was the loan origination fee?

Your strategy

Your solution

Solution on p. S16

OBJECTIVE B   To calculate the ongoing expenses of owning a home

Point of Interest

The number-one response of adults when asked what they would spend money on first if they suddenly became wealthy (for example, by winning the lottery) was a house; 31% gave this response. (*Source:* Yankelovich Partners for Lutheran Brotherhood)

Integrating Technology

In general, when a problem requests a monetary payment, the answer is rounded to the nearest cent. For the example at the right, enter

160000 **X** 0.0080462 **=**

The display reads 1287.392. Round this number to the nearest hundredth: 1287.39. The answer is $1287.39.

Besides the initial expenses of buying a house, there are continuing monthly expenses involved in owning a home. The **monthly mortgage payment** (one of 12 payments due each year to the lender of money to buy real estate), utilities, insurance, and **property tax** (a tax based on the value of real estate) are some of these ongoing expenses. Of these expenses, the largest one is normally the monthly mortgage payment.

For a **fixed-rate mortgage,** the monthly mortgage payment remains the same throughout the life of the loan. The calculation of the monthly mortgage payment is based on the amount of the loan, the interest rate on the loan, and the number of years required to pay back the loan. Calculating the monthly mortgage payment is fairly difficult, so tables such as the one in the Appendix are used to simplify these calculations.

HOW TO • 2   Find the monthly mortgage payment on a 30-year, $160,000 mortgage at an interest rate of 9%. Use the Monthly Payment Table in the Appendix.

$$160,000 \times \underline{0.0080462} \approx 1287.39$$

↑

From the table

The monthly mortgage payment is $1287.39.

The monthly mortgage payment includes the payment of both principal and interest on the mortgage. The interest charged during any one month is charged on the unpaid balance of the loan. Therefore, during the early years of the mortgage, when the unpaid balance is high, most of the monthly mortgage payment is interest charged on the loan. During the last few years of a mortgage, when the unpaid balance is low, most of the monthly mortgage payment goes toward paying off the loan.

Point of Interest

Home buyers rated the following characteristics "extremely important" in their purchase decision.

Natural, open space: 77%
Walking and biking paths: 74%
Gardens with native plants: 56%
Clustered retail stores: 55%
Wilderness area: 52%
Outdoor pool: 52%
Community recreation center: 52%
Interesting little parks: 50%

(*Sources:* American Lives, Inc; Intercommunications, Inc.)

HOW TO 3 Find the interest paid on a mortgage during a month in which the monthly mortgage payment is $886.26 and $358.08 of that amount goes toward paying off the principal.

Monthly mortgage payment	−	principal	=	interest
886.26	−	358.08	=	528.18

The interest paid on the mortgage is $528.18.

Property tax is another ongoing expense of owning a house. Property tax is normally an annual expense that may be paid on a monthly basis. The monthly property tax, which is determined by dividing the annual property tax by 12, is usually added to the monthly mortgage payment.

HOW TO 4 A homeowner must pay $3120 in property tax annually. Find the property tax that must be added each month to the homeowner's monthly mortgage payment.

$3120 \div 12 = 260$

Each month, $260 must be added to the monthly mortgage payment for property tax.

EXAMPLE 3

Serge purchased some land for $120,000 and made a down payment of $25,000. The savings and loan association charges an annual interest rate of 8% on Serge's 25-year mortgage. Find the monthly mortgage payment.

Strategy
To find the monthly mortgage payment:
• Subtract the down payment from the purchase price to find the mortgage.
• Multiply the mortgage by the factor found in the Monthly Payment Table in the Appendix.

Solution

Purchase price	−	down payment	=	mortgage
120,000	−	25,000	=	95,000

$95,000 \times 0.0077182 \approx 733.23$

↑
From the table

The monthly mortgage payment is $733.23.

YOU TRY IT 3

A new condominium project is selling townhouses for $175,000. A down payment of $17,500 is required, and a 20-year mortgage at an annual interest rate of 9% is available. Find the monthly mortgage payment.

Your strategy

Your solution

Solution on p. S16

EXAMPLE · 4

A home has a mortgage of $134,000 for 25 years at an annual interest rate of 7%. During a month in which $375.88 of the monthly mortgage payment is principal, how much of the payment is interest?

Strategy

To find the interest:

- Multiply the mortgage by the factor found in the Monthly Payment Table in the Appendix to find the monthly mortgage payment.
- Subtract the principal from the monthly mortgage payment.

Solution

$134,000 \times 0.0070678 \approx 947.09$

↑ From the table   ↑ Monthly mortgage payment

Monthly mortgage payment	−	principal	=	interest
947.09	−	375.88	=	571.21

$571.21 of the payment is interest on the mortgage.

YOU TRY IT · 4

An office building has a mortgage of $625,000 for 25 years at an annual interest rate of 7%. During a month in which $2516.08 of the monthly mortgage payment is principal, how much of the payment is interest?

Your strategy

Your solution

EXAMPLE · 5

The monthly mortgage payment for a home is $998.75. The annual property tax is $4020. Find the total monthly payment for the mortgage and property tax.

Strategy

To find the monthly payment:

- Divide the annual property tax by 12 to find the monthly property tax.
- Add the monthly property tax to the monthly mortgage payment.

Solution

$4020 \div 12 = 335$ • **Monthly property tax**
$998.75 + 335 = 1333.75$

The total monthly payment is $1333.75.

YOU TRY IT · 5

The monthly mortgage payment for a home is $815.20. The annual property tax is $3000. Find the total monthly payment for the mortgage and property tax.

Your strategy

Your solution

Solutions on p. S16

6.4 EXERCISES

OBJECTIVE A   To calculate the initial expenses of buying a home

Paul Conklin/PhotoEdit, Inc.

1. A condominium at Mt. Baldy Ski Resort was purchased for $197,000, and a down payment of $24,550 was made. Find the mortgage.

2. An insurance business was purchased for $173,000, and a down payment of $34,600 was made. Find the mortgage.

3. Brian Stedman made a down payment of 25% of the $850,000 purchase price of an apartment building. How much was the down payment?

4. A clothing store was purchased for $625,000, and a down payment that was 25% of the purchase price was made. Find the down payment.

5. A loan of $150,000 is obtained to purchase a home. The loan origination fee is $2\frac{1}{2}$ points. Find the amount of the loan origination fee.

6. Security Savings & Loan requires a borrower to pay $3\frac{1}{2}$ points for a loan. Find the amount of the loan origination fee for a loan of $90,000.

7. Baja Construction Inc. is selling homes for $350,000. A down payment of 10% is required. Find the mortgage.

8. A cattle rancher purchased some land for $240,000. The bank requires a down payment of 15% of the purchase price. Find the mortgage.

9. Vivian Tom purchased a home for $210,000. Find the mortgage if the down payment Vivian made is 10% of the purchase price.

10. A mortgage lender requires a down payment of 5% of the $180,000 purchase price of a condominium. How much is the mortgage?

 11. A home is purchased for $435,000. The mortgage lender requires a 10% down payment. Which expression below represents the mortgage?
   (i) $0.10 \times 435,000$ (ii) $0.10 \times 435,000 - 435,000$
   (iii) $435,000 - 0.10 \times 435,000$ (iv) $435,000 + 0.10 \times 435,000$

OBJECTIVE B   To calculate the ongoing expenses of owning a home

For Exercises 12 to 23, solve. Use the Monthly Payment Table in the Appendix. Round to the nearest cent.

12. An investor obtained a loan of $850,000 to buy a car wash business. The monthly mortgage payment was based on 25 years at 8%. Find the monthly mortgage payment.

13. A beautician obtained a 20-year mortgage of $90,000 to expand the business. The credit union charges an annual interest rate of 6%. Find the monthly mortgage payment.

14. A couple interested in buying a home determines that they can afford a monthly mortgage payment of $800. Can they afford to buy a home with a 30-year, $110,000 mortgage at 8% interest?

15. A lawyer is considering purchasing a new office building with a 15-year, $400,000 mortgage at 6% interest. The lawyer can afford a monthly mortgage payment of $3500. Can the lawyer afford the monthly mortgage payment on the new office building?

16. The county tax assessor has determined that the annual property tax on a $325,000 house is $3032.40. Find the monthly property tax.

17. The annual property tax on a $155,000 home is $1992. Find the monthly property tax.

18. Abacus Imports Inc. has a warehouse with a 25-year mortgage of $200,000 at an annual interest rate of 9%.
   a. Find the monthly mortgage payment.
   b. During a month in which $941.72 of the monthly mortgage payment is principal, how much of the payment is interest?

19. A vacation home has a mortgage of $135,000 for 30 years at an annual interest rate of 7%.
   a. Find the monthly mortgage payment.
   b. During a month in which $392.47 of the monthly mortgage payment is principal, how much of the payment is interest?

20. The annual mortgage payment on a duplex is $20,844.40. The owner must pay an annual property tax of $1944. Find the total monthly payment for the mortgage and property tax.

21. The monthly mortgage payment on a home is $716.40, and the homeowner pays an annual property tax of $1512. Find the total monthly payment for the mortgage and property tax.

22. Maria Hernandez purchased a home for $210,000 and made a down payment of $15,000. The balance was financed for 15 years at an annual interest rate of 6%. Find the monthly mortgage payment.

23. A customer of a savings and loan purchased a $385,000 home and made a down payment of $40,000. The savings and loan charges its customers an annual interest rate of 7% for 30 years for a home mortgage. Find the monthly mortgage payment.

24. The monthly mortgage payment for a home is $623.57. The annual property tax is $1400. Which expression below represents the total monthly payment for the mortgage and property tax? Which expression represents the total amount of money the owner will spend on the mortgage and property tax in one year?
   (i) $623.57 + 1400$ (ii) $12 \times 623.57 + 1400$
   (iii) $\dfrac{623.57 + 1400}{12}$ (iv) $623.57 + \dfrac{1400}{12}$

Applying the Concepts

25. **Mortgages** A couple considering a mortgage of $100,000 have a choice of loans. One loan is an 8% loan for 20 years, and the other loan is at 8% for 30 years. Find the amount of interest that the couple can save by choosing the 20-year loan.

SECTION

6.5   Car Expenses

OBJECTIVE A   **To calculate the initial expenses of buying a car**

The initial expenses in the purchase of a car usually include the down payment, the **license fees** (fees charged for authorization to operate a vehicle), and the **sales tax** (a tax levied by a state or municipality on purchases). The down payment may be very small or as much as 25% or 30% of the purchase price of the car, depending on the lending institution. License fees and sales tax are regulated by each state, so these expenses vary from state to state.

EXAMPLE • 1

A car is purchased for $38,500, and the lender requires a down payment of 15% of the purchase price. Find the amount financed.

Strategy

To find the amount financed:

• Find the down payment by solving the basic percent equation for *amount*.
• Subtract the down payment from the purchase price.

Solution

Percent   ×   base   =   amount

$$\boxed{\text{Percent}} \times \boxed{\begin{array}{c}\text{purchase}\\\text{price}\end{array}} = \boxed{\begin{array}{c}\text{down}\\\text{payment}\end{array}}$$

0.15   ×   38,500   = n
5775 = n

38,500 − 5775 = 32,725

The amount financed is $32,725.

YOU TRY IT • 1

A down payment of 20% of the $19,200 purchase price of a new car is made. Find the amount financed.

Your strategy

Your solution

EXAMPLE • 2

A sales clerk purchases a used car for $16,500 and pays a sales tax that is 5% of the purchase price. How much is the sales tax?

Strategy

To find the sales tax, solve the basic percent equation for *amount*.

Solution

Percent   ×   base   =   amount

$$\boxed{\text{Percent}} \times \boxed{\begin{array}{c}\text{purchase}\\\text{price}\end{array}} = \boxed{\begin{array}{c}\text{sales}\\\text{tax}\end{array}}$$

0.05   ×   16,500   = n
825 = n

The sales tax is $825.

YOU TRY IT • 2

A car is purchased for $27,350. The car license fee is 1.5% of the purchase price. How much is the license fee?

Your strategy

Your solution

Solutions on pp. S16–S17

OBJECTIVE B   To calculate the ongoing expenses of owning a car

 Take Note

The same formula that is used to calculate a monthly mortgage payment is used to calculate a monthly car payment.

Besides the initial expenses of buying a car, there are continuing expenses involved in owning a car. These ongoing expenses include car insurance, gas and oil, general maintenance, and the monthly car payment. The monthly car payment is calculated in the same manner as the monthly mortgage payment on a home loan. A monthly payment table, such as the one in the Appendix, is used to simplify the calculation of monthly car payments.

EXAMPLE · 3

At a cost of $.38 per mile, how much does it cost to operate a car during a year in which the car is driven 15,000 miles?

Strategy

To find the cost, multiply the cost per mile (0.38) by the number of miles driven (15,000).

Solution

$15,000 \times 0.38 = 5700$

The cost is $5700.

YOU TRY IT · 3

At a cost of $.41 per mile, how much does it cost to operate a car during a year in which the car is driven 23,000 miles?

Your strategy

Your solution

EXAMPLE · 4

During one month, your total gasoline bill was $252 and the car was driven 1200 miles. What was the cost per mile for gasoline?

Strategy

To find the cost per mile, divide the cost for gasoline (252) by the number of miles driven (1200).

Solution

$252 \div 1200 = 0.21$

The cost per mile was $.21.

YOU TRY IT · 4

In a year in which your total car insurance bill was $360 and the car was driven 15,000 miles, what was the cost per mile for car insurance?

Your strategy

Your solution

EXAMPLE · 5

A car is purchased for $18,500 with a down payment of $3700. The balance is financed for 3 years at an annual interest rate of 6%. Find the monthly car payment.

Strategy

To find the monthly payment:

• Subtract the down payment from the purchase price to find the amount financed.
• Multiply the amount financed by the factor found in the Monthly Payment Table in the Appendix.

Solution

$18,500 - 3700 = 14,800$

$14,800 \times 0.0304219 \approx 450.24$

The monthly payment is $450.24.

YOU TRY IT · 5

A truck is purchased for $25,900 with a down payment of $6475. The balance is financed for 4 years at an annual interest rate of 8%. Find the monthly car payment.

Your strategy

Your solution

Solutions on p. S17

6.5 EXERCISES

OBJECTIVE A   To calculate the initial expenses of buying a car

1. Amanda has saved $780 to make a down payment on a used minivan that costs $7100. The car dealer requires a down payment of 12% of the purchase price. Has she saved enough money to make the down payment?

2. A sedan was purchased for $23,500. A down payment of 15% of the purchase price was required. How much was the down payment?

3. A drapery installer bought a van to carry drapery samples. The purchase price of the van was $26,500, and a 4.5% sales tax was paid. How much was the sales tax?

4. A & L Lumber Company purchased a delivery truck for $28,500. A sales tax of 4% of the purchase price was paid. Find the sales tax.

5. A license fee of 2% of the purchase price is paid on a pickup truck costing $32,500. Find the license fee for the truck.

6. Your state charges a license fee of 1.5% on the purchase price of a car. How much is the license fee for a car that costs $16,998?

7. An electrician bought a $32,000 flatbed truck. A state license fee of $275 and a sales tax of 3.5% of the purchase price are required.
   a. Find the sales tax.
   b. Find the total cost of the sales tax and the license fee.

8. A physical therapist bought a used car for $9375 and made a down payment of $1875. The sales tax is 5% of the purchase price.
   a. Find the sales tax.
   b. Find the total cost of the sales tax and the down payment.

9. Martin bought a motorcycle for $16,200 and made a down payment of 25% of the purchase price. Find the amount financed.

10. A carpenter bought a utility van for $24,900 and made a down payment of 15% of the purchase price. Find the amount financed.

11. An author bought a sports car for $45,000 and made a down payment of 20% of the purchase price. Find the amount financed.

12. Tania purchased a used car for $13,500 and made a down payment of 25% of the cost. Find the amount financed.

13. The purchase price of a car is $25,700. The car dealer requires a down payment of 15% of the purchase price. There is a license fee of 2.5% of the purchase price and sales tax of 6% of the purchase price. What does the following expression represent?
   25,700 + 0.025 × 25,700 + 0.06 × 25,700

OBJECTIVE B   To calculate the ongoing expenses of owning a car

14. A driver had $1100 in car expenses and drove his car 8500 miles. Would you use *multiplication* or *division* to find the cost per mile to operate the car?

15. A car costs $.36 per mile to operate. Would you use *multiplication* or *division* to find the cost of driving the car 23,000 miles?

For Exercises 16 to 25, solve. Use the Monthly Payment Table in the Appendix. Round to the nearest cent.

16. A rancher financed $24,000 for the purchase of a truck through a credit union at 5% interest for 4 years. Find the monthly truck payment.

17. A car loan of $18,000 is financed for 3 years at an annual interest rate of 4%. Find the monthly car payment.

18. An estimate of the cost of owning a compact car is $.38 per mile. Using this estimate, find how much it costs to operate a car during a year in which the car is driven 16,000 miles.

19. An estimate of the cost of care and maintenance of automobile tires is $.018 per mile. Using this estimate, find how much it costs for care and maintenance of tires during a year in which the car is driven 14,000 miles.

20. A family spent $2600 on gas, oil, and car insurance during a period in which the car was driven 14,000 miles. Find the cost per mile for gas, oil, and car insurance.

21. Last year you spent $2400 for gasoline for your car. The car was driven 15,000 miles. What was your cost per mile for gasoline?

22. The city of Colton purchased a fire truck for $164,000 and made a down payment of $10,800. The balance is financed for 5 years at an annual interest rate of 6%.
a. Find the amount financed.
b. Find the monthly truck payment.

23. A used car is purchased for $14,999, and a down payment of $2999 is made. The balance is financed for 3 years at an annual interest rate of 5%.
a. Find the amount financed.
b. Find the monthly car payment.

24. An artist purchased a new car costing $27,500 and made a down payment of $5500. The balance is financed for 3 years at an annual interest rate of 4%. Find the monthly car payment.

25. A camper is purchased for $39,500, and a down payment of $5000 is made. The balance is financed for 4 years at an annual interest rate of 6%. Find the monthly payment.

Ulrich Mueller/Flickr/Getty Images

Applying the Concepts

26. Car Loans One bank offers a 4-year car loan at an annual interest rate of 7% plus a loan application fee of $45. A second bank offers 4-year car loans at an annual interest rate of 8% but charges no loan application fee. If you need to borrow $5800 to purchase a car, which of the two bank loans has the lesser loan costs? Assume you keep the car for 4 years.

27. Car Loans How much interest is paid on a 5-year car loan of $19,000 if the interest rate is 9%? Round to the nearest dollar.

SECTION

6.6 Wages

OBJECTIVE A To calculate commissions, total hourly wages, and salaries

Commissions, hourly wage, and salary are three ways to receive payment for doing work.

Commissions are usually paid to salespersons and are calculated as a percent of total sales.

> **HOW TO · 1** As a real estate broker, Emma Smith receives a commission of 4.5% of the selling price of a house. Find the commission she earned for selling a home for $275,000.
>
> To find the commission Emma earned, solve the basic percent equation for *amount*.
>
> Percent $\times$ base $=$ amount
>
> | Commission rate | $\times$ | total sales | $=$ | commission |
>
> $0.045 \times 275,000 = 12,375$
>
> The commission is $12,375.

An employee who receives an **hourly wage** is paid a certain amount for each hour worked.

> **HOW TO · 2** A plumber receives an hourly wage of $28.25. Find the plumber's total wages for working 37 hours.
>
> To find the plumber's total wages, multiply the hourly wage by the number of hours worked.
>
> | Hourly wage | $\times$ | number of hours worked | $=$ | total wages |
>
> $28.25 \times 37 = 1045.25$
>
> The plumber's total wages for working 37 hours are $1045.25.

An employee who is paid a **salary** receives payment based on a weekly, biweekly (every other week), monthly, or annual time schedule. Unlike the employee who receives an hourly wage, the salaried worker does not receive additional pay for working more than the regularly scheduled workday.

> **HOW TO · 3** Ravi Basar is a computer operator who receives a weekly salary of $895. Find his salary for 1 month (4 weeks).
>
> To find Ravi's salary for 1 month, multiply the salary per pay period by the number of pay periods.
>
> | Salary per pay period | $\times$ | number of pay periods | $=$ | total salary |
>
> $895 \times 4 = 3580$
>
> Ravi's total salary for 1 month is $3580.

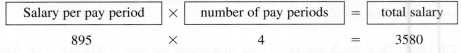

EXAMPLE · 1

A pharmacist's hourly wage is $48. On Saturday, the pharmacist earns time and a half (1.5 times the regular hourly wage). How much does the pharmacist earn for working 6 hours on Saturday?

Strategy
To find the pharmacist's earnings:

- Find the hourly wage for working on Saturday by multiplying the hourly wage by 1.5.
- Multiply the hourly wage by the number of hours worked.

Solution
$48 \times 1.5 = 72 \qquad 72 \times 6 = 432$

The pharmacist earns $432.

YOU TRY IT · 1

A construction worker, whose hourly wage is $28.50, earns double time (2 times the regular hourly wage) for working overtime. Find the worker's wages for working 8 hours of overtime.

Your strategy

Your solution

EXAMPLE · 2

An efficiency expert received a contract for $3000. The expert spent 75 hours on the project. Find the consultant's hourly wage.

Strategy
To find the hourly wage, divide the total earnings by the number of hours worked.

Solution
$3000 \div 75 = 40$

The hourly wage was $40.

YOU TRY IT · 2

A contractor for a bridge project receives an annual salary of $70,980. What is the contractor's salary per month?

Your strategy

Your solution

EXAMPLE · 3

Dani Greene earns $38,500 per year plus a 5.5% commission on sales over $100,000. During one year, Dani sold $150,000 worth of computers. Find Dani's total earnings for the year.

Strategy
To find the total earnings:

- Find the sales over $100,000.
- Multiply the commission rate by sales over $100,000.
- Add the commission to the annual pay.

Solution
$150,000 - 100,000 = 50,000$
$50,000 \times 0.055 = 2750$ • Commission
$38,500 + 2750 = 41,250$

Dani earned $41,250.

YOU TRY IT · 3

An insurance agent earns $37,000 per year plus a 9.5% commission on sales over $50,000. During one year, the agent's sales totaled $175,000. Find the agent's total earnings for the year.

Your strategy

Your solution

Solutions on p. S17

6.6 EXERCISES

OBJECTIVE A   To calculate commissions, total hourly wages, and salaries

1. Lewis works in a clothing store and earns $11.50 per hour. How much does he earn in a 40-hour work week?

2. Sasha pays a gardener an hourly wage of $11. How much does she pay the gardener for working 25 hours?

3. A real estate agent receives a 3% commission for selling a house. Find the commission that the agent earned for selling a house for $131,000.

4. Ron Caruso works as an insurance agent and receives a commission of 40% of the first year's premium. Find Ron's commission for selling a life insurance policy with a first-year premium of $1050.

5. A stockbroker receives a commission of 1.5% of the price of stock that is bought or sold. Find the commission on 100 shares of stock that were bought for $5600.

Jeff Greenberg/PhotoEdit, Inc.

6. The owner of the Carousel Art Gallery receives a commission of 20% on paintings that are sold on consignment. Find the commission on a painting that sold for $22,500.

7. Keisha Brown receives an annual salary of $38,928 as a teacher of Italian. How much does Keisha receive each month?

8. An apprentice plumber receives an annual salary of $27,900. How much does the plumber receive per month?

9. Carlos receives a commission of 12% of his weekly sales as a sales representative for a medical supply company. Find the commission he earned during a week in which sales were $4500.

10. A golf pro receives a commission of 25% for selling a golf set. Find the commission the pro earned for selling a golf set costing $450.

11. Steven receives $5.75 per square yard to install carpet. How much does he receive for installing 160 square yards of carpet?

12. A typist charges $3.75 per page for typing technical material. How much does the typist earn for typing a 225-page book?

13. A nuclear chemist received $15,000 in consulting fees while working on a nuclear power plant. The chemist worked 120 hours on the project. Find the chemist's hourly wage.

14. Maxine received $3400 for working on a project as a computer consultant for 40 hours. Find her hourly wage.

15. Gil Stratton's hourly wage is $10.78. For working overtime, he receives double time.
a. What is Gil's hourly wage for working overtime?
b. How much does he earn for working 16 hours of overtime?

16. Mark is a lathe operator and receives an hourly wage of $15.90. When working on Saturday, he receives time and a half.
a. What is Mark's hourly wage on Saturday?
b. How much does he earn for working 8 hours on Saturday?

17. A stock clerk at a supermarket earns $8.20 an hour. For working the night shift, the clerk's wage increases by 15%.
a. What is the increase in hourly pay for working the night shift?
b. What is the clerk's hourly wage for working the night shift?

18. A nurse earns $31.50 an hour. For working the night shift, the nurse receives a 10% increase in pay.
a. What is the increase in hourly pay for working the night shift?
b. What is the hourly pay for working the night shift?

19. Nicole Tobin, a salesperson, receives a salary of $250 per week plus a commission of 15% on all sales over $1500. Find her earnings during a week in which sales totaled $3000.

20. A veterinarian's assistant works 35 hours a week at $20 an hour. The assistant is paid time and a half for overtime hours. Which expression represents the assistant's earnings for a week in which the assistant worked 41 hours?
(i) 41 × 20
(ii) (35 × 20) + (41 × 30)
(iii) (35 × 20) + (6 × 30)
(iv) 41 × 30

Applying the Concepts

Compensation The table at the right shows the average starting salaries for recent college graduates. Use this table for Exercises 21 to 24. Round to the nearest dollar.

21. What was the starting salary in the previous year for an accountant?

22. How much did the starting salary for a chemical engineer increase over that of the previous year?

23. What was the starting salary in the previous year for a computer science major?

Average Starting Salaries		
Bachelor's Degree	**Average Starting Salary**	**Change from Previous Year**
Chemical Engineering	$52,169	1.8% increase
Electrical Engineering	$50,566	0.4% increase
Computer Science	$46,536	7.6% decrease
Accounting	$41,360	2.6% increase
Business	$36,515	3.7% increase
Biology	$29,554	1.0% decrease
Political Science	$28,546	12.6% decrease
Psychology	$26,738	10.7% decrease

Source: National Association of Colleges

24. How much did the starting salary for a political science major decrease from that of the previous year?

SECTION

6.7   Bank Statements

OBJECTIVE A   To calculate checkbook balances

✓ Take Note

A **checking account** is a bank account that enables you to withdraw money or make payments to other people using checks. A **check** is a printed form that, when filled out and signed, instructs a bank to pay a specified sum of money to the person named on it.

A **deposit slip** is a form for depositing money in a checking account.

A checking account can be opened at most banks and savings and loan associations by depositing an amount of money in the bank. A checkbook contains checks and deposit slips and a checkbook register in which to record checks written and amounts deposited in the checking account. A sample check is shown below.

Payee   Date Check is Written

East Phoenix Rental Equipment		NO. 2023	Check Number
3011 N.W. Ventura Street		68 - 461	
Phoenix, Arizona 85280		1052	

Date October 11, 2011

PAY TO THE
ORDER OF _Tellas Manufacturing Co._ $ 827 00/100   Amount of Check

Eight Hundred Twenty-Seven and 00/100 —— DOLLARS

MEYERS' NATIONAL BANK
11 N.W. Nova Street
Phoenix, Arizona 85215

Memo _____
I: 1052 III 0461 I: 5008 2023 III•

Eugene L. Madison

Amount of Check in Words   Depositor's Signature

⊙ Point of Interest

There are a number of computer programs that serve as "electronic" checkbooks. With these programs, you can pay your bills by using a computer to write the check and then transmit the check over telephone lines using a modem.

Each time a check is written, the amount of the check is subtracted from the amount in the account. When a deposit is made, the amount deposited is added to the amount in the account.

A portion of a checkbook register is shown below. The account holder had a balance of $587.93 before writing two checks, one for $286.87 and the other for $202.38, and making one deposit of $345.00.

		RECORD ALL CHARGES OR CREDITS THAT AFFECT YOUR ACCOUNT				BALANCE	
NUMBER	DATE	DESCRIPTION OF TRANSACTION	PAYMENT/DEBIT (–)	√ T	FEE (IF ANY) (–)	DEPOSIT/CREDIT (+)	$ 587 93
108	8/4	Plumber	$286 87	$	$		301 06
109	8/10	Car Payment	202 38				98 68
	8/14	Deposit				345 00	443 68

To find the current checking account balance, subtract the amount of each check from the previous balance. Then add the amount of the deposit.

The current checking account balance is $443.68.

EXAMPLE · 1

EXAMPLE · 1

A mail carrier had a checking account balance of $485.93 before writing two checks, one for $18.98 and another for $35.72, and making a deposit of $250. Find the current checking account balance.

YOU TRY IT · 1

A cement mason had a checking account balance of $302.46 before writing a check for $20.59 and making two deposits, one in the amount of $176.86 and another in the amount of $94.73. Find the current checking account balance.

Strategy

To find the current balance:

• Subtract the amount of each check from the old balance.
• Add the amount of the deposit.

Your strategy

Solution

$$485.93$$
$$\underline{-\ \ 18.98} \quad \text{first check}$$
$$466.95$$
$$\underline{-\ \ 35.72} \quad \text{second check}$$
$$431.23$$
$$\underline{+\ 250.00} \quad \text{deposit}$$
$$681.23$$

The current checking account balance is $681.23.

Your solution

Solution on p. S17

OBJECTIVE B To balance a checkbook

Each month a bank statement is sent to the account holder. A **bank statement** is a document showing all the transactions in a bank account during the month. It shows the checks that the bank has paid, the deposits received, and the current bank balance.

A bank statement and checkbook register are shown on the next page.

Balancing a checkbook, or determining whether the checking account balance is accurate, requires a number of steps.

1. In the checkbook register, put a check mark (✓) by each check paid by the bank and by each deposit recorded by the bank.

RECORD ALL CHARGES OR CREDITS THAT AFFECT YOUR ACCOUNT

NUMBER	DATE	DESCRIPTION OF TRANSACTION	PAYMENT/DEBIT (−)		√ T	FEE (IF ANY) (−)	DEPOSIT/CREDIT (+)		BALANCE $ 840	27
263	5/20	Dentist	$ 75	00	√	$	$		765	27
264	5/22	Post Office	33	61	√				731	66
265	5/22	Gas Company	67	14					664	52
	5/29	Deposit			√		192	00	856	52
266	5/29	Pharmacy	38	95	√				817	57
267	5/30	Telephone	63	85					753	72
268	6/2	Groceries	73	19	√				680	53
	6/3	Deposit			√		215	00	895	53
269	6/7	Insurance	103	00	√				792	53
	6/10	Deposit					225	00	1017	53
270	6/15	Photo Shop	16	63	√				1000	90
271	6/18	Newspaper	27	00					973	90

CHECKING ACCOUNT Monthly Statement			Account Number: 924-297-8
Date	**Transaction**	**Amount**	**Balance**
5/20	OPENING BALANCE		840.27
5/21	CHECK	75.00	765.27
5/23	CHECK	33.61	731.66
5/29	DEPOSIT	192.00	923.66
6/1	CHECK	38.95	884.71
6/1	INTEREST	4.47	889.18
6/3	CHECK	73.19	815.99
6/3	DEPOSIT	215.00	1030.99
6/9	CHECK	103.00	927.99
6/16	CHECK	16.63	911.36
6/20	SERVICE CHARGE	3.00	908.36
6/20	CLOSING BALANCE		908.36

2. Add to the current checkbook balance all checks that have been written but have not yet been paid by the bank and any interest paid on the account.

Current checkbook balance:		973.90
Checks: 265		67.14
267		63.85
271		27.00
Interest:	+	4.47
		1136.36

3. Subtract any service charges and any deposits not yet recorded by the bank. This is the checkbook balance.

Service charge:	−	3.00
		1133.36
Deposit:	−	225.00
Checkbook balance:		908.36

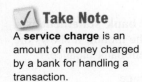
✓ Take Note

A **service charge** is an amount of money charged by a bank for handling a transaction.

4. Compare the balance with the bank balance listed on the bank statement. If the two numbers are equal, the bank statement and the checkbook "balance."

Closing bank balance from bank statement $908.36 = Checkbook balance $908.36

The bank statement and checkbook balance.

HOW TO · 1

RECORD ALL CHARGES OR CREDITS THAT AFFECT YOUR ACCOUNT

NUMBER	DATE	DESCRIPTION OF TRANSACTION	PAYMENT/DEBIT (−)		√ T	FEE (IF ANY) (−)	DEPOSIT/CREDIT (+)		BALANCE $ 1620	42
413	3/2	Car Payment	$232	15	√	$	$		1388	27
414	3/2	Utilities	67	14	√				1321	13
415	3/5	Restaurant	78	14					1242	99
	3/8	Deposit			√		1842	66	3085	65
416	3/10	House Payment	672	14	√				2413	51
417	3/14	Insurance	177	10					2236	41

PPT

CHECKING ACCOUNT Monthly Statement

Account Number: 924-297-8

Date	Transaction	Amount	Balance
3/1	OPENING BALANCE		1620.42
3/4	CHECK	232.15	1388.27
3/5	CHECK	67.14	1321.13
3/8	DEPOSIT	1842.66	3163.79
3/10	INTEREST	6.77	3170.56
3/12	CHECK	672.14	2498.42
3/25	SERVICE CHARGE	2.00	2496.42
3/30	CLOSING BALANCE		2496.42

PPT

Balance the checkbook shown above.

1. In the checkbook register, put a check mark (✓) by each check paid by the bank and by each deposit recorded by the bank.

2. Add to the current checkbook balance all checks that have been written but have not yet been paid by the bank and any interest paid on the account.

Current checkbook balance:	2236.41
Checks: 415	78.14
417	177.10
Interest:	+ 6.77
	2498.42

3. Subtract any service charges and any deposits not yet recorded by the bank. This is the checkbook balance.

Service charge:	− 2.00
Checkbook balance:	2496.42

4. Compare the balance with the bank balance listed on the bank statement. If the two numbers are equal, the bank statement and the checkbook "balance."

Closing bank balance from bank statement	Checkbook balance
$2496.42	= $2496.42

The bank statement and checkbook balance.

EXAMPLE · 2

Balance the checkbook shown below.

		RECORD ALL CHARGES OR CREDITS THAT AFFECT YOUR ACCOUNT			√ T	FEE (IF ANY) (−)	DEPOSIT/CREDIT (+)		BALANCE $ 412	64
NUMBER	DATE	DESCRIPTION OF TRANSACTION	PAYMENT/DEBIT (−)					$		
345	1/14	Phone Bill	$ 54	75	√	$	$		357	89
346	1/19	News Shop	18	98	√				338	91
347	1/23	Theater Tickets	95	00					243	91
	1/31	Deposit			√		947	00	1190	91
348	2/5	Cash	250	00	√				940	91
349	2/12	Rent	840	00					100	91

CHECKING ACCOUNT Monthly Statement		Account Number: 924-297-8	
Date	Transaction	Amount	Balance
1/10	OPENING BALANCE		412.64
1/18	CHECK	54.75	357.89
1/23	CHECK	18.98	338.91
1/31	DEPOSIT	947.00	1285.91
2/1	INTEREST	4.52	1290.43
2/10	CHECK	250.00	1040.43
2/10	CLOSING BALANCE		1040.43

Solution

Current checkbook balance:	100.91
Checks: 347	95.00
349	840.00
Interest:	+ 4.52
	1040.43
Service charge:	− 0.00
	1040.43
Deposit:	− 0.00
Checkbook balance:	1040.43

Closing bank balance from bank statement: $1040.43

Checkbook balance: $1040.43

The bank statement and the checkbook balance.

YOU TRY IT • 2

Balance the checkbook shown below.

RECORD ALL CHARGES OR CREDITS THAT AFFECT YOUR ACCOUNT

NUMBER	DATE	DESCRIPTION OF TRANSACTION	PAYMENT/DEBIT (−)	√T	FEE (IF ANY) (−)	DEPOSIT/CREDIT (+)	BALANCE	
	2/15	Deposit	$		$	$ 523 84	$ 903	17
							1427	01
234	2/20	Mortgage	773 21				653	80
235	2/27	Cash	200 00				453	80
	3/1	Deposit				523 84	977	64
236	3/12	Insurance	275 50				702	14
237	3/12	Telephone	78 73				623	41

CHECKING ACCOUNT Monthly Statement			Account Number: 314-271-4
Date	Transaction	Amount	Balance
2/14	OPENING BALANCE		903.17
2/15	DEPOSIT	523.84	1427.01
2/21	CHECK	773.21	653.80
2/28	CHECK	200.00	453.80
3/1	INTEREST	2.11	455.91
3/14	CHECK	275.50	180.41
3/14	CLOSING BALANCE		180.41

Your solution

Solution on p. S17

6.7 EXERCISES

OBJECTIVE A   To calculate checkbook balances

1. You had a checking account balance of $342.51 before making a deposit of $143.81. What is your new checking account balance?

2. The business checking account for R and R Tires showed a balance of $1536.97. What is the balance in this account after a deposit of $439.21 has been made?

3. A nutritionist had a checking account balance of $1204.63 before writing one check for $119.27 and another check for $260.09. Find the current checkbook balance.

4. Sam had a checking account balance of $3046.93 before writing a check for $1027.33 and making a deposit of $150.00. Find the current checkbook balance.

5. The business checking account for Rachael's Dry Cleaning had a balance of $3476.85 before a deposit of $1048.53 was made. The store manager then wrote two checks, one for $848.37 and another for $676.19. Find the current checkbook balance.

6. Joel had a checking account balance of $427.38 before a deposit of $127.29 was made. Joel then wrote two checks, one for $43.52 and one for $249.78. Find the current checkbook balance.

7. A carpenter had a checkbook balance of $404.96 before making a deposit of $350 and writing a check for $71.29. Is there enough money in the account to purchase a refrigerator for $675?

8. A taxi driver had a checkbook balance of $149.85 before making a deposit of $245 and writing a check for $387.68. Is there enough money in the account for the bank to pay the check?

9. A sporting goods store has the opportunity to buy downhill skis and cross-country skis at a manufacturer's closeout sale. The downhill skis will cost $3500, and the cross-country skis will cost $2050. There is currently $5625.42 in the sporting goods store's checking account. Is there enough money in the account to make both purchases by check?

10. A lathe operator's current checkbook balance is $1143.42. The operator wants to purchase a utility trailer for $525 and a used piano for $650. Is there enough money in the account to make the two purchases?

 For Exercises 11 and 12, suppose the given transactions take place on an account in one day. State whether the account's ending balance on that day *must be less than, might be less than,* or *cannot be less than* its starting balance on that day.

11. Two deposits and one check written   12. Three checks written

OBJECTIVE B To balance a checkbook

13. Balance the checkbook.

RECORD ALL CHARGES OR CREDITS THAT AFFECT YOUR ACCOUNT

NUMBER	DATE	DESCRIPTION OF TRANSACTION	PAYMENT/DEBIT (–)		√T	FEE (IF ANY) (–)	DEPOSIT/CREDIT (+)		BALANCE $	
									2466	79
223	3/2	Groceries	$ 167	32		$	$		2299	47
	3/5	Deposit					960	70	3260	17
224	3/5	Rent	860	00					2400	17
225	3/7	Gas & Electric	142	35					2257	82
226	3/7	Cash	300	00					1957	82
227	3/7	Insurance	218	44					1739	38
228	3/7	Credit Card	419	32					1320	06
229	3/12	Dentist	92	00					1228	06
230	3/13	Drug Store	47	03					1181	03
	3/19	Deposit					960	70	2141	73
231	3/22	Car Payment	241	35					1900	38
232	3/25	Cash	300	00					1600	38
233	3/25	Oil Company	166	40					1433	98
234	3/28	Plumber	155	73					1278	25
235	3/29	Department Store	288	39					989	86

CHECKING ACCOUNT Monthly Statement		Account Number: 122-345-1	
Date	Transaction	Amount	Balance
3/1	OPENING BALANCE		2466.79
3/5	DEPOSIT	960.70	3427.49
3/7	CHECK	167.32	3260.17
3/8	CHECK	860.00	2400.17
3/8	CHECK	300.00	2100.17
3/9	CHECK	142.35	1957.82
3/12	CHECK	218.44	1739.38
3/14	CHECK	92.00	1647.38
3/18	CHECK	47.03	1600.35
3/19	DEPOSIT	960.70	2561.05
3/25	CHECK	241.35	2319.70
3/27	CHECK	300.00	2019.70
3/29	CHECK	155.73	1863.97
3/30	INTEREST	13.22	1877.19
4/1	CLOSING BALANCE		1877.19

14. Balance the checkbook.

RECORD ALL CHARGES OR CREDITS THAT AFFECT YOUR ACCOUNT

NUMBER	DATE	DESCRIPTION OF TRANSACTION	PAYMENT/DEBIT (−)		√ T	FEE (IF ANY) (−)	DEPOSIT/CREDIT (+)		BALANCE $ 1219	43
	5/1	Deposit	$			$	$ 619	14	1838	57
515	5/2	Electric Bill	42	35					1796	22
516	5/2	Groceries	95	14					1701	08
517	5/4	Insurance	122	17					1578	91
518	5/5	Theatre Tickets	84	50					1494	41
	5/8	Deposit					619	14	2113	55
519	5/10	Telephone	37	39					2076	16
520	5/12	Newspaper	22	50					2053	66
	5/15	Deposit					619	14	2672	80
521	5/20	Computer Store	172	90					2499	90
522	5/21	Credit Card	313	44					2186	46
523	5/22	Eye Exam	82	00					2104	46
524	5/24	Groceries	107	14					1997	32
525	5/24	Deposit					619	14	2616	46
526	5/25	Oil Company	144	16					2472	30
527	5/30	Car Payment	288	62					2183	68
528	5/30	Mortgage Payment	877	42					1306	26

CHECKING ACCOUNT Monthly Statement		Account Number: 122-345-1	
Date	**Transaction**	**Amount**	**Balance**
5/1	OPENING BALANCE		1219.43
5/1	DEPOSIT	619.14	1838.57
5/3	CHECK	95.14	1743.43
5/4	CHECK	42.35	1701.08
5/6	CHECK	84.50	1616.58
5/8	CHECK	122.17	1494.41
5/8	DEPOSIT	619.14	2113.55
5/15	INTEREST	7.82	2121.37
5/15	CHECK	37.39	2083.98
5/15	DEPOSIT	619.14	2703.12
5/23	CHECK	82.00	2621.12
5/23	CHECK	172.90	2448.22
5/24	CHECK	107.14	2341.08
5/24	DEPOSIT	619.14	2960.22
5/30	CHECK	288.62	2671.60
6/1	CLOSING BALANCE		2671.60

15. Balance the checkbook.

NUMBER	DATE	DESCRIPTION OF TRANSACTION	PAYMENT/DEBIT (−)		√ T	FEE (IF ANY) (−)	DEPOSIT/CREDIT (+)		BALANCE $	
									2035	18
218	7/2	Mortgage	$984	60		$	$		1050	58
219	7/4	Telephone	63	36					987	22
220	7/7	Cash	200	00					787	22
	7/12	Deposit					792	60	1579	82
221	7/15	Insurance	292	30					1287	52
222	7/18	Investment	500	00					787	52
223	7/20	Credit Card	414	83					372	69
	7/26	Deposit					792	60	1165	29
224	7/27	Department Store	113	37					1051	92

CHECKING ACCOUNT Monthly Statement Account Number: 122-345-1

Date	Transaction	Amount	Balance
7/1	OPENING BALANCE		2035.18
7/1	INTEREST	5.15	2040.33
7/4	CHECK	984.60	1055.73
7/6	CHECK	63.36	992.37
7/12	DEPOSIT	792.60	1784.97
7/20	CHECK	292.30	1492.67
7/24	CHECK	500.00	992.67
7/26	DEPOSIT	792.60	1785.27
7/28	CHECK	200.00	1585.27
7/30	CLOSING BALANCE		1585.27

16. The ending balance on a monthly bank statement is greater than the beginning balance, and the bank did not include a service charge. Was the total of all deposits recorded *less than* or *greater than* the total of all checks paid?

17. When balancing your checkbook, you find that all the deposits in your checkbook register have been recorded by the bank, four checks in the register have not yet been paid by the bank, and the bank did not include a service charge. Is the ending balance on the monthly bank statement *less than* or *greater than* the ending balance on the check register?

Applying the Concepts

18. Define the words *credit* and *debit* as they apply to checkbooks.

FOCUS ON PROBLEM SOLVING

Counterexamples

An example that is given to show that a statement is not true is called a **counterexample.** For instance, suppose someone makes the statement "All colors are red." A counterexample to that statement would be to show someone the color blue or some other color.

If a statement is *always* true, there are no counterexamples. The statement "All even numbers are divisible by 2" is always true. It is not possible to give an example of an even number that is not divisible by 2.

In mathematics, statements that are always true are called *theorems,* and mathematicians are always searching for theorems. Sometimes a conjecture by a mathematician appears to be a theorem. That is, the statement appears to be always true, but later on someone finds a counterexample.

One example of this occurred when the French mathematician Pierre de Fermat (1601–1665) conjectured that $2^{(2^n)} + 1$ is always a prime number for any natural number n. For instance, when $n = 3$, we have $2^{(2^3)} + 1 = 2^8 + 1 = 257$, and 257 is a prime number. However, in 1732 Leonhard Euler (1707–1783) showed that when $n = 5$, $2^{(2^5)} + 1 = 4,294,967,297$, and that $4,294,967,297 = 641 \cdot 6,700,417$—without a calculator! Because 4,294,967,297 is the product of two numbers (other than itself and 1), it is not a prime number. This counterexample showed that Fermat's conjecture is not a theorem.

> **✓ Take Note**
>
> Recall that a prime number is a natural number greater than 1 that can be divided by only itself and 1. For instance, 17 is a prime number. 12 is not a prime number because 12 is divisible by numbers other than 1 and 12—for example, 4.

For Exercises 1 to 5, find at least one counterexample.

1. All composite numbers are divisible by 2.

2. All prime numbers are odd numbers.

3. The square of any number is always bigger than the number.

4. The reciprocal of a number is always less than 1.

5. A number ending in 9 is always larger than a number ending in 3.

When a problem is posed, it may not be known whether the problem statement is true or false. For instance, Christian Goldbach (1690–1764) stated that every even number greater than 2 can be written as the sum of two prime numbers. For example,

$$12 = 5 + 7 \qquad 32 = 3 + 29$$

Although this problem is approximately 250 years old, mathematicians have not been able to prove it is a theorem, nor have they been able to find a counterexample.

For Exercises 6 to 9, answer true if the statement is always true. If there is an instance in which the statement is false, give a counterexample.

6. The sum of two positive numbers is always larger than either of the two numbers.

7. The product of two positive numbers is always larger than either of the two numbers.

8. Percents always represent a number less than or equal to 1.

9. It is never possible to divide by zero.

PROJECTS AND GROUP ACTIVITIES

Buying a Car   Suppose a student has an after-school job to earn money to buy and maintain a car. We will make assumptions about the monthly costs in several categories in order to determine how many hours per week the student must work to support the car. Assume the student earns $10.50 per hour.

1. Monthly payment

   Assume that the car cost $18,500 with a down payment of $2220. The remainder is financed for 3 years at an annual simple interest rate of 9%.

   Monthly payment = _____

2. Insurance

   Assume that insurance costs $3000 per year.

   Monthly insurance payment = _____

3. Gasoline

   Assume that the student travels 750 miles per month, that the car travels 25 miles per gallon of gasoline, and that gasoline costs $3.50 per gallon.

   Number of gallons of gasoline purchased per month = _____

   Monthly cost for gasoline = _____

4. Miscellaneous

   Assume $.42 per mile for upkeep.

   Monthly expense for upkeep = _____

5. Total monthly expenses for the monthly payment, insurance, gasoline, and miscellaneous
   = _____

6. To find the number of hours per month that the student must work to finance the car, divide the total monthly expenses by the hourly rate.

   Number of hours per month = _____

7. To find the number of hours per week that the student must work, divide the number of hours per month by 4.

   Number of hours per week = _____

   The student has to work _____ hours per week to pay the monthly car expenses.

If you own a car, make out your own expense record. If you do not own a car, make assumptions about the kind of car that you would like to purchase, and calculate the total monthly expenses that you would have. An insurance company will give you rates on different kinds of insurance. An automobile club can give you approximations of miscellaneous expenses.

CHAPTER 6

SUMMARY

KEY WORDS	EXAMPLES
The *unit cost* is the cost of one item. [6.1A, p. 234]	Three paperback books cost $36. The unit cost is the cost of one paperback book, $12.
Percent increase is used to show how much a quantity has increased over its original value. [6.2A, p. 238]	The city's population increased 5%, from 10,000 people to 10,500 people.
Cost is the price a business pays for a product. *Selling price* is the price at which a business sells a product to a customer. *Markup* is the difference between selling price and cost. *Markup rate* is the markup expressed as a percent of the product's cost. [6.2B, p. 239]	A business pays $90 for a pair of cross trainers; the cost is $90. The business sells the cross trainers for $135; the selling price is $135. The markup is $135 − $90 = $45.
Percent decrease is used to show how much a quantity has decreased from its original value. [6.2C, p. 241]	Sales decreased 10%, from 10,000 units in the third quarter to 9000 units in the fourth quarter.
Sale price is the price after a reduction from the regular price. *Discount* is the difference between the regular price and the sale price. *Discount rate* is the discount as a percent of the product's regular price. [6.2D, p. 242]	A skateboard deck that regularly sells for $50 is on sale for $40. The regular price is $50. The sale price is $40. The discount is $50 − $40 = $10.
Interest is the amount paid for the privilege of using someone else's money. *Principal* is the amount of money originally deposited or borrowed. The percent used to determine the amount of interest is the *interest rate*. Interest computed on the original amount is called *simple interest*. The principal plus the interest owed on a loan is called the *maturity value*. [6.3A, p. 248]	Consider a 1-year loan of $5000 at an annual simple interest rate of 8%. The principal is $5000. The interest rate is 8%. The interest paid on the loan is $5000 × 0.08 = $400. The maturity value is $5000 + $400 = $5400.
The interest charged on purchases made with a credit card is called a *finance charge*. [6.3B, p. 250]	A credit card company charges 1.5% per month on any unpaid balance. The finance charge on an unpaid balance of $1000 is $1000 × 0.015 × 1 = $15.
Compound interest is computed not only on the original principal but also on the interest already earned. [6.3C, p. 251]	$10,000 is invested at 5% annual interest, compounded monthly. The value of the investment after 5 years can be found by multiplying 10,000 by the factor found in the Compound Interest Table in the Appendix. $10,000 × 1.283359 = $12,833.59
A *mortgage* is an amount that is borrowed to buy real estate. The *loan origination fee* is usually a percent of the mortgage and is expressed as *points*. [6.4A, p. 258]	The loan origination fee of 3 points paid on a mortgage of $200,000 is 0.03 × $200,000 = $6000.
A *commission* is usually paid to a salesperson and is calculated as a percent of sales. [6.6A, p. 268]	A commission of 5% on sales of $50,000 is 0.05 × $50,000 = $2500.

An employee who receives an *hourly wage* is paid a certain amount for each hour worked.   [6.6A, p. 268]

An employee is paid an hourly wage of $15. The employee's wages for working 10 hours are $15 × 10 = $150.

An employee who is paid a *salary* receives payment based on a weekly, biweekly, monthly, or annual time schedule. [6.6A, p. 268]

An employee paid an annual salary of $60,000 is paid $60,000 ÷ 12 = $5000 per month.

Balancing a checkbook is determining whether the checkbook balance is accurate.   [6.7B, pp. 273–274]

To balance a checkbook:

(1) Put a check mark in the checkbook register by each check paid by the bank and by each deposit recorded by the bank.

(2) Add to the current checkbook balance all checks that have been written but have not yet been paid by the bank and any interest paid on the account.

(3) Subtract any charges and any deposits not yet recorded by the bank. This is the checkbook balance.

(4) Compare the balance with the bank balance listed on the bank statement. If the two numbers are equal, the bank statement and the checkbook "balance."

ESSENTIAL RULES AND PROCEDURES

EXAMPLES

To find unit cost, divide the total cost by the number of units. [6.1A, p. 234]

Three paperback books cost $36. The unit cost is $36 ÷ 3 = $12 per book.

To find total cost, multiply the unit cost by the number of units purchased.   [6.1C, p. 235]

One melon costs $3. The total cost for 5 melons is $3 × 5 = $15.

Basic Markup Equations   [6.2B, p. 239]
Selling price − cost = markup
Cost + markup = selling price
Markup rate × cost = markup

A pair of cross trainers that cost a business $90 has a 50% markup rate. The markup is 0.50 × $90 = $45. The selling price is $90 + $45 = $135.

Basic Discount Equations   [6.2D, p. 242]
Regular price − sale price = discount
Regular price − discount = sale price
Discount rate × regular price = discount

A movie DVD is on sale for 20% off the regular price of $50. The discount is 0.20 × $50 = $10. The sale price is $50 − $10 = $40.

Simple Interest Formula for Annual Interest Rates   [6.3A, p. 248]
Principal × annual interest rate × time (in years) = interest

The simple interest due on a 2-year loan of $5000 that has an annual interest rate of 5% is $5000 × 0.05 × 2 = $500.

Maturity Value Formula for a Simple Interest Loan   [6.3A, p. 248]
Principal + interest = maturity value

The interest to be paid on a 2-year loan of $5000 is $500. The maturity value of the loan is $5000 + $500 = $5500.

Monthly Payment on a Simple Interest Loan   [6.3A, p. 249]
Maturity value ÷ length of the loan in months = monthly payment

The maturity value of a simple interest 8-month loan is $8000. The monthly payment is $8000 ÷ 8 = $1000.

CHAPTER 6

CONCEPT REVIEW

Test your knowledge of the concepts presented in this chapter. Answer each question.
Then check your answers against the ones provided in the Answer Section.

1.　Find the unit cost if 4 cans cost $2.96.

2.　Find the total cost of 3.4 pounds of apples if apples cost $.85 per pound.

3.　How do you find the selling price if you know the cost and the markup?

4.　How do you use the markup rate to find the markup?

5.　How do you find the amount of decrease if you know the percent decrease?

6.　How do you find the discount if you know the regular price and the sale price?

7.　How do you find the discount rate?

8.　How do you find simple interest?

9.　How do you find the maturity value for a simple interest loan?

10.　What is the principal?

11.　How do you find the monthly payment for a loan of 18 months if you know the
maturity value of the loan?

12.　What is compound interest?

13.　What is a fixed-rate mortgage?

14.　What expenses are involved in owning a car?

15.　How do you balance a checkbook?

CHAPTER 6

REVIEW EXERCISES

1. **Consumerism** A 20-ounce box of cereal costs $3.90. Find the unit cost.

2. **Car Expenses** An account executive had car expenses of $1025.58 for insurance, $1805.82 for gas, $37.92 for oil, and $288.27 for maintenance during a year in which 11,320 miles were driven. Find the cost per mile for these four items taken as a group. Round to the nearest tenth of a cent.

3. **Investments** An oil stock was bought for $42.375 per share. Six months later, the stock was selling for $55.25 per share. Find the percent increase in the price of the stock over the 6 months. Round to the nearest tenth of a percent.

4. **Markup** A sporting goods store uses a markup rate of 40%. What is the markup on a ski suit that costs the store $180?

5. **Simple Interest** A contractor borrowed $100,000 from a credit union for 9 months at an annual interest rate of 4%. What is the simple interest due on the loan?

6. **Compound Interest** A computer programmer invested $25,000 in a retirement account that pays 6% interest, compounded daily. What is the value of the investment in 10 years? Use the Compound Interest Table in the Appendix. Round to the nearest cent.

7. **Investments** Last year an oil company had earnings of $4.12 per share. This year the earnings are $4.73 per share. What is the percent increase in earnings per share? Round to the nearest percent.

8. **Real Estate** The monthly mortgage payment for a condominium is $923.67. The owner must pay an annual property tax of $2582.76. Find the total monthly payment for the mortgage and property tax.

9. **Car Expenses** A used pickup truck is purchased for $24,450. A down payment of 8% is made, and the remaining cost is financed for 4 years at an annual interest rate of 5%. Find the monthly payment. Use the Monthly Payment Table in the Appendix. Round to the nearest cent.

Car Culture/Getty Images

10. **Compound Interest** A fast-food restaurant invested $50,000 in an account that pays 7% annual interest, compounded quarterly. What is the value of the investment in 1 year? Use the Compound Interest Table in the Appendix.

11. **Real Estate** Paula Mason purchased a home for $195,000. The lender requires a down payment of 15%. Find the amount of the down payment.

12. **Car Expenses** A plumber bought a truck for $28,500. A state license fee of $315 and a sales tax of 6.25% of the purchase price are required. Find the total cost of the sales tax and the license fee.

13. **Markup** Techno-Center uses a markup rate of 35% on all computer systems. Find the selling price of a computer system that costs the store $1540.

14. **Car Expenses** Mien pays a monthly car payment of $222.78. During a month in which $65.45 is principal, how much of the payment is interest?

15. **Compensation** The manager of the retail store at a ski resort receives a commission of 3% on all sales at the alpine shop. Find the total commission received during a month in which the shop had $108,000 in sales.

16. **Discount** A suit that regularly costs $235 is on sale for 40% off the regular price. Find the sale price.

17. **Banking** Luke had a checking account balance of $1568.45 before writing checks for $123.76, $756.45, and $88.77. He then deposited a check for $344.21. Find Luke's current checkbook balance.

18. **Simple Interest** Pros' Sporting Goods borrowed $30,000 at an annual interest rate of 8% for 6 months. Find the maturity value of the loan.

19. **Real Estate** A credit union requires a borrower to pay $2\frac{1}{2}$ points for a loan. Find the origination fee for a $75,000 loan.

20. **Consumerism** Sixteen ounces of mouthwash cost $3.49. A 33-ounce container of the same brand of mouthwash costs $6.99. Which is the better buy?

21. **Real Estate** The Sweeneys bought a home for $356,000. The family made a 10% down payment and financed the remainder with a 30-year loan at an annual interest rate of 7%. Find the monthly mortgage payment. Use the Monthly Payment Table in the Appendix. Round to the nearest cent.

22. **Compensation** Richard Valdez receives $12.60 per hour for working 40 hours a week and time and a half for working over 40 hours. Find his total income during a week in which he worked 48 hours.

23. **Banking** The business checking account of a donut shop showed a balance of $9567.44 before checks of $1023.55, $345.44, and $23.67 were written and checks of $555.89 and $135.91 were deposited. Find the current checkbook balance.

24. **Simple Interest** The simple interest due on a 4-month loan of $55,000 is $1375. Find the monthly payment on the loan.

25. **Simple Interest** A credit card company charges a customer 1.25% per month on the unpaid balance of charges on the card. What is the finance charge in a month in which the customer has an unpaid balance of $576?

CHAPTER 6

TEST

1. **Consumerism** Twenty feet of lumber cost $138.40. What is the cost per foot?

2. **Consumerism** Which is the more economical purchase: 3 pounds of tomatoes for $7.49 or 5 pounds of tomatoes for $12.59?

3. **Consumerism** Red snapper costs $4.15 per pound. Find the cost of $3\frac{1}{2}$ pounds. Round to the nearest cent.

4. **Business** An exercise bicycle increased in price from $415 to $498. Find the percent increase in the cost of the exercise bicycle.

5. **Markup** A department store uses a 40% markup rate. Find the selling price of a blu-ray disc player that the store purchased for $315.

6. **Investments** The price of gold rose from $790 per ounce to $860 per ounce. What percent increase does this amount represent? Round to the nearest tenth of a percent.

7. **Consumerism** The price of a video camera dropped from $1120 to $896. What percent decrease does this price drop represent?

8. **Discount** A corner hutch with a regular price of $299 is on sale for 30% off the regular price. Find the sale price.

9. **Discount** A box of stationery that regularly sells for $9.50 is on sale for $5.70. Find the discount rate.

10. **Simple Interest** A construction company borrowed $75,000 for 4 months at an annual interest rate of 8%. Find the simple interest due on the loan.

11. **Simple Interest** Craig Allen borrowed $25,000 at an annual interest rate of 9.2% for 9 months. Find the maturity value of the loan.

12. **Simple Interest** A credit card company charges a customer 1.2% per month on the unpaid balance of charges on the card. What is the finance charge in a month in which the customer has an unpaid balance of $374.95?

13. **Compound Interest** Jorge, who is self-employed, placed $30,000 in an account that pays 6% annual interest, compounded quarterly. How much interest was earned in 10 years? Use the Compound Interest Table in the Appendix.

14. **Real Estate** A savings and loan institution is offering mortgage loans that have a loan origination fee of $2\frac{1}{2}$ points. Find the loan origination fee when a home is purchased with a loan of $134,000.

15. **Real Estate** A new housing development offers homes with a mortgage of $222,000 for 25 years at an annual interest rate of 8%. Find the monthly mortgage payment. Use the Monthly Payment Table in the Appendix.

16. **Car Expenses** A Chevrolet was purchased for $23,750, and a 20% down payment was made. Find the amount financed.

17. **Car Expenses** A rancher purchased an SUV for $33,714 and made a down payment of 15% of the cost. The balance was financed for 4 years at an annual interest rate of 7%. Find the monthly truck payment. Use the Monthly Payment Table in the Appendix.

18. **Compensation** Shaney receives an hourly wage of $30.40 an hour as an emergency room nurse. When called in at night, she receives time and a half. How much does Shaney earn in a week when she works 30 hours at normal rates and 15 hours during the night?

19. **Banking** The business checking account for a pottery store had a balance of $7349.44 before checks for $1349.67 and $344.12 were written. The store manager then made a deposit of $956.60. Find the current checkbook balance.

20. **Banking** Balance the checkbook shown.

		RECORD ALL CHARGES OR CREDITS THAT AFFECT YOUR ACCOUNT					BALANCE	
NUMBER	DATE	DESCRIPTION OF TRANSACTION	PAYMENT/DEBIT (−)	√T	FEE (IF ANY) (−)	DEPOSIT/CREDIT (+)	$ 1422	13
843	8/1	House Payment	$ 713 72		$	$	708	41
	8/4	Deposit				852 60	1561	01
844	8/5	Loan Payment	162 40				1398	61
845	8/6	Groceries	166 44				1232	17
846	8/10	Car Payment	322 37				909	80
	8/15	Deposit				852 60	1762	40
847	8/16	Credit Card	413 45				1348	95
848	8/18	Pharmacy	92 14				1256	81
849	8/22	Utilities	72 30				1184	51
850	8/28	Telephone	78 20				1106	31

CHECKING ACCOUNT Monthly Statement		Account Number: 122-345-1	
Date	Transaction	Amount	Balance
8/1	OPENING BALANCE		1422.13
8/3	CHECK	713.72	708.41
8/4	DEPOSIT	852.60	1561.01
8/8	CHECK	166.44	1394.57
8/8	CHECK	162.40	1232.17
8/15	DEPOSIT	852.60	2084.77
8/23	CHECK	72.30	2012.47
8/24	CHECK	92.14	1920.33
9/1	CLOSING BALANCE		1920.33

CUMULATIVE REVIEW EXERCISES

1. Simplify: $12 - (10 - 8)^2 \div 2 + 3$

2. Add: $3\frac{1}{3} + 4\frac{1}{8} + 1\frac{1}{12}$

3. Find the difference between $12\frac{3}{16}$ and $9\frac{5}{12}$.

4. Find the product of $5\frac{5}{8}$ and $1\frac{9}{15}$.

5. Divide: $3\frac{1}{2} \div 1\frac{3}{4}$

6. Simplify: $\left(\frac{3}{4}\right)^2 \div \left(\frac{3}{8} - \frac{1}{4}\right) + \frac{1}{2}$

7. Divide: $0.059\overline{)3.0792}$
    Round to the nearest tenth.

8. Convert $\frac{17}{12}$ to a decimal. Round to the nearest thousandth.

9. Write "$410 in 8 hours" as a unit rate.

10. Solve the proportion $\frac{5}{n} = \frac{16}{35}$.
    Round to the nearest hundredth.

11. Write $\frac{5}{8}$ as a percent.

12. Find 6.5% of 420.

13. Write 18.2% as a decimal.

14. What percent of 20 is 8.4?

15. 30 is 12% of what?

16. 65 is 42% of what? Round to the nearest hundredth.

17. **Meteorology** A series of late-summer storms produced rainfall of $3\frac{3}{4}$, $8\frac{1}{2}$, and $1\frac{2}{3}$ inches during a 3-week period. Find the total rainfall during the 3 weeks.

18. **Taxes** The Homer family pays $\frac{1}{5}$ of its total monthly income for taxes. The family has a total monthly income of $4850. Find the amount of their monthly income that the Homers pay in taxes.

19. **Consumerism** In 5 years, the cost of a scientific calculator went from $75 to $30. What is the ratio of the decrease in price to the original price?

20. **Fuel Efficiencies** A compact car was driven 417.5 miles on 12.5 gallons of gasoline. Find the number of miles driven per gallon of gasoline.

21. **Consumerism** A 14-pound turkey costs $15.40. Find the unit cost. Round to the nearest cent.

22. **Investments** Eighty shares of a stock paid a dividend of $112. At the same rate, find the dividend on 200 shares of the stock.

23. **Discount** A laptop computer that regularly sells for $900 is on sale for 20% off the regular price. What is the sale price?

24. **Markup** A pro skate shop bought a grinding rail for $85 and used a markup rate of 40%. Find the selling price of the grinding rail.

25. **Compensation** Sook Kim, an elementary school teacher, received an increase in salary from $2800 per month to $3024 per month. Find the percent increase in her salary.

26. **Simple Interest** A contractor borrowed $120,000 for 6 months at an annual interest rate of 4.5%. How much simple interest is due on the loan?

27. **Car Expenses** A red Ford Mustang was purchased for $26,900, and a down payment of $2000 was made. The balance is financed for 3 years at an annual interest rate of 9%. Find the monthly payment. Use the Monthly Payment Table in the Appendix. Round to the nearest cent.

28. **Banking** A family had a checking account balance of $1846.78. A check of $568.30 was deposited into the account, and checks of $123.98 and $47.33 were written. Find the new checking account balance.

29. **Car Expenses** During 1 year, Anna Gonzalez spent $1840 on gasoline and oil, $820 on insurance, $185 on tires, and $432 on repairs. Find the cost per mile to drive the car 10,000 miles during the year. Round to the nearest cent.

30. **Real Estate** A house has a mortgage of $172,000 for 20 years at an annual interest rate of 6%. Find the monthly mortgage payment. Use the Monthly Payment Table in the Appendix. Round to the nearest cent.

Appendix

Equations and Properties

Equations

Basic Percent Equation

Percent $\times$ base = amount

Proportion Method of Solving Percent Equations

$$\frac{\text{Percent}}{100} = \frac{\text{amount}}{\text{base}}$$

Basic Markup Equations

Selling price = cost + markup
Markup = markup rate $\times$ cost

Basic Discount Equations

Sale price = regular price − discount
Discount = discount rate $\times$ regular price

Annual Simple Interest Formula

Principle $\times$ annual simple interest rate $\times$ time in years = interest

Maturity Value Formula for Simple Interest

Principle + interest = maturity value

Monthly Payment on a Simple Interest Loan

Maturilty value $\div$ length of the loan in months = monthly payment

Properties

Addition Property of Zero

Zero added to a number does not change the number.

Commutative Property of Addition

Two numbers can be added in either order; the sum will be the same.

Associative Property of Addition

Grouping addition in any order gives the same result.

Multiplication Property of Zero

The product of a number and zero is zero.

Multiplication Property of One

The product of a number and one is the number.

Commutative Property of Multiplication

Two numbers can be multiplied in either order. The product will be the same.

Associative Property of Multiplication

Grouping the numbers to be multiplied in any order gives the same result.

Properties of Zero and One in Division

Any number divided by 1 is the number.
Any number other than zero divided by itself is 1.
Division by zero is not allowed.
Zero divided by any number other than zero is zero.

Compound Interest Table

Compounded Annually							
	4%	**5%**	**6%**	**7%**	**8%**	**9%**	**10%**
1 year	1.04000	1.05000	1.06000	1.07000	1.08000	1.09000	1.10000
5 years	1.21665	1.27628	1.33823	1.40255	1.46933	1.53862	1.61051
10 years	1.48024	1.62890	1.79085	1.96715	2.15893	2.36736	2.59374
15 years	1.80094	2.07893	2.39656	2.75903	3.17217	3.64248	4.17725
20 years	2.19112	2.65330	3.20714	3.86968	4.66095	5.60441	6.72750

Compounded Semiannually							
	4%	**5%**	**6%**	**7%**	**8%**	**9%**	**10%**
1 year	1.04040	1.05062	1.06090	1.07123	1.08160	1.09203	1.10250
5 years	1.21899	1.28008	1.34392	1.41060	1.48024	1.55297	1.62890
10 years	1.48595	1.63862	1.80611	1.98979	2.19112	2.41171	2.65330
15 years	1.81136	2.09757	2.42726	2.80679	3.24340	3.74531	4.32194
20 years	2.20804	2.68506	3.26204	3.95926	4.80102	5.81634	7.03999

Compounded Quarterly							
	4%	**5%**	**6%**	**7%**	**8%**	**9%**	**10%**
1 year	1.04060	1.05094	1.06136	1.07186	1.08243	1.09308	1.10381
5 years	1.22019	1.28204	1.34686	1.41478	1.48595	1.56051	1.63862
10 years	1.48886	1.64362	1.81402	2.00160	2.20804	2.43519	2.68506
15 years	1.81670	2.10718	2.44322	2.83182	3.28103	3.80013	4.39979
20 years	2.21672	2.70148	3.29066	4.00639	4.87544	5.93015	7.20957

Compounded Monthly							
	4%	**5%**	**6%**	**7%**	**8%**	**9%**	**10%**
1 year	1.04074	1.051162	1.061678	1.072290	1.083000	1.093807	1.104713
5 years	1.220997	1.283359	1.348850	1.417625	1.489846	1.565681	1.645309
10 years	1.490833	1.647009	1.819397	2.009661	2.219640	2.451357	2.707041
15 years	1.820302	2.113704	2.454094	2.848947	3.306921	3.838043	4.453920
20 years	2.222582	2.712640	3.310204	4.038739	4.926803	6.009152	7.328074

Compounded Daily							
	4%	**5%**	**6%**	**7%**	**8%**	**9%**	**10%**
1 year	1.04080	1.05127	1.06183	1.07250	1.08328	1.09416	1.10516
5 years	1.22139	1.28400	1.34983	1.41902	1.49176	1.56823	1.64861
10 years	1.49179	1.64866	1.82203	2.01362	2.22535	2.45933	2.71791
15 years	1.82206	2.11689	2.45942	2.85736	3.31968	3.85678	4.48077
20 years	2.22544	2.71810	3.31979	4.05466	4.95217	6.04830	7.38703

To use this table:
1. Locate the section that gives the desired compounding period.
2. Locate the interest rate in the top row of that section.
3. Locate the number of years in the left-hand column of that section.
4. Locate the number where the interest-rate column and the number-of-years row meet. This is the compound interest factor.

Example An investment yields an annual interest rate of 10% compounded quarterly for 5 years.
The compounding period is "compounded quarterly."
The interest rate is 10%.
The number of years is 5.
The number where the row and column meet is 1.63862. This is the compound interest factor.

Compound Interest Table

Compounded Annually							
	11%	**12%**	**13%**	**14%**	**15%**	**16%**	**17%**
1 year	1.11000	1.12000	1.13000	1.14000	1.15000	1.16000	1.17000
5 years	1.68506	1.76234	1.84244	1.92542	2.01136	2.10034	2.19245
10 years	2.83942	3.10585	3.39457	3.70722	4.04556	4.41144	4.80683
15 years	4.78459	5.47357	6.25427	7.13794	8.13706	9.26552	10.53872
20 years	8.06239	9.64629	11.52309	13.74349	16.36654	19.46076	23.10560

Compounded Semiannually							
	11%	**12%**	**13%**	**14%**	**15%**	**16%**	**17%**
1 year	1.11303	1.12360	1.13423	1.14490	1.15563	1.16640	1.17723
5 years	1.70814	1.79085	1.87714	1.96715	2.06103	2.15893	2.26098
10 years	2.91776	3.20714	3.52365	3.86968	4.24785	4.66096	5.11205
15 years	4.98395	5.74349	6.61437	7.61226	8.75496	10.06266	11.55825
20 years	8.51331	10.28572	12.41607	14.97446	18.04424	21.72452	26.13302

Compounded Quarterly							
	11%	**12%**	**13%**	**14%**	**15%**	**16%**	**17%**
1 year	1.11462	1.12551	1.13648	1.14752	1.15865	1.16986	1.18115
5 years	1.72043	1.80611	1.89584	1.98979	2.08815	2.19112	2.29891
10 years	2.95987	3.26204	3.59420	3.95926	4.36038	4.80102	5.28497
15 years	5.09225	5.89160	6.81402	7.87809	9.10513	10.51963	12.14965
20 years	8.76085	10.64089	12.91828	15.67574	19.01290	23.04980	27.93091

Compounded Monthly							
	11%	**12%**	**13%**	**14%**	**15%**	**16%**	**17%**
1 year	1.115719	1.126825	1.138032	1.149342	1.160755	1.172271	1.183892
5 years	1.728916	1.816697	1.908857	2.005610	2.107181	2.213807	2.325733
10 years	2.989150	3.300387	3.643733	4.022471	4.440213	4.900941	5.409036
15 years	5.167988	5.995802	6.955364	8.067507	9.356334	10.849737	12.579975
20 years	8.935015	10.892554	13.276792	16.180270	19.715494	24.019222	29.257669

Compounded Daily							
	11%	**12%**	**13%**	**14%**	**15%**	**16%**	**17%**
1 year	1.11626	1.12747	1.13880	1.15024	1.16180	1.17347	1.18526
5 years	1.73311	1.82194	1.91532	2.01348	2.11667	2.22515	2.33918
10 years	3.00367	3.31946	3.66845	4.05411	4.48031	4.95130	5.47178
15 years	5.20569	6.04786	7.02625	8.16288	9.48335	11.01738	12.79950
20 years	9.02203	11.01883	13.45751	16.43582	20.07316	24.51534	29.94039

Monthly Payment Table

	4%	5%	6%	7%	8%	9%
1 year	0.0851499	0.0856075	0.0860664	0.0865267	0.0869884	0.0874515
2 years	0.0434249	0.0438714	0.0443206	0.0447726	0.0452273	0.0456847
3 years	0.0295240	0.0299709	0.0304219	0.0308771	0.0313364	0.0317997
4 years	0.0225791	0.0230293	0.0234850	0.0239462	0.0244129	0.0248850
5 years	0.0184165	0.0188712	0.0193328	0.0198012	0.0202764	0.0207584
15 years	0.0073969	0.0079079	0.0084386	0.0089883	0.0095565	0.0101427
20 years	0.0060598	0.0065996	0.0071643	0.0077530	0.0083644	0.0089973
25 years	0.0052784	0.0058459	0.0064430	0.0070678	0.0077182	0.0083920
30 years	0.0047742	0.0053682	0.0059955	0.0066530	0.0073376	0.0080462

	10%	11%	12%	13%		
1 year	0.0879159	0.0883817	0.0888488	0.0893173		
2 years	0.0461449	0.0466078	0.0470735	0.0475418		
3 years	0.0322672	0.0327387	0.0332143	0.0336940		
4 years	0.0253626	0.0258455	0.0263338	0.0268275		
5 years	0.0212470	0.0217424	0.0222445	0.0227531		
15 years	0.0107461	0.0113660	0.0120017	0.0126524		
20 years	0.0096502	0.0103219	0.0110109	0.0117158		
25 years	0.0090870	0.0098011	0.0105322	0.0112784		
30 years	0.0087757	0.0095232	0.0102861	0.0110620		

To use this table:
1. Locate the desired interest rate in the top row.
2. Locate the number of years in the left-hand column.
3. Locate the number where the interest-rate column and the number-of-years row meet. This is the monthly payment factor.

Example A home has a 30-year mortgage at an annual interest rate of 12%.
The interest rate is 12%.
The number of years is 30.
The number where the row and column meet is 0.0102861. This is the monthly payment factor.

Solutions to "You Try It"

SOLUTIONS TO CHAPTER 1 "YOU TRY IT"

SECTION 1.1

You Try It 1

You Try It 2   **a.** $45 > 29$ **b.** $27 > 0$

You Try It 3   Thirty-six million four hundred sixty-two thousand seventy-five

You Try It 4   452,007

You Try It 5 $60,000 + 8000 + 200 + 80 + 1$

You Try It 6 $100,000 + 9000 + 200 + 7$

You Try It 7

```
        ┌── Given place value
368,492
        └── 8 > 5
```

368,492 rounded to the nearest ten-thousand is 370,000.

You Try It 8

```
        ┌── Given place value
3962
        └── 6 > 5
```

3962 rounded to the nearest hundred is 4000.

SECTION 1.2

You Try It 1

```
   1 1
   347     • 7 + 3 = 10
+12,453      Write the 0 in the ones
 12,800      column. Carry the 1 to
             the tens column.
             1 + 4 + 5 = 10
             Write the 0 in the tens
             column. Carry the 1 to
             the hundreds column.
             1 + 3 + 4 = 8
```

347 increased by 12,453 is 12,800.

You Try It 2

```
  2
  95    • 5 + 8 + 7 = 20
  88      Write the 0 in the ones column.
+ 67      Carry the 2 to the tens column.
 250
```

You Try It 3

```
  1 1 2 1
    392
  4,079
 89,035
+ 4,992
 98,498
```

You Try It 4

Strategy   To find the total square footage of Wal-Mart stores in the United States, read the table to find the square footage of discount stores, Supercenters, Sam's Clubs, and neighborhood markets. Then add the four numbers.

Solution
```
  105
  457
   78
+   5
  645
```

The total square footage of Wal-Mart stores in the United States is 645 million square feet.

SECTION 1.3

You Try It 1
```
  8925    Check:   6413
- 6413           + 2512
  2512             8925
```

You Try It 2
```
 17,504   Check:   9,302
- 9,302          + 8,202
  8,202           17,504
```

You Try It 3
```
 2 14 7 11   Check:    865
 3̶ 4̶ 8̶ 1̶           + 2616
-    8 6 5            3481
  2 6 1 6
```

You Try It 4
```
         15
       4 8̶ 12
  5 4, 8̶ 6̶ 2̶    Check:   14,485
- 1 4, 4 8 5           + 40,077
  4 0, 0 7 7            54,562
```

You Try It 5

$$\begin{array}{r} \overset{3}{\cancel{6}}\,\overset{10}{\cancel{4}},\overset{}{0}\ 0\ 3 \\ -\ 5\ 4,9\ 3\ 6 \end{array}$$

• There are two zeros in the minuend. Borrow 1 thousand from the thousands column and write 10 in the hundreds column.

$$\begin{array}{r} \overset{3}{\cancel{6}}\ \overset{9}{\cancel{4}},\overset{10\ 10}{\cancel{0}\ \cancel{0}}\ 3 \\ -\ 5\ 4,9\ 3\ 6 \end{array}$$

• Borrow 1 hundred from the hundreds column and write 10 in the tens column.

$$\begin{array}{r} \overset{5}{\cancel{6}}\ \overset{13}{\cancel{3}}\ \overset{9}{\cancel{4}},\overset{9}{\cancel{0}}\ \overset{9}{\cancel{0}}\ \overset{13}{\cancel{3}} \\ 6\ 4,0\ 0\ 3 \\ -\ 5\ 4,9\ 3\ 6 \\ \hline 9,0\ 6\ 7 \end{array}$$

• Borrow 1 ten from the tens column and add 10 to the 3 in the ones column.

Check:
$$\begin{array}{r} 54,936 \\ +\ 9,067 \\ \hline 64,003 \end{array}$$

You Try It 6

Strategy   To find the difference, subtract the number of personnel on active duty in the Air Force in 1945 (2,282,259) from the number of personnel on active duty in the Navy in 1945 (3,380,817).

Solution
$$\begin{array}{r} 3,380,817 \\ -\ 2,282,259 \\ \hline 1,098,558 \end{array}$$

The difference was 1,098,558 personnel.

You Try It 7

Strategy   To find your take-home pay:
- Add to find the total of the deductions (127 + 18 + 35).
- Subtract the total of the deductions from your total salary (638).

Solution
$$\begin{array}{r} 127 \\ 18 \\ +\ 35 \\ \hline 180 \end{array} \qquad \begin{array}{r} 638 \\ -\ 180 \\ \hline 458 \end{array}$$

180 in deductions

Your take-home pay is $458.

SECTION 1.4

You Try It 1

$$\begin{array}{r} \overset{3\ 5}{} \\ 648 \\ \times\ \ 7 \\ \hline 4536 \end{array}$$

• 7 × 8 = 56
Write the 6 in the ones column. Carry the 5 to the tens column.
7 × 4 = 28, 28 + 5 = 33
7 × 6 = 42, 42 + 3 = 45

You Try It 2

$$\begin{array}{r} 756 \\ \times\ 305 \\ \hline 3780 \\ 22680 \\ \hline 230,580 \end{array}$$

• 5 × 756 = 3780
Write a zero in the tens column for 0 × 756.
3 × 756 = 2268

You Try It 3

Strategy   To find the number of cars the dealer will receive in 12 months, multiply the number of months (12) by the number of cars received each month (37).

Solution
$$\begin{array}{r} 37 \\ \times\ 12 \\ \hline 74 \\ 37 \\ \hline 444 \end{array}$$

The dealer will receive 444 cars in 12 months.

You Try It 4

Strategy   To find the total cost of the order:
- Find the cost of the sports jackets by multiplying the number of jackets (25) by the cost for each jacket (23).
- Add the product to the cost for the suits (4800).

Solution
$$\begin{array}{r} 23 \\ \times\ 25 \\ \hline 115 \\ 46 \\ \hline 575 \end{array} \qquad \begin{array}{r} 4800 \\ +\ 575 \\ \hline 5375 \end{array}$$

575 cost for jackets

The total cost of the order is $5375.

SECTION 1.5

You Try It 1

$$9\overline{)63} \quad \overset{7}{}$$

Check: 7 × 9 = 63

You Try It 2

$$\begin{array}{r} 453 \\ 9\overline{)4077} \\ -36 \\ \hline 47 \\ -45 \\ \hline 27 \\ -27 \\ \hline 0 \end{array}$$

Check: 453 × 9 = 4077

You Try It 3

$$
\begin{array}{r}
705 \\
9{\overline{\smash{\big)}\,6345}} \\
-63 \\
\hline
04 \\
-\ 0 \\
\hline
45 \\
-45 \\
\hline
0
\end{array}
$$

• Think $9{\overline{)}\,4}$. Place 0 in quotient.
• Subtract 0×9.
• Bring down the 5.

Check: $705 \times 9 = 6345$

You Try It 4

$$
\begin{array}{r}
870 \text{ r}5 \\
6{\overline{\smash{\big)}\,5225}} \\
-48 \\
\hline
42 \\
-42 \\
\hline
05 \\
-\ 0 \\
\hline
5
\end{array}
$$

• Think $6{\overline{)}\,5}$. Place 0 in quotient.
• Subtract 0×6.

Check: $(870 \times 6) + 5 =$
$5220 + 5 = 5225$

You Try It 5

$$
\begin{array}{r}
3{,}058 \text{ r}3 \\
7{\overline{\smash{\big)}\,21{,}409}} \\
-21 \\
\hline
04 \\
-\ 0 \\
\hline
40 \\
-35 \\
\hline
59 \\
-56 \\
\hline
3
\end{array}
$$

• Think $7{\overline{)}\,4}$. Place 0 in quotient.
• Subtract 0×7.

Check: $(3058 \times 7) + 3 =$
$21{,}406 + 3 = 21{,}409$

You Try It 6

$$
\begin{array}{r}
109 \\
42{\overline{\smash{\big)}\,4578}} \\
-42 \\
\hline
37 \\
-\ 0 \\
\hline
378 \\
-378 \\
\hline
0
\end{array}
$$

• Think $42{\overline{)}\,37}$. Place 0 in quotient.
• Subtract 0×42.

Check: $(109 \times 42) = 4578$

You Try It 7

$$
\begin{array}{r}
470 \text{ r}29 \\
39{\overline{\smash{\big)}\,18{,}359}} \\
-15\ 6 \\
\hline
2\ 75 \\
-2\ 73 \\
\hline
29 \\
-\ 0 \\
\hline
29
\end{array}
$$

• Think $3{\overline{)}\,18}$. 6×39 is too large. Try 5. 5×39 is too large. Try 4.

Check: $(470 \times 39) + 29 =$
$18{,}330 + 29 = 18{,}359$

You Try It 8

$$
\begin{array}{r}
62 \text{ r}111 \\
534{\overline{\smash{\big)}\,33{,}219}} \\
-32\ 04 \\
\hline
1\ 179 \\
-1\ 068 \\
\hline
111
\end{array}
$$

Check: $(62 \times 534) + 111 =$
$33{,}108 + 111 = 33{,}219$

You Try It 9

$$
\begin{array}{r}
421 \text{ r}33 \\
515{\overline{\smash{\big)}\,216{,}848}} \\
-206\ 0 \\
\hline
10\ 84 \\
-10\ 30 \\
\hline
548 \\
-515 \\
\hline
33
\end{array}
$$

Check: $(421 \times 515) + 33 =$
$216{,}815 + 33 = 216{,}848$

You Try It 10

Strategy

To find the number of tires that can be stored on each shelf, divide the number of tires (270) by the number of shelves (15).

Solution

$$
\begin{array}{r}
18 \\
15{\overline{\smash{\big)}\,270}} \\
-15 \\
\hline
120 \\
-120 \\
\hline
0
\end{array}
$$

Each shelf can store 18 tires.

You Try It 11

Strategy

To find the number of cases produced in 8 hours:

• Find the number of cases produced in 1 hour by dividing the number of cans produced (12,600) by the number of cans to a case (24).
• Multiply the number of cases produced in 1 hour by 8.

Solution

$$
\begin{array}{r}
525 \\
24{\overline{\smash{\big)}\,12{,}600}} \\
-12\ 0 \\
\hline
60 \\
-48 \\
\hline
120 \\
-120 \\
\hline
0
\end{array}
$$
cases produced in 1 hour

$$
\begin{array}{r}
525 \\
\times\ \ \ 8 \\
\hline
4200
\end{array}
$$

In 8 hours, 4200 cases are produced.

SECTION 1.6

You Try It 1 $2 \cdot 2 \cdot 2 \cdot 2 \cdot 3 \cdot 3 \cdot 3 = 2^4 \cdot 3^3$

You Try It 2 $10 \cdot 10 \cdot 10 \cdot 10 \cdot 10 \cdot 10 \cdot 10 = 10^7$

You Try It 3 $2^3 \cdot 5^2 = (2 \cdot 2 \cdot 2) \cdot (5 \cdot 5) = 8 \cdot 25$
$$= 200$$

You Try It 4

$5 \cdot (8 - 4)^2 \div 4 - 2$ • Parentheses
$= 5 \cdot 4^2 \div 4 - 2$ • Exponents
$= 5 \cdot 16 \div 4 - 2$ • Multiplication
$= 80 \div 4 - 2$ and division
$= 20 - 2$ • Subtraction
$= 18$

SECTION 1.7

You Try It 1

$40 \div 1 = 40$
$40 \div 2 = 20$
$40 \div 3$ Will not divide evenly
$40 \div 4 = 10$
$40 \div 5 = 8$
$40 \div 6$ Will not divide evenly
$40 \div 7$ Will not divide evenly
$40 \div 8 = 5$

1, 2, 4, 5, 8, 10, 20, and 40 are factors of 40.

You Try It 2

$$
\begin{array}{r|r}
44 & \\
\hline
2 & 22 \\
2 & 11 \\
11 & 1 \\
\end{array}
$$
   • $44 \div 2 = 22$
   • $22 \div 2 = 11$
   • $11 \div 11 = 1$

$44 = 2 \cdot 2 \cdot 11$

You Try It 3

$$
\begin{array}{r|r}
177 & \\
\hline
3 & 59 \\
59 & 1 \\
\end{array}
$$
   • Try only 2, 3, 4, 7, and 11 because $11^2 > 59$.

$177 = 3 \cdot 59$

SOLUTIONS TO CHAPTER 2 "YOU TRY IT"

SECTION 2.1

You Try It 1

	2	3	5
12 =	(2 · 2)	3	
27 =		(3 · 3 · 3)	
50 =	2		(5 · 5)

The LCM $= 2 \cdot 2 \cdot 3 \cdot 3 \cdot 3 \cdot 5 \cdot 5 = 2700$

You Try It 2

	2	3	5
36 =	(2 · 2)	3 · 3	
60 =	2 · 2	(3)	5
72 =	2 · 2 · 2	3 · 3	

The GCF $= 2 \cdot 2 \cdot 3 = 12$.

You Try It 3

	2	3	5	11
11 =				11
24 =	2 · 2 · 2	3		
30 =	2	3	5	

Because no numbers are circled, the GCF = 1.

SECTION 2.2

You Try It 1 $4\dfrac{1}{4}$

You Try It 2 $\dfrac{17}{4}$

You Try It 3

$$
\begin{array}{r}
4 \\
5{\overline{\smash{)}22}} \\
-20 \\
\hline
2
\end{array}
$$
$\dfrac{22}{5} = 4\dfrac{2}{5}$

You Try It 4

$$
\begin{array}{r}
4 \\
7{\overline{\smash{)}28}} \\
-28 \\
\hline
0
\end{array}
$$
$\dfrac{28}{7} = 4$

You Try It 5 $14\dfrac{5}{8} = \dfrac{112 + 5}{8} = \dfrac{117}{8}$

SECTION 2.3

You Try It 1 $45 \div 5 = 9$ $\dfrac{3}{5} = \dfrac{3 \cdot 9}{5 \cdot 9} = \dfrac{27}{45}$

$\dfrac{27}{45}$ is equivalent to $\dfrac{3}{5}$.

You Try It 2 Write 6 as $\dfrac{6}{1}$.

$18 \div 1 = 18$ $6 = \dfrac{6 \cdot 18}{1 \cdot 18} = \dfrac{108}{18}$

$\dfrac{108}{18}$ is equivalent to 6.

You Try It 3 $\dfrac{16}{24} = \dfrac{\cancel{2} \cdot \cancel{2} \cdot \cancel{2} \cdot 2}{\cancel{2} \cdot \cancel{2} \cdot \cancel{2} \cdot 3} = \dfrac{2}{3}$

You Try It 4 $\dfrac{8}{56} = \dfrac{\cancel{2} \cdot \cancel{2} \cdot \cancel{2}}{\cancel{2} \cdot \cancel{2} \cdot \cancel{2} \cdot 7} = \dfrac{1}{7}$

You Try It 5 $\dfrac{15}{32} = \dfrac{3 \cdot 5}{2 \cdot 2 \cdot 2 \cdot 2 \cdot 2} = \dfrac{15}{32}$

You Try It 6 $\dfrac{48}{36} = \dfrac{\cancel{2} \cdot \cancel{2} \cdot 2 \cdot 2 \cdot \cancel{3}}{\cancel{2} \cdot \cancel{2} \cdot \cancel{3} \cdot 3} = \dfrac{4}{3} = 1\dfrac{1}{3}$

SECTION 2.4

You Try It 1

$$\begin{array}{r} \frac{3}{8} \\ + \frac{7}{8} \\ \hline \frac{10}{8} = \frac{5}{4} = 1\frac{1}{4} \end{array}$$

- The denominators are the same. Add the numerators. Place the sum over the common denominator.

You Try It 2

$$\begin{array}{r} \frac{5}{12} = \frac{20}{48} \\ + \frac{9}{16} = \frac{27}{48} \\ \hline \frac{47}{48} \end{array}$$

- The LCM of 12 and 16 is 48.

You Try It 3

$$\begin{array}{r} \frac{7}{8} = \frac{105}{120} \\ + \frac{11}{15} = \frac{88}{120} \\ \hline \frac{193}{120} = 1\frac{73}{120} \end{array}$$

- The LCM of 8 and 15 is 120.

You Try It 4

$$\begin{array}{r} \frac{3}{4} = \frac{30}{40} \\ \frac{4}{5} = \frac{32}{40} \\ + \frac{5}{8} = \frac{25}{40} \\ \hline \frac{87}{40} = 2\frac{7}{40} \end{array}$$

- The LCM of 4, 5, and 8 is 40.

You Try It 5

$$7 + \frac{6}{11} = 7\frac{6}{11}$$

You Try It 6

$$\begin{array}{r} 29 \\ + 17\frac{5}{12} \\ \hline 46\frac{5}{12} \end{array}$$

You Try It 7

$$\begin{array}{r} 7\frac{4}{5} = 7\frac{24}{30} \\ 6\frac{7}{10} = 6\frac{21}{30} \\ + 13\frac{11}{15} = 13\frac{22}{30} \\ \hline 26\frac{67}{30} = 28\frac{7}{30} \end{array}$$

- LCM = 30

You Try It 8

$$\begin{array}{r} 9\frac{3}{8} = 9\frac{45}{120} \\ 17\frac{7}{12} = 17\frac{70}{120} \\ + 10\frac{14}{15} = 10\frac{112}{120} \\ \hline 36\frac{227}{120} = 37\frac{107}{120} \end{array}$$

- LCM = 120

You Try It 9

Strategy To find the total time spent on the activities, add the three times $\left(4\frac{1}{2}, 3\frac{3}{4}, 1\frac{1}{3}\right)$.

Solution

$$\begin{array}{r} 4\frac{1}{2} = 4\frac{6}{12} \\ 3\frac{3}{4} = 3\frac{9}{12} \\ + 1\frac{1}{3} = 1\frac{4}{12} \\ \hline 8\frac{19}{12} = 9\frac{7}{12} \end{array}$$

The total time spent on the three activities was $9\frac{7}{12}$ hours.

You Try It 10

Strategy To find the overtime pay:
- Find the total number of overtime hours $\left(1\frac{2}{3} + 3\frac{1}{3} + 2\right)$.
- Multiply the total number of hours by the overtime hourly wage (36).

Solution

$$\begin{array}{r} 1\frac{2}{3} \\ 3\frac{1}{3} \\ + 2 \\ \hline 6\frac{3}{3} = 7 \text{ hours} \end{array} \qquad \begin{array}{r} 36 \\ \times\ 7 \\ \hline 252 \end{array}$$

Jeff earned \$252 in overtime pay.

SECTION 2.5

You Try It 1

$$\begin{array}{r} \frac{16}{27} \\ - \frac{7}{27} \\ \hline \frac{9}{27} = \frac{1}{3} \end{array}$$

- The denominators are the same. Subtract the numerators. Place the difference over the common denominator.

You Try It 2

$$\begin{array}{r} \frac{13}{18} = \frac{52}{72} \\ - \frac{7}{24} = \frac{21}{72} \\ \hline \frac{31}{72} \end{array}$$

- LCM = 72

You Try It 3

$$17\frac{5}{9} = 17\frac{20}{36} \qquad \bullet \text{ LCM} = 36$$
$$- 11\frac{5}{12} = 11\frac{15}{36}$$
$$\rule{3cm}{0.4pt}$$
$$6\frac{5}{36}$$

You Try It 4

$$8 = 7\frac{13}{13} \qquad \bullet \text{ LCM} = 13$$
$$- 2\frac{4}{13} = 2\frac{4}{13}$$
$$\rule{3cm}{0.4pt}$$
$$5\frac{9}{13}$$

You Try It 5

$$21\frac{7}{9} = 21\frac{28}{36} = 20\frac{64}{36} \bullet \text{ LCM} = 36$$
$$- 7\frac{11}{12} = 7\frac{33}{36} = 7\frac{33}{36}$$
$$\rule{5cm}{0.4pt}$$
$$13\frac{31}{36}$$

You Try It 6

Strategy
To find the time remaining before the plane lands, subtract the number of hours already in the air $\left(2\frac{3}{4}\right)$ from the total time of the trip $\left(5\frac{1}{2}\right)$.

Solution
$$5\frac{1}{2} = 5\frac{2}{4} = 4\frac{6}{4}$$
$$- 2\frac{3}{4} = 2\frac{3}{4} = 2\frac{3}{4}$$
$$\rule{5cm}{0.4pt}$$
$$2\frac{3}{4} \text{ hours}$$

The plane will land in $2\frac{3}{4}$ hours.

You Try It 7

Strategy
To find the amount of weight to be lost during the third month:

● Find the total weight loss during the first two months $\left(7\frac{1}{2} + 5\frac{3}{4}\right)$.

● Subtract the total weight loss from the goal (24 pounds).

Solution
$$7\frac{1}{2} = 7\frac{2}{4}$$
$$+ 5\frac{3}{4} = 5\frac{3}{4}$$
$$\rule{3cm}{0.4pt}$$
$$12\frac{5}{4} = 13\frac{1}{4} \text{ pounds lost}$$

$$24 = 23\frac{4}{4}$$
$$- 13\frac{1}{4} = 13\frac{1}{4}$$
$$\rule{3cm}{0.4pt}$$
$$10\frac{3}{4} \text{ pounds}$$

The patient must lose $10\frac{3}{4}$ pounds to achieve the goal.

SECTION 2.6

You Try It 1

$$\frac{4}{21} \times \frac{7}{44} = \frac{4 \cdot 7}{21 \cdot 44}$$
$$= \frac{\overset{1}{\cancel{2}} \cdot \overset{1}{\cancel{2}} \cdot \overset{1}{\cancel{7}}}{3 \cdot \underset{1}{\cancel{7}} \cdot \underset{1}{\cancel{2}} \cdot \underset{1}{\cancel{2}} \cdot 11} = \frac{1}{33}$$

You Try It 2

$$\frac{2}{21} \times \frac{10}{33} = \frac{2 \cdot 10}{21 \cdot 33}$$
$$= \frac{2 \cdot 2 \cdot 5}{3 \cdot 7 \cdot 3 \cdot 11} = \frac{20}{693}$$

You Try It 3

$$\frac{16}{5} \times \frac{15}{24} = \frac{16 \cdot 15}{5 \cdot 24}$$
$$= \frac{\overset{1}{\cancel{2}} \cdot \overset{1}{\cancel{2}} \cdot \overset{1}{\cancel{2}} \cdot 2 \cdot 3 \cdot \overset{1}{\cancel{5}}}{\underset{1}{\cancel{5}} \cdot \underset{1}{\cancel{2}} \cdot \underset{1}{\cancel{2}} \cdot \underset{1}{\cancel{2}} \cdot \underset{1}{\cancel{3}}} = \frac{2}{1} = 2$$

You Try It 4

$$5\frac{2}{5} \times \frac{5}{9} = \frac{27}{5} \times \frac{5}{9} = \frac{27 \cdot 5}{5 \cdot 9}$$
$$= \frac{\overset{1}{\cancel{3}} \cdot \overset{1}{\cancel{3}} \cdot 3 \cdot \overset{1}{\cancel{5}}}{\underset{1}{\cancel{5}} \cdot \underset{1}{\cancel{3}} \cdot \underset{1}{\cancel{3}}} = \frac{3}{1} = 3$$

You Try It 5

$$3\frac{2}{5} \times 6\frac{1}{4} = \frac{17}{5} \times \frac{25}{4} = \frac{17 \cdot 25}{5 \cdot 4}$$
$$= \frac{17 \cdot \overset{1}{\cancel{5}} \cdot 5}{\underset{1}{\cancel{5}} \cdot 2 \cdot 2} = \frac{85}{4} = 21\frac{1}{4}$$

You Try It 6

$$3\frac{2}{7} \times 6 = \frac{23}{7} \times \frac{6}{1} = \frac{23 \cdot 6}{7 \cdot 1}$$
$$= \frac{23 \cdot 2 \cdot 3}{7 \cdot 1} = \frac{138}{7} = 19\frac{5}{7}$$

You Try It 7

Strategy
To find the value of the house today, multiply the old value of the house (170,000) by $2\frac{1}{2}$.

Solution
$$170,000 \times 2\frac{1}{2} = \frac{170,000}{1} \times \frac{5}{2}$$
$$= \frac{170,000 \cdot 5}{1 \cdot 2}$$
$$= 425,000$$

The value of the house today is $425,000.

You Try It 8

Strategy To find the cost of the air compressor:

- Multiply to find the value of the drying chamber $\left(\dfrac{4}{5} \times 160{,}000\right)$.

- Subtract the value of the drying chamber from the total value of the two items (160,000).

Solution
$$\frac{4}{5} \times \frac{160{,}000}{1} = \frac{640{,}000}{5}$$
$$= 128{,}000 \quad \bullet \text{ Value of the drying chamber}$$

$$160{,}000 - 128{,}000 = 32{,}000$$

The cost of the air compressor was $32,000.

SECTION 2.7

You Try It 1 $\dfrac{3}{7} \div \dfrac{2}{3} = \dfrac{3}{7} \times \dfrac{3}{2} = \dfrac{3 \cdot 3}{7 \cdot 2} = \dfrac{9}{14}$

You Try It 2 $\dfrac{3}{4} \div \dfrac{9}{10} = \dfrac{3}{4} \times \dfrac{10}{9}$

$$= \frac{3 \cdot 10}{4 \cdot 9} = \frac{\overset{1}{\cancel{3}} \cdot \overset{1}{\cancel{2}} \cdot 5}{\underset{1}{\cancel{2}} \cdot 2 \cdot \underset{1}{\cancel{3}} \cdot 3} = \frac{5}{6}$$

You Try It 3 $\dfrac{5}{7} \div 6 = \dfrac{5}{7} \div \dfrac{6}{1}$ • $6 = \dfrac{6}{1}$. The reciprocal of $\dfrac{6}{1}$ is $\dfrac{1}{6}$.

$$= \frac{5}{7} \times \frac{1}{6} = \frac{5 \cdot 1}{7 \cdot 6}$$
$$= \frac{5}{7 \cdot 2 \cdot 3} = \frac{5}{42}$$

You Try It 4 $12\dfrac{3}{5} \div 7 = \dfrac{63}{5} \div \dfrac{7}{1} = \dfrac{63}{5} \times \dfrac{1}{7}$

$$= \frac{63 \cdot 1}{5 \cdot 7} = \frac{3 \cdot 3 \cdot \overset{1}{\cancel{7}}}{5 \cdot \underset{1}{\cancel{7}}} = \frac{9}{5} = 1\frac{4}{5}$$

You Try It 5 $3\dfrac{2}{3} \div 2\dfrac{2}{5} = \dfrac{11}{3} \div \dfrac{12}{5}$

$$= \frac{11}{3} \times \frac{5}{12} = \frac{11 \cdot 5}{3 \cdot 12}$$
$$= \frac{11 \cdot 5}{3 \cdot 2 \cdot 2 \cdot 3} = \frac{55}{36} = 1\frac{19}{36}$$

You Try it 6 $2\dfrac{5}{6} \div 8\dfrac{1}{2} = \dfrac{17}{6} \div \dfrac{17}{2}$

$$= \frac{17}{6} \times \frac{2}{17} = \frac{17 \cdot 2}{6 \cdot 17}$$
$$= \frac{\overset{1}{\cancel{17}} \cdot \overset{1}{\cancel{2}}}{\underset{1}{\cancel{2}} \cdot 3 \cdot \underset{1}{\cancel{17}}} = \frac{1}{3}$$

You Try It 7 $6\dfrac{2}{5} \div 4 = \dfrac{32}{5} \div \dfrac{4}{1}$

$$= \frac{32}{5} \times \frac{1}{4} = \frac{32 \cdot 1}{5 \cdot 4}$$
$$= \frac{2 \cdot 2 \cdot 2 \cdot \overset{1}{\cancel{2}} \cdot \overset{1}{\cancel{2}}}{5 \cdot \underset{1}{\cancel{2}} \cdot \underset{1}{\cancel{2}}} = \frac{8}{5} = 1\frac{3}{5}$$

You Try It 8

Strategy To find the number of products, divide the number of minutes in 1 hour (60) by the time to assemble one product $\left(7\dfrac{1}{2}\right)$.

Solution $60 \div 7\dfrac{1}{2} = \dfrac{60}{1} \div \dfrac{15}{2} = \dfrac{60}{1} \cdot \dfrac{2}{15}$

$$= \frac{60 \cdot 2}{1 \cdot 15} = 8$$

The factory worker can assemble 8 products in 1 hour.

You Try It 9

Strategy To find the length of the remaining piece:

- Divide the total length of the board (16) by the length of each shelf $\left(3\dfrac{1}{3}\right)$. This will give you the number of shelves cut, with a certain fraction of a shelf left over.

- Multiply the fractional part of the result in step 1 by the length of one shelf to determine the length of the remaining piece.

Solution $16 \div 3\dfrac{1}{3} = \dfrac{16}{1} \div \dfrac{10}{3}$

$$= \frac{16}{1} \times \frac{3}{10} = \frac{16 \cdot 3}{1 \cdot 10}$$
$$= \frac{\overset{1}{\cancel{2}} \cdot 2 \cdot 2 \cdot 2 \cdot 3}{\underset{1}{\cancel{2}} \cdot 5} = \frac{24}{5}$$
$$= 4\frac{4}{5}$$

There are 4 pieces that are each $3\dfrac{1}{3}$ feet long. There is 1 piece that is $\dfrac{4}{5}$ of $3\dfrac{1}{3}$ feet long.

$$\frac{4}{5} \times 3\frac{1}{3} = \frac{4}{5} \times \frac{10}{3}$$
$$= \frac{4 \cdot 10}{5 \cdot 3} = \frac{2 \cdot 2 \cdot 2 \cdot \overset{1}{\cancel{5}}}{\underset{1}{\cancel{5}} \cdot 3}$$
$$= \frac{8}{3} = 2\frac{2}{3}$$

The length of the piece remaining is $2\dfrac{2}{3}$ feet.

SECTION 2.8

You Try It 1 $\dfrac{9}{14} = \dfrac{27}{42}$ $\dfrac{13}{21} = \dfrac{26}{42}$ $\dfrac{9}{14} > \dfrac{13}{21}$

You Try It 2

$$\left(\frac{7}{11}\right)^2 \cdot \left(\frac{2}{7}\right) = \left(\frac{7}{11} \cdot \frac{7}{11}\right) \cdot \left(\frac{2}{7}\right)$$

$$= \frac{\overset{1}{\cancel{7}} \cdot 7 \cdot 2}{11 \cdot 11 \cdot \underset{1}{\cancel{7}}} = \frac{14}{121}$$

You Try It 3

$$\left(\frac{1}{13}\right)^2 \cdot \left(\frac{1}{4} + \frac{1}{6}\right) \div \frac{5}{13} =$$

$$\left(\frac{1}{13}\right)^2 \cdot \left(\frac{5}{12}\right) \div \frac{5}{13} =$$

$$\left(\frac{1}{169}\right) \cdot \left(\frac{5}{12}\right) \div \frac{5}{13} =$$

$$\left(\frac{1 \cdot 5}{13 \cdot 13 \cdot 12}\right) \div \frac{5}{13} =$$

$$\left(\frac{1 \cdot 5}{13 \cdot 13 \cdot 12}\right) \times \frac{13}{5} =$$

$$\frac{1 \cdot \overset{1}{\cancel{5}} \cdot \overset{1}{\cancel{13}}}{\underset{1}{\cancel{13}} \cdot 13 \cdot 12 \cdot \underset{1}{\cancel{5}}} = \frac{1}{156}$$

SOLUTIONS TO CHAPTER 3 "YOU TRY IT"

SECTION 3.1

You Try It 1   The digit 4 is in the thousandths place.

You Try It 2 $\dfrac{501}{1000} = 0.501$
(five hundred one thousandths)

You Try It 3 $0.67 = \dfrac{67}{100}$ (sixty-seven hundredths)

You Try It 4   Fifty-five and six thousand eighty-three ten-thousandths

You Try It 5   806.00491   • 1 is in the hundred-thousandths place.

You Try It 6

┌── Given place value
3.675849
    └── 4 < 5

3.675849 rounded to the nearest ten-thousandth is 3.6758.

You Try It 7

┌── Given place value
48.907
    └── 0 < 5

48.907 rounded to the nearest tenth is 48.9.

You Try It 8

┌── Given place value
31.8652
    └── 8 > 5

31.8652 rounded to the nearest whole number is 32.

You Try It 9   2.65 rounded to the nearest whole number is 3.

To the nearest inch, the average annual precipitation in Yuma is 3 inches.

SECTION 3.2

You Try It 1

```
  1 2
   4.62        • Place the decimal
  27.9           points on a vertical
+  0.62054       line.
─────────
  33.14054
```

You Try It 2

```
      1
   6.05
  12.
+  0.374
─────────
  18.424
```

You Try It 3

Strategy   To determine the number, add the numbers of hearing-impaired Americans ages 45 to 54, 55 to 64, 65 to 74, and 75 and over.

Solution

```
   4.48
   4.31
   5.41
+  3.80
───────
  18.00
```

18 million Americans ages 45 and older are hearing-impaired.

You Try It 4

Strategy   To find the total income, add the four commissions (985.80, 791.46, 829.75, and 635.42) to the salary (875).

Solution $875 + 985.80 + 791.46 + 829.75 + 635.42 = 4117.43$

Anita's total income was $4117.43.

SECTION 3.3

You Try It 1

```
   11 9
  6 ⁷ ⁰13
  7 2.0 3 9        Check:     1  11
 −    8.4 7                   8.47
 ──────────               + 63.569
  6 3.5 6 9               ─────────
                           72.039
```

You Try It 2

$$\begin{array}{r}{\scriptstyle 14\ 9}\\ {\scriptstyle 2\ \ 4\ 10\ 10}\\ \cancel{3}\,\cancel{5}.\cancel{0}\,\cancel{0}\\ -\ \ 9.6\,7\\ \hline 2\,5.3\,3\end{array}$$ *Check:* $\begin{array}{r}{\scriptstyle 1\ 1\ 1}\\ 9.67\\ +\ 25.33\\ \hline 35.00\end{array}$

You Try It 3

$$\begin{array}{r}{\scriptstyle 16}\\ {\scriptstyle 2\ 6\ 9\ 9\ 10}\\ \cancel{3}.\cancel{7}\,\cancel{0}\,\cancel{0}\,\cancel{0}\\ -\ 1.9\,7\,1\,5\\ \hline 1.7\,2\,8\,5\end{array}$$ *Check:* $\begin{array}{r}{\scriptstyle 1\ 1\ 1\ 1}\\ 1.9715\\ +\ 1.7285\\ \hline 3.7000\end{array}$

You Try It 4

Strategy   To find the amount of change, subtract the amount paid (6.85) from 10.00.

Solution
$$\begin{array}{r}10.00\\ -\ 6.85\\ \hline 3.15\end{array}$$

Your change was $3.15.

You Try It 5

Strategy   To find the new balance:
- Add to find the total of the three checks (1025.60 + 79.85 + 162.47).
- Subtract the total from the previous balance (2472.69).

Solution
$$\begin{array}{r}1025.60\\ 79.85\\ +\ 162.47\\ \hline 1267.92\end{array} \qquad \begin{array}{r}2472.69\\ -\ 1267.92\\ \hline 1204.77\end{array}$$

The new balance is $1204.77.

SECTION 3.4

You Try It 1
$$\begin{array}{r}870\\ \times\ \ 4.6\\ \hline 522\,0\\ 3480\\ \hline 4002.0\end{array}$$ • 1 decimal place
• 1 decimal place

You Try It 2
$$\begin{array}{r}0.000086\\ \times\ \ \ 0.057\\ \hline 602\\ 430\\ \hline 0.000004902\end{array}$$ • 6 decimal places
• 3 decimal places
• 9 decimal places

You Try It 3
$$\begin{array}{r}4.68\\ \times\ 6.03\\ \hline 1404\\ 28\,080\\ \hline 28.2204\end{array}$$ • 2 decimal places
• 2 decimal places
• 4 decimal places

You Try It 4 $6.9 \times 1000 = 6900$

You Try It 5 $4.0273 \times 10^2 = 402.73$

You Try It 6

Strategy   To find the total bill:
- Find the number of gallons of water used by multiplying the number of gallons used per day (5000) by the number of days (62).
- Find the cost of water by multiplying the cost per 1000 gallons (1.39) by the number of 1000-gallon units used.
- Add the cost of the water to the meter fee (133.70).

Solution   Number of gallons = 5000(62) = 310,000

Cost of water = $\frac{310,000}{1000} \times 1.39 = 430.90$

Total cost = 430.90 + 133.70 = 564.60

The total bill is $564.60.

You Try It 7

Strategy   To find the cost of running the freezer for 210 hours, multiply the hourly cost (0.035) by the number of hours the freezer has run (210).

Solution
$$\begin{array}{r}0.035\\ \times\ \ 210\\ \hline 350\\ 7\,0\\ \hline 7.350\end{array}$$

The cost of running the freezer for 210 hours is $7.35.

You Try It 8

Strategy   To find the total cost of the electronic drum kit:
- Multiply the monthly payment (37.18) by the number of months (18).
- Add that product to the down payment (175.00).

Solution
$$\begin{array}{r}37.18\\ \times\ \ 18\\ \hline 29744\\ 3718\\ \hline 669.24\end{array} \qquad \begin{array}{r}175.00\\ +\ 669.24\\ \hline 844.24\end{array}$$

The total cost of the electronic drum kit is $844.24.

SECTION 3.5

You Try It 1
$$\begin{array}{r}2.7\\ 0.052.\overline{)0.140.4}\\ -\underline{104}\\ 36\,4\\ -\underline{36\,4}\\ 0\end{array}$$
• Move the decimal point 3 places to the right in the divisor and the dividend. Write the decimal point in the quotient directly over the decimal point in the dividend.

You Try It 2

$$0.4873 \approx 0.487$$

$$76\overline{)37.0420}$$
$$\underline{-30\ 4}$$
$$6\ 64$$
$$\underline{-6\ 08}$$
$$562$$
$$\underline{-532}$$
$$300$$
$$\underline{-228}$$

• Write the decimal point in the quotient directly over the decimal point in the dividend.

You Try It 3

$$72.73 \approx 72.7$$

$$5.09.\overline{)370.20.00}$$
$$\underline{-356\ 3}$$
$$13\ 90$$
$$10\ 18$$
$$3\ 720$$
$$\underline{-3\ 563}$$
$$1570$$
$$\underline{-1527}$$

You Try It 4 $\quad 309.21 \div 10,000 = 0.030921$

You Try It 5 $\quad 42.93 \div 10^4 = 0.004293$

You Try It 6

Strategy To find how many times greater the average hourly earnings were, divide the 1998 average hourly earnings (12.88) by the 1978 average hourly earnings (5.70).

Solution $12.88 \div 5.70 \approx 2.3$
The average hourly earnings in 1998 were about 2.3 times greater than in 1978.

You Try It 7

Strategy To find the average number of people watching TV:
- Add the numbers of people watching each day of the week.
- Divide the total number of people watching by 7.

Solution $91.9 + 89.8 + 90.6 + 93.9 + 78.0$
$\qquad\qquad + 77.1 + 87.7 = 609$

$$\frac{609}{7} = 87$$

An average of 87 million people watch television per day.

SECTION 3.6

You Try It 1

$$0.56 \approx 0.6$$
$$16\overline{)9.00}$$

You Try It 2 $\quad 4\dfrac{1}{6} = \dfrac{25}{6}$

$$4.166 \approx 4.17$$
$$6\overline{)25.000}$$

You Try It 3

$$0.56 = \frac{56}{100} = \frac{14}{25}$$

$$5.35 = 5\frac{35}{100} = 5\frac{7}{20}$$

You Try It 4

$$0.12\frac{7}{8} = \frac{12\frac{7}{8}}{100} = 12\frac{7}{8} \div 100$$

$$= \frac{103}{8} \times \frac{1}{100} = \frac{103}{800}$$

You Try It 5

$$\frac{5}{8} = 0.625$$

$$0.630 > 0.625$$

$$0.63 > \frac{5}{8}$$

• Convert the fraction $\dfrac{5}{8}$ to a decimal.
• Compare the two decimals.
• Convert 0.625 back to a fraction.

SOLUTIONS TO CHAPTER 4 "YOU TRY IT"

SECTION 4.1

You Try It 1

$$\frac{20 \text{ pounds}}{24 \text{ pounds}} = \frac{20}{24} = \frac{5}{6}$$

20 pounds : 24 pounds = 20 : 24 = 5 : 6

20 pounds to 24 pounds = 20 to 24
$\qquad\qquad\qquad\qquad\qquad = 5 \text{ to } 6$

You Try It 2

$$\frac{64 \text{ miles}}{8 \text{ miles}} = \frac{64}{8} = \frac{8}{1}$$

64 miles : 8 miles = 64 : 8 = 8 : 1

64 miles to 8 miles = 64 to 8 = 8 to 1

You Try It 3

Strategy To find the ratio, write the ratio of board feet of cedar (12,000) to board feet of ash (18,000) in simplest form.

Solution $\dfrac{12,000}{18,000} = \dfrac{2}{3}$

The ratio is $\dfrac{2}{3}$.

You Try It 4

Strategy To find the ratio, write the ratio of the amount spent on radio advertising (450,000) to the amount spent on radio and television advertising (450,000 + 600,000) in simplest form.

Solution $\dfrac{\$450,000}{\$450,000 + \$600,000} = \dfrac{450,000}{1,050,000} = \dfrac{3}{7}$

The ratio is $\dfrac{3}{7}$.

SECTION 4.2

You Try It 1 $\dfrac{15 \text{ pounds}}{12 \text{ trees}} = \dfrac{5 \text{ pounds}}{4 \text{ trees}}$

You Try It 2 $\dfrac{260 \text{ miles}}{8 \text{ hours}}$

$8)\overline{260.0}$: 32.5

32.5 miles/hour

You Try It 3

Strategy   To find Erik's profit per ounce:

- Find the total profit by subtracting the cost ($1625) from the selling price ($1720).
- Divide the total profit by the number of ounces (5).

Solution $1720 - 1625 = 95$
$95 \div 5 = 19$

The profit was $19/ounce.

SECTION 4.3

You Try It 1 $\dfrac{6}{10} \quad \dfrac{9}{15}$ $10 \times 9 = 90$
$6 \times 15 = 90$

The cross products are equal. The proportion is true.

You Try It 2 $\dfrac{32}{6} \quad \dfrac{90}{8}$ $6 \times 90 = 540$
$32 \times 8 = 256$

The cross products are not equal. The proportion is not true.

You Try It 3 $\dfrac{n}{14} = \dfrac{3}{7}$ • Find the cross products. Then solve for n.
$n \times 7 = 14 \times 3$
$n \times 7 = 42$
$n = 42 \div 7$
$n = 6$

Check: $\dfrac{6}{14} \quad \dfrac{3}{7}$ $14 \times 3 = 42$
$6 \times 7 = 42$

You Try It 4 $\dfrac{5}{7} = \dfrac{n}{20}$ • Find the cross products. Then solve for n.
$5 \times 20 = 7 \times n$
$100 = 7 \times n$
$100 \div 7 = n$
$14.3 \approx n$

You Try It 5 $\dfrac{15}{20} = \dfrac{12}{n}$ • Find the cross products. Then solve for n.
$15 \times n = 20 \times 12$
$15 \times n = 240$
$n = 240 \div 15$
$n = 16$

Check: $\dfrac{15}{20} \quad \dfrac{12}{16}$ $20 \times 12 = 240$
$15 \times 16 = 240$

You Try It 6

$7.5 = n$
$12 \times 4 = n \times 7$
$48 = n \times 7$
$48 \div 7 = n$
$6.86 \approx n$

You Try It 7 $\dfrac{n}{12} = \dfrac{4}{1}$
$n \times 1 = 12 \times 4$
$n \times 1 = 48$
$n = 48 \div 1$
$n = 48$

Check: $\dfrac{48}{12} \quad \dfrac{4}{1}$ $12 \times 4 = 48$
$48 \times 1 = 48$

You Try It 8

Strategy   To find the number of tablespoons of fertilizer needed, write and solve a proportion using n to represent the number of tablespoons of fertilizer.

Solution

$\dfrac{3 \text{ tablespoons}}{4 \text{ gallons}} = \dfrac{n \text{ tablespoons}}{10 \text{ gallons}}$ • The unit "tablespoons" is in the numerator. The unit "gallons" is in the denominator.
$3 \times 10 = 4 \times n$
$30 = 4 \times n$
$30 \div 4 = n$
$7.5 = n$

For 10 gallons of water, 7.5 tablespoons of fertilizer are required.

You Try It 9

Strategy   To find the number of jars that can be packed in 15 boxes, write and solve a proportion using n to represent the number of jars.

Solution $\dfrac{24 \text{ jars}}{6 \text{ boxes}} = \dfrac{n \text{ jars}}{15 \text{ boxes}}$
$24 \times 15 = 6 \times n$
$360 = 6 \times n$
$360 \div 6 = n$
$60 = n$

60 jars can be packed in 15 boxes.

SOLUTIONS TO CHAPTER 5 "YOU TRY IT"

SECTION 5.1

You Try It 1   **a.** $125\% = 125 \times \dfrac{1}{100} = \dfrac{125}{100} = 1\dfrac{1}{4}$

b. $125\% = 125 \times 0.01 = 1.25$

You Try It 2

$$33\frac{1}{3}\% = 33\frac{1}{3} \times \frac{1}{100}$$

$$= \frac{100}{3} \times \frac{1}{100}$$

$$= \frac{100}{300} = \frac{1}{3}$$

You Try It 3 $0.25\% = 0.25 \times 0.01 = 0.0025$

You Try It 4

$$0.048 = 0.048 \times 100\% = 4.8\%$$

$$3.67 = 3.67 \times 100\% = 367\%$$

$$0.62\frac{1}{2} = 0.62\frac{1}{2} \times 100\%$$

$$= 62\frac{1}{2}\%$$

You Try It 5 $\dfrac{5}{6} = \dfrac{5}{6} \times 100\% = \dfrac{500}{6}\% = 83\dfrac{1}{3}\%$

You Try It 6 $1\dfrac{4}{9} = \dfrac{13}{9} = \dfrac{13}{9} \times 100\%$

$$= \frac{1300}{9}\% \approx 144.4\%$$

SECTION 5.2

You Try It 1   Percent × base = amount

$$0.063 \times 150 = n$$

$$9.45 = n$$

You Try It 2   Percent × base = amount

$$\frac{1}{6} \times 66 = n \qquad \bullet\ 16\frac{2}{3}\% = \frac{1}{6}$$

$$11 = n$$

You Try It 3

Strategy   To determine the amount that came from corporations, write and solve the basic percent equation using n to represent the amount. The percent is 4%. The base is $212 billion.

Solution   Percent × base = amount

$$4\% \times 212 = n$$

$$0.04 \times 212 = n$$

$$8.48 = n$$

Corporations gave $8.48 billion to charities.

You Try It 4

Strategy   To find the new hourly wage:
- Find the amount of the raise. Write and solve the basic percent equation using n to represent the amount of the raise (amount). The percent is 8%. The base is $33.50.
- Add the amount of the raise to the old wage (33.50).

Solution
$$8\% \times 33.50 = n \qquad\qquad 33.50$$
$$0.08 \times 33.50 = n \qquad\qquad +\ 2.68$$
$$2.68 = n \qquad\qquad\quad 36.18$$

The new hourly wage is $36.18.

SECTION 5.3

You Try It 1   Percent × base = amount

$$n \times 32 = 16$$

$$n = 16 \div 32$$

$$n = 0.50$$

$$n = 50\%$$

You Try It 2   Percent × base = amount

$$n \times 15 = 48$$

$$n = 48 \div 15$$

$$n = 3.2$$

$$n = 320\%$$

You Try It 3   Percent × base = amount

$$n \times 45 = 30$$

$$n = 30 \div 45$$

$$n = \frac{2}{3} = 66\frac{2}{3}\%$$

You Try It 4

Strategy   To find what percent of the income the income tax is, write and solve the basic percent equation using n to represent the percent. The base is $33,500, and the amount is $5025.

Solution
$$n \times 33{,}500 = 5025$$
$$n = 5025 \div 33{,}500$$
$$n = 0.15 = 15\%$$

The income tax is 15% of the income.

You Try It 5

Strategy   To find the percent who were women:
- Subtract to find the number of enlisted personnel who were women (518,921 − 512,370).
- Write and solve the basic percent equation using n to represent the percent. The base is 518,921, and the amount is the number of enlisted personnel who were women.

Solution $518{,}921 - 512{,}370 = 6551$

$$n \times 518{,}921 = 6551$$
$$n = 6551 \div 518{,}921$$
$$n \approx 0.013 = 1.3\%$$

In 1950, 1.3% of the enlisted personnel in the U.S. Army were women.

SECTION 5.4

You Try It 1

$$\text{Percent} \times \text{base} = \text{amount}$$
$$0.86 \times n = 215$$
$$n = 215 \div 0.86$$
$$n = 250$$

You Try It 2

$$\text{Percent} \times \text{base} = \text{amount}$$
$$0.025 \times n = 15$$
$$n = 15 \div 0.025$$
$$n = 600$$

You Try It 3

$$\text{Percent} \times \text{base} = \text{amount}$$
$$\frac{1}{6} \times n = 5 \qquad \cdot \ 16\frac{2}{3}\% = \frac{1}{6}$$
$$n = 5 \div \frac{1}{6}$$
$$n = 30$$

You Try It 4

Strategy   To find the original value of the car, write and solve the basic percent equation using n to represent the original value (base). The percent is 42%, and the amount is $10,458.

Solution $42\% \times n = 10,458$
$0.42 \times n = 10,458$
$n = 10,458 \div 0.42$
$n = 24,900$

The original value of the car was $24,900.

You Try It 5

Strategy   To find the difference between the original price and the sale price:

- Find the original price. Write and solve the basic percent equation using n to represent the original price (base). The percent is 80%, and the amount is $89.60.
- Subtract the sale price (89.60) from the original price.

Solution $80\% \times n = 89.60$
$0.80 \times n = 89.60$
$n = 89.60 \div 0.80$
$n = 112.00$ (original price)

$112.00 - 89.60 = 22.40$

The difference between the original price and the sale price is $22.40.

SECTION 5.5

You Try It 1

$$\frac{26}{100} = \frac{22}{n}$$
$$26 \times n = 100 \times 22$$
$$26 \times n = 2200$$
$$n = 2200 \div 26$$
$$n \approx 84.62$$

You Try It 2

$$\frac{16}{100} = \frac{n}{132}$$
$$16 \times 132 = 100 \times n$$
$$2112 = 100 \times n$$
$$2112 \div 100 = n$$
$$21.12 = n$$

You Try It 3

Strategy   To find the number of days it snowed, write and solve a proportion using n to represent the number of days it snowed (amount). The percent is 64%, and the base is 150.

Solution

$$\frac{64}{100} = \frac{n}{150}$$
$$64 \times 150 = 100 \times n$$
$$9600 = 100 \times n$$
$$9600 \div 100 = n$$
$$96 = n$$

It snowed 96 days.

You Try It 4

Strategy   To find the percent of alarms that were not false alarms:

- Subtract to find the number of alarms that were not false alarms ($200 - 24$).
- Write and solve a proportion using n to represent the percent of alarms that were not false. The base is 200, and the amount is the number of alarms that were not false.

Solution $200 - 24 = 176$

$$\frac{n}{100} = \frac{176}{200}$$
$$n \times 200 = 100 \times 176$$
$$n \times 200 = 17,600$$
$$n = 17,600 \div 200$$
$$n = 88$$

88% of the alarms were not false alarms.

SOLUTIONS TO CHAPTER 6 "YOU TRY IT"

SECTION 6.1

You Try It 1

Strategy   To find the unit cost, divide the total cost by the number of units.

Solution   **a.** $7.67 \div 8 = 0.95875$
$.959 per battery
b. $2.29 \div 15 \approx 0.153$
$.153 per ounce

You Try It 2

Strategy   To find the more economical purchase, compare the unit costs.

Solution $8.70 \div 6 = 1.45$
$6.96 \div 4 = 1.74$
$\$1.45 < \$1.74$

The more economical purchase is 6 cans for $8.70.

You Try It 3

Strategy   To find the total cost, multiply the unit cost (9.96) by the number of units (7).

Solution

Unit cost	×	number of units	=	total cost
9.96	×	7	=	69.72

The total cost is $69.72.

SECTION 6.2

You Try It 1

Strategy   To find the percent increase:

- Find the amount of the increase.
- Solve the basic percent equation for *percent*.

Solution

New value	−	original value	=	amount of increase
3.83	−	3.46	=	0.37

Percent × base = amount
$$n \times 3.46 = 0.37$$
$$n = 0.37 \div 3.46$$
$$n \approx 0.11 = 11\%$$

The percent increase was 11%.

You Try It 2

Strategy   To find the new hourly wage:

- Solve the basic percent equation for *amount*.
- Add the amount of the increase to the original wage.

Solution   Percent × base = amount
$$0.14 \times 12.50 = n$$
$$1.75 = n$$

$$12.50 + 1.75 = 14.25$$

The new hourly wage is $14.25.

You Try It 3

Strategy   To find the markup, solve the basic percent equation for *amount*.

Solution

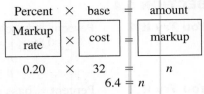

$$0.20 \times 32 = n$$
$$6.4 = n$$

The markup is $6.40.

You Try It 4

Strategy   To find the selling price:

- Find the markup by solving the basic percent equation for *amount*.
- Add the markup to the cost.

Solution

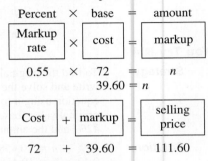

$$0.55 \times 72 = n$$
$$39.60 = n$$

Cost	+	markup	=	selling price
72	+	39.60	=	111.60

The selling price is $111.60.

You Try It 5

Strategy   To find the percent decrease:

- Find the amount of the decrease.
- Solve the basic percent equation for *percent*.

Solution

Original value	−	new value	=	amount of decrease
261,000	−	215,000	=	46,000

Percent × base = amount
$$n \times 261,000 = 46,000$$
$$n = 46,000 \div 261,000$$
$$n \approx 0.176$$

The percent decrease is 17.6%.

You Try It 6

Strategy   To find the visibility:

- Find the amount of decrease by solving the basic percent equation for *amount*.
- Subtract the amount of decrease from the original visibility.

Solution   Percent × base = amount
$$0.40 \times 5 = n$$
$$2 = n$$

$$5 - 2 = 3$$

The visibility was 3 miles.

You Try It 7

Strategy   To find the discount rate:

- Find the discount.
- Solve the basic percent equation for *percent*.

Solution

Regular price	−	sale price	=	discount

| 12.50 | − | 10.99 | = | 1.51 |

Percent	×	base	=	amount

| Discount rate | × | regular price | = | discount |

$$n \times 12.50 = 1.51$$
$$n = 1.51 \div 12.50$$
$$n = 0.1208$$

The discount rate is 12.1%.

You Try It 8

Strategy   To find the sale price:

- Find the discount by solving the basic percent equation for *amount*.
- Subtract to find the sale price.

Solution

Percent	×	base	=	amount

| Discount rate | × | regular price | = | discount |

$$0.15 \times 225 = n$$
$$33.75 = n$$

Regular price	−	discount	=	sale price

| 225 | − | 33.75 | = | 191.25 |

The sale price is $191.25.

SECTION 6.3

You Try It 1

Strategy   To find the simple interest due, multiply the principal (15,000) times the annual interest rate (8% = 0.08) times the time in years
$\left(18 \text{ months} = \dfrac{18}{12} \text{ years} = 1.5 \text{ years}\right)$.

Solution

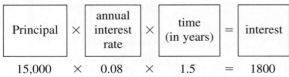

Principal	×	annual interest rate	×	time (in years)	=	interest

| 15,000 | × | 0.08 | × | 1.5 | = | 1800 |

The interest due is $1800.

You Try It 2

Strategy   To find the maturity value:

- Use the simple interest formula to find the simple interest due.
- Find the maturity value by adding the principal and the interest.

Solution

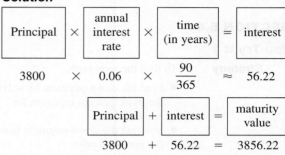

Principal	×	annual interest rate	×	time (in years)	=	interest

| 3800 | × | 0.06 | × | $\dfrac{90}{365}$ | ≈ | 56.22 |

Principal	+	interest	=	maturity value

| 3800 | + | 56.22 | = | 3856.22 |

The maturity value is $3856.22.

You Try It 3

Strategy   To find the monthly payment:

- Find the maturity value by adding the principal and the interest.
- Divide the maturity value by the length of the loan in months (12).

Solution

Principal + interest = maturity value

| 1900 | + | 152 | = | 2052 |

Maturity value ÷ length of the loan = payment

| 2052 | ÷ | 12 | = | 171 |

The monthly payment is $171.

You Try It 4

Strategy   To find the finance charge, multiply the principal, or unpaid balance (1250), times the monthly interest rate (1.6%) times the number of months (1).

Solution

Principal	×	monthly interest rate	×	time (in months)

| 1250 | × | 0.016 | × | 1 = 20 |

The finance charge is $20.

You Try It 5

Strategy   To find the interest earned:

- Find the new principal by multiplying the original principal (1000) by the factor found in the Compound Interest Table (3.29066).
- Subtract the original principal from the new principal.

(Continued)

(Continued)

Solution $1000 \times 3.29066 = 3290.66$

The new principal is $3290.66.

$3290.66 - 1000 = 2290.66$

The interest earned is $2290.66.

SECTION 6.4
You Try It 1

Strategy To find the mortgage:

- Find the down payment by solving the basic percent equation for *amount*.
- Subtract the down payment from the purchase price.

Solution

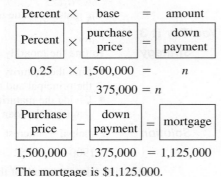

Percent $\times$ base $=$ amount

Percent	$\times$	purchase price	$=$	down payment

$$0.25 \times 1{,}500{,}000 = n$$
$$375{,}000 = n$$

Purchase price	$-$	down payment	$=$	mortgage

$$1{,}500{,}000 - 375{,}000 = 1{,}125{,}000$$

The mortgage is $1,125,000.

You Try It 2

Strategy To find the loan origination fee, solve the basic percent equation for *amount*.

Solution Percent $\times$ base $=$ amount

Points	$\times$	mortgage	$=$	fee

$$0.045 \times 180{,}000 = n$$
$$8100 = n$$

The loan origination fee was $8100.

You Try It 3

Strategy To find the monthly mortgage payment:

- Subtract the down payment from the purchase price to find the mortgage.
- Multiply the mortgage by the factor found in the Monthly Payment Table.

Solution

Purchase price	$-$	down payment	$=$	mortgage

$$175{,}000 - 17{,}500 = 157{,}500$$
$$157{,}500 \times 0.0089973 \approx 1417.08$$
↑
From the table

The monthly mortgage payment is $1417.08.

You Try It 4

Strategy To find the interest:

- Multiply the mortgage by the factor found in the Monthly Payment Table to find the monthly mortgage payment.
- Subtract the principal from the monthly mortgage payment.

Solution $625{,}000 \times 0.0070678 \approx 4417.38$

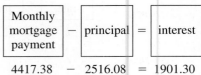

From the   Monthly mortgage
table   payment

Monthly mortgage payment	$-$	principal	$=$	interest

$$4417.38 - 2516.08 = 1901.30$$

The interest on the mortgage is $1901.30.

You Try It 5

Strategy To find the monthly payment:

- Divide the annual property tax by 12 to find the monthly property tax.
- Add the monthly property tax to the monthly mortgage payment.

Solution $3000 \div 12 = 250$

The monthly property tax is $250.

$$815.20 + 250 = 1065.20$$

The total monthly payment is $1065.20.

SECTION 6.5
You Try It 1

Strategy To find the amount financed:

- Find the down payment by solving the basic percent equation for *amount*.
- Subtract the down payment from the purchase price.

Solution Percent $\times$ base $=$ amount

Percent	$\times$	purchase price	$=$	down payment

$$0.20 \times 19{,}200 = n$$
$$3840 = n$$

The down payment is $3840.

$$19{,}200 - 3840 = 15{,}360$$

The amount financed is $15,360.

You Try It 2

Strategy To find the license fee, solve the basic percent equation for *amount*.

Solution

| Percent | × | base | = | amount |

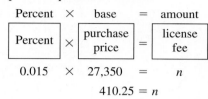

$$0.015 \times 27{,}350 = n$$
$$410.25 = n$$

The license fee is $410.25.

You Try It 3

Strategy To find the cost, multiply the cost per mile (0.41) by the number of miles driven (23,000).

Solution $23{,}000 \times 0.41 = 9430$

The cost is $9430.

You Try It 4

Strategy To find the cost per mile for car insurance, divide the cost for insurance (360) by the number of miles driven (15,000).

Solution $360 \div 15{,}000 = 0.024$

The cost per mile for insurance is $.024.

You Try It 5

Strategy To find the monthly payment:
- Subtract the down payment from the purchase price to find the amount financed.
- Multiply the amount financed by the factor found in the Monthly Payment Table.

Solution $25{,}900 - 6475 = 19{,}425$

$19{,}425 \times 0.0244129 \approx 474.22$

The monthly payment is $474.22.

SECTION 6.6

You Try It 1

Strategy To find the worker's earnings:
- Find the worker's overtime wage by multiplying the hourly wage by 2.
- Multiply the number of overtime hours worked by the overtime wage.

Solution $28.50 \times 2 = 57$

The hourly wage for overtime is $57.

$57 \times 8 = 456$

The construction worker earns $456.

You Try It 2

Strategy To find the salary per month, divide the annual salary by the number of months in a year (12).

Solution $70{,}980 \div 12 = 5915$

The contractor's monthly salary is $5915.

You Try It 3

Strategy To find the total earnings:
- Find the sales over $50,000.
- Multiply the commission rate by sales over $50,000.
- Add the commission to the annual salary.

Solution $175{,}000 - 50{,}000 = 125{,}000$

Sales over $50,000 totaled $125,000.

$125{,}000 \times 0.095 = 11{,}875$

Earnings from commissions totaled $11,875.

$37{,}000 + 11{,}875 = 48{,}875$

The insurance agent earned $48,875.

SECTION 6.7

You Try It 1

Strategy To find the current balance:
- Subtract the amount of the check from the old balance.
- Add the amount of each deposit.

Solution

$$
\begin{array}{rl}
302.46 & \\
-\;\;20.59 & \text{check} \\
\hline
281.87 & \\
176.86 & \text{first deposit} \\
+\;\;94.73 & \text{second deposit} \\
\hline
553.46 &
\end{array}
$$

The current checking account balance is $553.46.

You Try It 2

$$
\begin{array}{llr}
\text{Current checkbook} & & \\
\text{balance:} & & 623.41 \\
\text{Check: 237} & + & 78.73 \\
\hline
& & 702.14 \\
\text{Interest:} & + & 2.11 \\
\hline
& & 704.25 \\
\text{Deposit:} & & -523.84 \\
\hline
& & 180.41
\end{array}
$$

Closing bank balance from bank statement: $180.41

Checkbook balance: $180.41

The bank statement and the checkbook balance.

Answers to Selected Exercises

ANSWERS TO CHAPTER 1 SELECTED EXERCISES

PREP TEST

1. 8 **2.** 1 2 3 4 5 6 7 8 9 10 **3.** a and D; b and E; c and A; d and B; e and F; f and C

SECTION 1.1

1.
```
+--+--+--+--•--+--+--+--+--+--+--+--+-->
0  1  2  3  4  5  6  7  8  9 10 11 12
```
3.
```
+--+--+--+--+--+--+--+--+--•--+--+--+-->
0  1  2  3  4  5  6  7  8  9 10 11 12
```
5. $37 < 49$ **7.** $101 > 87$
9. $2701 > 2071$ **11.** $107 > 0$ **13.** Yes **15.** Millions **17.** Hundred-thousands **19.** Three thousand seven hundred ninety **21.** Fifty-eight thousand four hundred seventy-three **23.** Four hundred ninety-eight thousand five hundred twelve **25.** Six million eight hundred forty-two thousand seven hundred fifteen **27.** 357 **29.** 63,780 **31.** 7,024,709 **33.** $5000 + 200 + 80 + 7$ **35.** $50,000 + 8000 + 900 + 40 + 3$ **37.** $200,000 + 500 + 80 + 3$ **39.** $400,000 + 3000 + 700 + 5$ **41.** No **43.** 850 **45.** 4000 **47.** 53,000 **49.** 630,000 **51.** 250,000 **53.** 72,000,000 **55.** No. Round 3846 to the nearest hundred.

SECTION 1.2

1. 28 **3.** 125 **5.** 102 **7.** 154 **9.** 1489 **11.** 828 **13.** 1584 **15.** 1219 **17.** 102,317 **19.** 79,326 **21.** 1804 **23.** 1579 **25.** 19,740 **27.** 7420 **29.** 120,570 **31.** 207,453 **33.** 24,218 **35.** 11,974 **37.** 9323 **39.** 77,139 **41.** 14,383 **43.** 9473 **45.** 33,247 **47.** 5058 **49.** 1992 **51.** 68,263 **53.** Cal.: 17,754 Est.: 17,700 **55.** Cal.: 2872 Est.: 2900 **57.** Cal.: 101,712 Est.: 101,000 **59.** Cal.: 158,763 Est.: 158,000 **61.** Cal.: 261,595 Est.: 260,000 **63.** Cal.: 946,718 Est.: 940,000 **65.** Commutative Property of Addition **67.** There were 118,295 multiple births during the year. **69.** The estimated income from the four *Star Wars* movies was $1,500,000,000. **71. a.** The income from the two movies with the lowest box-office returns is $599,300,000. **b.** Yes, this income exceeds the income from the 1977 *Star Wars* production. **73. a.** During the three days, 1285 miles will be driven. **b.** At the end of the trip, the odometer will read 69,977 miles. **75.** The total length of the trail is 740 miles. **77.** 11 different sums **79.** No. $0 + 0 = 0$ **81.** 10 numbers

SECTION 1.3

1. 4 **3.** 4 **5.** 10 **7.** 4 **9.** 9 **11.** 22 **13.** 60 **15.** 66 **17.** 31 **19.** 901 **21.** 791 **23.** 1125 **25.** 3131 **27.** 47 **29.** 925 **31.** 4561 **33.** 3205 **35.** 1222 **37.** 53 **39.** 29 **41.** 8 **43.** 37 **45.** 58 **47.** 574 **49.** 337 **51.** 1423 **53.** 754 **55.** 2179 **57.** 6489 **59.** 889 **61.** 71,129 **63.** 698 **65.** 29,405 **67.** 49,624 **69.** 628 **71.** 6532 **73.** 4286 **75.** 4042 **77.** 5209 **79.** 10,378 **81.** (ii) and (iii) **83.** 11,239 **85.** 8482 **87.** 625 **89.** 76,725 **91.** 23 **93.** 4648 **95.** Cal.: 29,837 Est.: 30,000 **97.** Cal.: 36,668 Est.: 40,000 **99.** Cal.: 101,998 Est.: 100,000

101. a. The honey bee has 91 more smell genes than the mosquito. **b.** The mosquito has eight more taste genes than the fruit fly. **c.** The honey bee has the best sense of smell. **d.** The honey bee has the worst sense of taste. **103.** The difference between the maximum eruption heights is 15 feet. **105.** 202,345 more women than men earned a bachelor's degree. **107. a.** The smallest expected increase occurs from 2010 to 2012. **b.** The greatest expected increase occurs from 2018 to 2020. **109.** Your new credit card balance is $360.

SECTION 1.4

1. 6×2 or $6 \cdot 2$ **3.** 4×7 or $4 \cdot 7$ **5.** 12   **7.** 35   **9.** 25   **11.** 0   **13.** 72   **15.** 198   **17.** 335   **19.** 2492   **21.** 5463   **23.** 4200   **25.** 6327   **27.** 1896   **29.** 5056   **31.** 1685   **33.** 46,963   **35.** 59,976   **37.** 19,120   **39.** 19,790   **41.** 140   **43.** 22,456   **45.** 18,630   **47.** 336   **49.** 910   **51.** 63,063   **53.** 33,520   **55.** 380,834   **57.** 541,164   **59.** 400,995   **61.** 105,315   **63.** 428,770   **65.** 260,000   **67.** 344,463   **69.** 41,808   **71.** 189,500   **73.** 401,880   **75.** 1,052,763   **77.** 4,198,388   **79.** For example, 5 and 20   **81.** 198,423   **83.** 18,834   **85.** 260,178   **87.** Cal.: 440,076   **89.** Cal.: 6,491,166   **91.** Cal.: 18,728,744   **93.** Cal.: 57,691,192
Est.: 450,000   Est.: 6,300,000   Est.: 18,000,000   Est.: 54,000,000
95. The car could travel 516 miles on 12 gallons of gas.   **97.** The perimeter is 64 miles.   **99. a.** eHarmony can take credit for 630 marriages a week.   **b.** eHarmony can take credit for 32,850 marriages a year.   **101.** The estimated cost of the electricians' labor is $5100.   **103.** The total cost is $2138.   **105.** There are 12 accidental deaths each hour; 288 deaths each day; and 105,120 deaths each year.

SECTION 1.5

1. 2   **3.** 6   **5.** 7   **7.** 16   **9.** 210   **11.** 44   **13.** 703   **15.** 910   **17.** 21,560   **19.** 3580   **21.** 1075   **23.** 1   **25.** 47   **27.** 23   **29.** 3 r1   **31.** 9 r7   **33.** 16 r1   **35.** 10 r4   **37.** 90 r3   **39.** 120 r5   **41.** 309 r3   **43.** 1160 r4   **45.** 708 r2   **47.** 3825 r1   **49.** 9044 r2   **51.** 11,430   **53.** 510   **55.** False   **57.** 1 r38   **59.** 1 r26   **61.** 21 r21   **63.** 30 r22   **65.** 5 r40   **67.** 9 r17   **69.** 200 r21   **71.** 303 r1   **73.** 67 r13   **75.** 176 r13   **77.** 1086 r7   **79.** 403   **81.** 12 r456   **83.** 4 r160   **85.** 160 r27   **87.** 1669 r14   **89.** 7950   **91.** Cal.: 5129   **93.** Cal.: 21,968   **95.** Cal.: 24,596   **97.** Cal.: 2836   **99.** Cal.: 3024   **101.** Cal.: 32,036
Est.: 5000   Est.: 20,000   Est.: 22,500   Est.: 3000   Est.: 3000   Est.: 30,000
103. The average monthly claim for theft is $25,000.   **105.** The average number of hours worked by employees in Britain is 35 hours.   **107.** 380 pennies are in circulation for each person.   **109.** On average, each household will receive 175 pieces of mail.   **111.** (i) and (iii)   **113.** The total of the deductions is $350.   **115.** 49,500,000 more cases of eggs were sold by retail stores.   **117.** The average monthly expense for housing is $976.   **119.** A major's annual pay is $75,024.   **121.** The total amount paid is $11,860.

SECTION 1.6

1. 2^3 **3.** $6^3 \cdot 7^4$ **5.** $2^3 \cdot 3^3$ **7.** $5 \cdot 7^5$ **9.** $3^3 \cdot 6^4$ **11.** $3^3 \cdot 5 \cdot 9^3$ **13.** 8   **15.** 400   **17.** 900   **19.** 972   **21.** 120   **23.** 360   **25.** 0   **27.** 90,000   **29.** 540   **31.** 4050   **33.** 11,025   **35.** 25,920   **37.** 4,320,000   **39.** 5   **41.** 4   **43.** 23   **45.** 5   **47.** 10   **49.** 7   **51.** 8   **53.** 6   **55.** 52   **57.** 26   **59.** 52   **61.** 42   **63.** 8   **65.** 16   **67.** 6   **69.** 8   **71.** 3   **73.** 4   **75.** 13   **77.** 0   **79.** $8 - 2 \cdot (3 + 1)$

SECTION 1.7

1. 1, 2, 4   **3.** 1, 2, 5, 10   **5.** 1, 7   **7.** 1, 3, 9   **9.** 1, 13   **11.** 1, 2, 3, 6, 9, 18   **13.** 1, 2, 4, 7, 8, 14, 28, 56   **15.** 1, 3, 5, 9, 15, 45   **17.** 1, 29   **19.** 1, 2, 11, 22   **21.** 1, 2, 4, 13, 26, 52   **23.** 1, 2, 41, 82   **25.** 1, 3, 19, 57   **27.** 1, 2, 3, 4, 6, 8, 12, 16, 24, 48   **29.** 1, 5, 19, 95   **31.** 1, 2, 3, 6, 9, 18, 27, 54   **33.** 1, 2, 3, 6, 11, 22, 33, 66   **35.** 1, 2, 4, 5, 8, 10, 16, 20, 40, 80   **37.** 1, 2, 3, 4, 6, 8, 12, 16, 24, 32, 48, 96   **39.** 1, 2, 3, 5, 6, 9, 10, 15, 18, 30, 45, 90   **41.** True   **43.** $2 \cdot 3$ **45.** Prime   **47.** $2 \cdot 2 \cdot 2 \cdot 3$ **49.** $3 \cdot 3 \cdot 3$ **51.** $2 \cdot 2 \cdot 3 \cdot 3$ **53.** Prime   **55.** $2 \cdot 3 \cdot 3 \cdot 5$ **57.** $5 \cdot 23$ **59.** $2 \cdot 3 \cdot 3$ **61.** $2 \cdot 2 \cdot 7$ **63.** Prime   **65.** $2 \cdot 31$ **67.** $2 \cdot 11$ **69.** Prime   **71.** $2 \cdot 3 \cdot 11$ **73.** $2 \cdot 37$ **75.** Prime   **77.** $5 \cdot 11$ **79.** $2 \cdot 2 \cdot 2 \cdot 3 \cdot 5$ **81.** $2 \cdot 2 \cdot 2 \cdot 2 \cdot 2 \cdot 5$ **83.** $2 \cdot 2 \cdot 2 \cdot 3 \cdot 3 \cdot 3$ **85.** $5 \cdot 5 \cdot 5 \cdot 5$ **87.** False

CHAPTER 1 CONCEPT REVIEW*

1. The symbol $<$ means "is less than." A number that appears to the left of a given number on the number line is less than ($<$) the given number. For example, $4 < 9$. The symbol $>$ means "is greater than." A number that appears to the right of a given number on the number line is greater than ($>$) the given number. For example, $5 > 2$.   [1.1A]

*Note: The numbers in brackets following the answers in the Concept Review are a reference to the objective that corresponds to that problem. For example, the reference [1.2A] stands for Section 1.2, Objective A. This notation will be used for all Prep Tests, Concept Reviews, Chapter Reviews, Chapter Tests, and Cumulative Reviews throughout the text.

2. To round a four-digit whole number to the nearest hundred, look at the digit in the tens place. If the digit in the tens place is less than 5, that digit and the digit in the ones place are replaced by zeros. If the digit in the tens place is greater than or equal to 5, increase the digit in the hundreds place by 1 and replace the digits in the tens place and the ones place by zeros.   [1.1D]

3. The Commutative Property of Addition states that two numbers can be added in either order; the sum is the same. For example, $3 + 5 = 5 + 3$. The Associative Property of Addition states that changing the grouping of three or more addends does not change their sum. For example, $3 + (4 + 5) = (3 + 4) + 5$. Note that in the Commutative Property of Addition, the order in which the numbers appear changes, while in the Associative Property of Addition, the order in which the numbers appear does not change.   [1.2A]

4. To estimate the sum of two numbers, round each number to the same place value. Then add the numbers. For example, to estimate the sum of 562,397 and 41,086, round the numbers to 560,000 and 40,000. Then add $560,000 + 40,000 = 600,000$.   [1.2A]

5. It is necessary to borrow when performing subtraction if, in any place value, the lower digit is larger than the upper digit.   [1.3B]

6. The Multiplication Property of Zero states that the product of a number and zero is zero. For example, $8 \times 0 = 0$. The Multiplication Property of One states that the product of a number and one is the number. For example, $8 \times 1 = 8$.   [1.4A]

7. To multiply a whole number by 100, write two zeros to the right of the number. For example, $64 \times 100 = 6400$.   [1.4B]

8. To estimate the product of two numbers, round each number so that it contains only one nonzero digit. Then multiply. For example, to estimate the product of 87 and 43, round the two numbers to 90 and 40; then multiply $90 \times 40 = 3600$.   [1.4B]

9. $0 \div 9 = 0$. Zero divided by any whole number except zero is zero. $9 \div 0$ is undefined. Division by zero is not allowed.   [1.5A]

10. To check the answer to a division problem that has a remainder, multiply the quotient by the divisor. Add the remainder to the product. The result should be the dividend. For example, $16 \div 5 = 3$ r1. Check: $(3 \times 5) + 1 = 16$, the dividend.   [1.5B]

11. The steps in the Order of Operations Agreement are:
1. Do all operations inside parentheses.
2. Simplify any number expressions containing exponents.
3. Do multiplication and division as they occur from left to right.
4. Do addition and subtraction as they occur from left to right.   [1.6B]

12. A number is a factor of another number if it divides that number evenly (there is no remainder). For example, 7 is a factor of 21 because $21 \div 7 = 3$, with a remainder of 0.   [1.7A]

13. Three is a factor of a number if the sum of the digits of the number is divisible by 3. For the number 285, $2 + 8 + 5 = 15$, which is divisible by 3. Thus 285 is divisible by 3.   [1.7A]

CHAPTER 1 REVIEW EXERCISES

1. 600   [1.6A]   **2.** $10,000 + 300 + 20 + 7$ [1.1C]   **3.** 1, 2, 3, 6, 9, 18   [1.7A]   **4.** 12,493   [1.2A]   **5.** 1749   [1.3B]
6. 2135   [1.5A]   **7.** $101 > 87$ [1.1A]   **8.** $5^2 \cdot 7^5$ [1.6A]   **9.** 619,833   [1.4B]   **10.** 5409   [1.3B]   **11.** 1081   [1.2A]
12. 2   [1.6B]   **13.** 45,700   [1.1D]   **14.** Two hundred seventy-six thousand fifty-seven   [1.1B]   **15.** 1306 r59   [1.5C]
16. 2,011,044   [1.1B]   **17.** 488 r2   [1.5B]   **18.** 17   [1.6B]   **19.** 32   [1.6B]   **20.** $2 \cdot 2 \cdot 2 \cdot 3 \cdot 3$ [1.7B]
21. 2133   [1.3A]   **22.** 22,761   [1.4B]   **23.** The total pay for last week's work is $768.   [1.4C]   **24.** He drove 27 miles per gallon of gasoline.   [1.5D]   **25.** Each monthly car payment is $560.   [1.5D]   **26.** The total income from commissions is $2567.   [1.2B]   **27.** The total amount deposited is $301. The new checking account balance is $817.   [1.2B]   **28.** The total of the car payments is $2952.   [1.4C]   **29.** More males were involved in college sports in 2005 than in 1972.   [1.1A]
30. The difference between the numbers of male and female athletes in 1972 was 140,407 students.   [1.3C]   **31.** The number of female athletes increased by 175,515 students from 1972 to 2005.   [1.3C]   **32.** 296,928 more students were involved in athletics in 2005 than in 1972.   [1.3C]

CHAPTER 1 TEST

1. 432   [1.6A, Example 3]   **2.** Two hundred seven thousand sixty-eight   [1.1B, Example 3]   **3.** 9333   [1.3B, Example 3]
4. 1, 2, 4, 5, 10, 20   [1.7A, Example 1]   **5.** 6,854,144   [1.4B, HOW TO 3]   **6.** 9   [1.6B, Example 4]
7. $900,000 + 6000 + 300 + 70 + 8$ [1.1C, Example 6]   **8.** 75,000   [1.1D, Example 8]   **9.** 1121 r27   [1.5C, Example 8]
10. $3^3 \cdot 7^2$ [1.6A, Example 1]   **11.** 54,915   [1.2A, Example 1]   **12.** $2 \cdot 2 \cdot 3 \cdot 7$ [1.7B, Example 2]   **13.** 4
[1.6B, Example 4]   **14.** 726,104   [1.4A, Example 1]   **15.** 1,204,006   [1.1B, Example 4]   **16.** 8710 r2   [1.5B, Example 5]
17. $21 > 19$ [1.1A, Example 2]   **18.** 703   [1.5A, Example 3]   **19.** 96,798   [1.2A, Example 3]   **20.** 19,922
[1.3B, Example 4]   **21.** The difference in projected total enrollment between 2016 and 2013 is 1,908,000 students.   [1.3C, Example 6]
22. The average projected enrollment in each of the grades 9 through 12 in 2016 is 4,171,000 students.   [1.5D, HOW TO 3]
23. 3000 boxes were needed to pack the lemons.   [1.5D, Example 10]   **24.** The investor receives $2844 over the 12-month period.
[1.4C, Example 3]   **25. a.** 855 miles were driven during the 3 days.   **b.** The odometer reading at the end of the 3 days is 48,481 miles.   [1.2B, Example 4]

ANSWERS TO CHAPTER 2 SELECTED EXERCISES

PREP TEST

1. 20 [1.4A]   **2.** 120 [1.4A]   **3.** 9 [1.4A]   **4.** 10 [1.2A]   **5.** 7 [1.3A]   **6.** 2 r3 [1.5C]

7. 1, 2, 3, 4, 6, 12 [1.7A]   **8.** 59 [1.6B]   **9.** 7 [1.3A]   **10.** $44 < 48$ [1.1A]

SECTION 2.1

1. 40   **3.** 24   **5.** 30   **7.** 12   **9.** 24   **11.** 60   **13.** 56   **15.** 9   **17.** 32   **19.** 36   **21.** 660   **23.** 9384

25. 24   **27.** 30   **29.** 24   **31.** 576   **33.** 1680   **35.** True   **37.** 1   **39.** 3   **41.** 5   **43.** 25   **45.** 1   **47.** 4

49. 4   **51.** 6   **53.** 4   **55.** 1   **57.** 7   **59.** 5   **61.** 8   **63.** 1   **65.** 25   **67.** 7   **69.** 8   **71.** True

73. They will have another day off together in 12 days.

SECTION 2.2

1. Improper fraction   **3.** Proper fraction   **5.** $\frac{3}{4}$ **7.** $\frac{7}{8}$ **9.** $1\frac{1}{2}$ **11.** $2\frac{5}{8}$ **13.** $3\frac{3}{5}$ **15.** $\frac{5}{4}$ **17.** $\frac{8}{3}$ **19.** $\frac{28}{8}$

21.   **23.**   **25.** False   **27.** $5\frac{1}{3}$ **29.** 2   **31.** $3\frac{1}{4}$ **33.** $14\frac{1}{2}$ **35.** 17   **37.** $1\frac{7}{9}$

39. $1\frac{4}{5}$ **41.** 23   **43.** $1\frac{15}{16}$ **45.** $6\frac{1}{3}$ **47.** 5   **49.** 1   **51.** $\frac{14}{3}$ **53.** $\frac{26}{3}$ **55.** $\frac{59}{8}$ **57.** $\frac{25}{4}$ **59.** $\frac{121}{8}$ **61.** $\frac{41}{12}$

63. $\frac{34}{9}$ **65.** $\frac{38}{3}$ **67.** $\frac{38}{7}$ **69.** $\frac{63}{5}$ **71.** $\frac{41}{9}$ **73.** $\frac{117}{14}$

SECTION 2.3

1. $\frac{5}{10}$ **3.** $\frac{9}{48}$ **5.** $\frac{12}{32}$ **7.** $\frac{9}{51}$ **9.** $\frac{12}{16}$ **11.** $\frac{27}{9}$ **13.** $\frac{20}{60}$ **15.** $\frac{44}{60}$ **17.** $\frac{12}{18}$ **19.** $\frac{35}{49}$ **21.** $\frac{10}{18}$ **23.** $\frac{21}{3}$

25. $\frac{35}{45}$ **27.** $\frac{60}{64}$ **29.** $\frac{21}{98}$ **31.** $\frac{30}{48}$ **33.** $\frac{15}{42}$ **35.** $\frac{102}{144}$ **37.** $\frac{1}{3}$ **39.** $\frac{1}{2}$ **41.** $\frac{1}{6}$ **43.** $1\frac{1}{9}$ **45.** 0   **47.** $\frac{9}{22}$

49. 3   **51.** $\frac{4}{21}$ **53.** $\frac{12}{35}$ **55.** $\frac{7}{11}$ **57.** $1\frac{1}{3}$ **59.** $\frac{3}{5}$ **61.** $\frac{1}{11}$ **63.** 4   **65.** $\frac{1}{3}$ **67.** $\frac{3}{5}$ **69.** $2\frac{1}{4}$ **71.** $\frac{1}{5}$

73. Answers will vary. For example, $\frac{4}{6}, \frac{6}{9}, \frac{8}{12}, \frac{10}{15}, \frac{12}{8}$.   **75. a.** $\frac{4}{25}$ **b.** $\frac{4}{25}$

SECTION 2.4

1. $\frac{3}{7}$ **3.** 1   **5.** $1\frac{4}{11}$ **7.** $3\frac{2}{5}$ **9.** $2\frac{4}{5}$ **11.** $2\frac{1}{4}$ **13.** $1\frac{3}{8}$ **15.** $1\frac{7}{15}$ **17.** $1\frac{5}{12}$ **19.** A whole number other than 1

21. The number 1   **23.** $1\frac{1}{6}$ **25.** $\frac{13}{14}$ **27.** $\frac{53}{60}$ **29.** $1\frac{1}{56}$ **31.** $\frac{23}{60}$ **33.** $1\frac{17}{18}$ **35.** $1\frac{11}{48}$ **37.** $1\frac{9}{20}$ **39.** $2\frac{17}{120}$

41. $2\frac{5}{72}$ **43.** $\frac{39}{40}$ **45.** $1\frac{19}{24}$ **47.** (ii)   **49.** $10\frac{1}{12}$ **51.** $9\frac{2}{7}$ **53.** $9\frac{47}{48}$ **55.** $8\frac{3}{13}$ **57.** $16\frac{29}{120}$ **59.** $24\frac{29}{40}$

61. $33\frac{7}{24}$ **63.** $10\frac{5}{36}$ **65.** $10\frac{5}{12}$ **67.** $14\frac{73}{90}$ **69.** $10\frac{13}{48}$ **71.** $9\frac{5}{24}$ **73.** $14\frac{1}{18}$ **75.** $11\frac{11}{12}$ **77.** No

79. The length of the shaft is $8\frac{9}{16}$ inches.   **81.** The sum represents the height of the table.   **83.** The total length of the course

is $10\frac{1}{2}$ miles.   **85.** The wall is $6\frac{5}{8}$ inches thick.   **87.** The minimum length of the bolt needed is $1\frac{7}{16}$ inches.

SECTION 2.5

1. $\frac{2}{17}$ **3.** $\frac{1}{3}$ **5.** $\frac{1}{10}$ **7.** $\frac{5}{13}$ **9.** $\frac{1}{3}$ **11.** $\frac{4}{7}$ **13.** $\frac{1}{4}$ **15.** Yes   **17.** $\frac{1}{2}$ **19.** $\frac{19}{56}$ **21.** $\frac{1}{2}$ **23.** $\frac{11}{60}$ **25.** $\frac{1}{32}$

27. $\frac{19}{60}$ **29.** $\frac{5}{72}$ **31.** $\frac{11}{60}$ **33.** $\frac{29}{60}$ **35.** (i)   **37.** $5\frac{1}{5}$ **39.** $4\frac{7}{8}$ **41.** $\frac{16}{21}$ **43.** $5\frac{1}{2}$ **45.** $5\frac{4}{7}$ **47.** $7\frac{5}{24}$ **49.** $1\frac{2}{5}$

51. $15\frac{11}{20}$ **53.** $4\frac{37}{45}$ **55.** No   **57.** The missing dimension is $9\frac{1}{2}$ inches.   **59.** The difference between Meyfarth's distance and

Coachman's distance was $9\frac{3}{8}$ inches. The difference between Kostadinova's distance and Meyfarth's distance was $5\frac{1}{4}$ inches.

61. a. The hikers plan to travel $17\frac{17}{24}$ miles the first two days.   **b.** There will be $9\frac{19}{24}$ miles left to travel on the third day.

63. The difference represents how much farther the hikers plan to travel on the second day than on the first day.   **65. a.** Yes

b. The wrestler needs to lose $3\frac{1}{4}$ pounds to reach the desired weight.   **67.** $\frac{11}{15}$ of the electrician's income is not spent for housing.

69. $6\frac{1}{8}$

SECTION 2.6

1. $\frac{7}{12}$ **3.** $\frac{7}{48}$ **5.** $\frac{1}{48}$ **7.** $\frac{11}{14}$ **9.** 6   **11.** $\frac{5}{12}$ **13.** 6   **15.** $\frac{2}{3}$ **17.** $\frac{3}{16}$ **19.** $\frac{3}{80}$ **21.** 10   **23.** $\frac{1}{15}$ **25.** $\frac{2}{3}$

27. $\frac{7}{26}$ **29.** 4   **31.** $\frac{100}{357}$ **33.** Answers will vary. For example, $\frac{3}{4}$ and $\frac{4}{3}$.   **35.** $1\frac{1}{3}$ **37.** $2\frac{1}{2}$ **39.** $\frac{9}{34}$ **41.** 10   **43.** $16\frac{2}{3}$

45. 1   **47.** $\frac{1}{2}$ **49.** 30   **51.** 42   **53.** $12\frac{2}{3}$ **55.** $1\frac{4}{5}$ **57.** $1\frac{2}{3}$ **59.** $1\frac{2}{3}$ **61.** 0   **63.** $27\frac{2}{3}$ **65.** $17\frac{85}{128}$ **67.** $\frac{2}{5}$

69. $8\frac{1}{16}$ **71.** 8   **73.** 9   **75.** $\frac{5}{8}$ **77.** $3\frac{1}{40}$ **79.** Less than   **81.** The cost is $11.   **83.** The length is $3\frac{1}{12}$ feet.

85. The area is $27\frac{9}{16}$ square feet.   **87.** Each year $5\frac{1}{2}$ billion bushels of corn are turned into ethanol.   **89.** The weight is

$54\frac{19}{36}$ pounds.   **91.** The total cost of the material is $363.   **93.** $\frac{1}{2}$ **95.** *A*

SECTION 2.7

1. $\frac{5}{6}$ **3.** 1   **5.** 0   **7.** $\frac{1}{2}$ **9.** $\frac{1}{6}$ **11.** $\frac{7}{10}$ **13.** 2   **15.** 2   **17.** $\frac{1}{6}$ **19.** 6   **21.** $\frac{1}{15}$ **23.** 2   **25.** $2\frac{1}{2}$

27. 3   **29.** $1\frac{1}{6}$ **31.** $3\frac{1}{3}$ **33.** True   **35.** 6   **37.** $\frac{1}{2}$ **39.** $\frac{1}{30}$ **41.** $1\frac{4}{5}$ **43.** 13   **45.** 3   **47.** $\frac{1}{5}$ **49.** $\frac{11}{28}$

51. 120   **53.** $\frac{11}{40}$ **55.** $\frac{33}{40}$ **57.** $4\frac{4}{9}$ **59.** $\frac{13}{32}$ **61.** $10\frac{2}{3}$ **63.** $\frac{12}{53}$ **65.** $4\frac{62}{191}$ **67.** 68   **69.** $8\frac{2}{7}$ **71.** $3\frac{13}{49}$

73. 4   **75.** $1\frac{3}{5}$ **77.** $\frac{9}{34}$ **79.** False   **81.** Less than   **83.** There are 12 servings in 16 ounces of cereal.   **85.** Each acre

costs $24,000.   **87.** The nut will make 12 turns in moving $1\frac{7}{8}$ inches.   **89. a.** The total weight of the fat and bone is $1\frac{5}{12}$ pounds.

b. The chef can cut 28 servings from the roast.   **91.** The distance between each post is $2\frac{3}{4}$ inches.   **93.** $\frac{31}{50}$ of the money borrowed

on home equity loans is spent on debt consolidation and home improvement.   **95.** $\frac{1}{6}$ of the puzzle is left to complete.

97. The difference was $\frac{3}{32}$ inch.   **99.** The dimensions of the board when it is closed are 14 inches by 7 inches by $1\frac{3}{4}$ inches.

101. The average teenage boy drinks 7 more cans of soda per week than the average teenage girl.   **103. a.** $\frac{2}{3}$ **b.** $2\frac{5}{8}$

SECTION 2.8

1. $\frac{11}{40} < \frac{19}{40}$ **3.** $\frac{2}{3} < \frac{5}{7}$ **5.** $\frac{5}{8} > \frac{7}{12}$ **7.** $\frac{7}{9} < \frac{11}{12}$ **9.** $\frac{13}{14} > \frac{19}{21}$ **11.** $\frac{7}{24} < \frac{11}{30}$ **13.** $\frac{4}{5}$ **15.** $\frac{25}{144}$ **17.** $\frac{2}{9}$ **19.** $\frac{3}{125}$

21. $\frac{4}{45}$ **23.** $\frac{16}{1225}$ **25.** $\frac{4}{49}$ **27.** $\frac{9}{125}$ **29.** $\frac{27}{88}$ **31.** $\frac{5}{6}$ **33.** $1\frac{5}{12}$ **35.** $\frac{7}{48}$ **37.** $\frac{29}{36}$ **39.** $\frac{55}{72}$ **41.** $\frac{35}{54}$ **43.** 2

45. $\frac{9}{19}$ **47.** $\frac{7}{32}$ **49.** $\frac{64}{75}$ **51. a.** More people choose a fast-food restaurant on the basis of its location.   **b.** Location was

the criterion cited by the most people.

CHAPTER 2 CONCEPT REVIEW*

1. To find the LCM of 75, 30, and 50, find the prime factorization of each number and write the factorization of each number in a
table. Circle the greatest product in each column. The LCM is the product of the circled numbers.

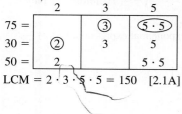

	2	3	5
75 =		③	5 · 5
30 =	②	3	5
50 =	2		5 · 5

LCM = 2 · 3 · 5 · 5 = 150   [2.1A]

---

*Note: The numbers in brackets following the answers in the Concept Review are a reference to the objective that corresponds to that
problem. For example, the reference [1.2A] stands for Section 1.2, Objective A. This notation will be used for all Prep Tests, Concept
Reviews, Chapter Reviews, Chapter Tests, and Cumulative Reviews throughout the text.

2. To find the GCF of 42, 14, and 21, find the prime factorization of each number and write the factorization of each number in a table. Circle the least product in each column that does not have a blank. The GCF is the product of the circled numbers.

	2	3	7
42 =	2	3	⑦
14 =	2		7
21 =		3	7

GCF = 7   [2.1B]

3. To write an improper fraction as a mixed number, divide the numerator by the denominator. The quotient without the remainder is the whole number part of the mixed number. To write the fractional part of the mixed number, write the remainder over the divisor. [2.2B]

4. A fraction is in simplest form when the numerator and denominator have no common factors other than 1. For example, $\frac{8}{12}$ is not in simplest form because 8 and 12 have a common factor of 4. $\frac{5}{7}$ is in simplest form because 5 and 7 have no common factors other than 1.   [2.3B]

5. When adding fractions, you have to convert to equivalent fractions with a common denominator. One way to explain this is that you can combine like things, but you cannot combine unlike things. You can combine 3 apples and 4 apples and get 7 apples. You cannot combine 4 apples and 3 oranges and get a sum consisting of just one item. In adding whole numbers, you add like things: ones, tens, hundreds, and so on. In adding fractions, you can combine 2 *ninths* and 5 *ninths* and get 7 *ninths*, but you cannot add 2 *ninths* and 3 *fifths*.   [2.4B]

6. To add mixed numbers, add the fractional parts and then add the whole number parts. Then reduce the sum to simplest form. [2.4C]

7. To subtract mixed numbers, the first step is to subtract the fractional parts. If we are subtracting a mixed number from a whole number, there is no fractional part in the whole number from which to subtract the fractional part of the mixed number. Therefore, we must borrow a 1 from the whole number and write 1 as an equivalent fraction with a denominator equal to the denominator of the fraction in the mixed number. Then we can subtract the fractional parts and then subtract the whole numbers.   [2.5C]

8. When multiplying two fractions, it is better to eliminate the common factors before multiplying the remaining factors in the numerator and denominator so that (1) we don't end up with very large products and (2) we don't have the added step of simplifying the resulting fraction.   [2.6A]

9. Let's look at an example, $\frac{1}{2} \times \frac{1}{3} = \frac{1}{6}$. The fractions $\frac{1}{2}$ and $\frac{1}{3}$ are less than 1. The product, $\frac{1}{6}$, is less than $\frac{1}{2}$ and less than $\frac{1}{3}$. Therefore, the product is less than the smaller number.   [2.6A]

10. Reciprocals are used to rewrite division problems as related multiplication problems. Since "divided by" means the same thing as "times the reciprocal of," we can change the division sign to a multiplication sign and change the divisor to its reciprocal. For example, $9 \div 3 = 9 \times \frac{1}{3}$.   [2.7A]

11. When a fraction is divided by a whole number, we write the whole number as a fraction before dividing so that we can easily determine the reciprocal of the whole number.   [2.7B]

12. When comparing fractions, we must first look at the denominators. If they are not the same, we must rewrite the fractions as equivalent fractions with a common denominator. If the denominators are the same, we must look at the numerators. The fraction that has the smaller numerator is the smaller fraction.   [2.8A]

13. We must follow the Order of Operations Agreement in simplifying the expression $\left(\frac{5}{6}\right)^2 - \left(\frac{3}{4} - \frac{2}{3}\right) \div \frac{1}{2}$. Therefore, we must first simplify the expression $\frac{3}{4} - \frac{2}{3}$ inside the parentheses: $\frac{3}{4} - \frac{2}{3} = \frac{1}{12}$. Then simplify the exponential expression: $\left(\frac{5}{6}\right)^2 = \frac{25}{36}$. Then perform the division: $\frac{1}{12} \div \frac{1}{2} = \frac{1}{6}$. Then perform the subtraction: $\frac{25}{36} - \frac{1}{6} = \frac{19}{36}$.   [2.8C]

CHAPTER 2 REVIEW EXERCISES

1. $\frac{2}{3}$ [2.3B]   2. $\frac{5}{16}$ [2.8B]   3. $\frac{13}{4}$ [2.2A]   4. $1\frac{13}{18}$ [2.4B]   5. $\frac{11}{18} < \frac{17}{24}$ [2.8A]   6. $14\frac{19}{42}$ [2.5C]   7. $\frac{5}{36}$ [2.8C]

8. $9\frac{1}{24}$ [2.6B]   9. 2 [2.7B]   10. $\frac{25}{48}$ [2.5B]   11. $3\frac{1}{3}$ [2.7B]   12. 4 [2.1B]   13. $\frac{24}{36}$ [2.3A]   14. $\frac{3}{4}$ [2.7A]

15. $\frac{32}{44}$ [2.3A]   16. $16\frac{1}{2}$ [2.6B]   17. 36 [2.1A]   18. $\frac{4}{11}$ [2.3B]   19. $1\frac{1}{8}$ [2.4A]   20. $10\frac{1}{8}$ [2.5C]

21. $18\frac{13}{54}$ [2.4C]   22. 5 [2.1B]   23. $3\frac{2}{5}$ [2.2B]   24. $\frac{1}{15}$ [2.8C]   25. $5\frac{7}{8}$ [2.4C]   26. 54 [2.1A]

27. $\frac{1}{3}$ [2.5A]   28. $\frac{19}{7}$ [2.2B]   29. 2 [2.7A]   30. $\frac{1}{15}$ [2.6A]   31. $\frac{1}{8}$ [2.6A]   32. $1\frac{7}{8}$ [2.2A]

33. The total rainfall for the 3 months was $21\frac{7}{24}$ inches.   [2.4D]   **34.** The cost per acre was \$36,000.   [2.7C]   **35.** The second checkpoint is $4\frac{3}{4}$ miles from the finish line.   [2.5D]   **36.** The car can travel 243 miles.   [2.6C]

CHAPTER 2 TEST

1. $\frac{4}{9}$ [2.6A, Example 1]   **2.** 8   [2.1B, Example 2]   **3.** $1\frac{3}{7}$ [2.7A, Example 2]   **4.** $\frac{7}{24}$ [2.8C, You Try It 3]

5. $\frac{49}{5}$ [2.2B, Example 5]   **6.** 8   [2.6B, Example 5]   **7.** $\frac{5}{8}$ [2.3B, Example 3]   **8.** $\frac{3}{8} < \frac{5}{12}$ [2.8A, Example 1]

9. $\frac{5}{6}$ [2.8C, Example 3]   **10.** 120   [2.1A, Example 1]   **11.** $\frac{1}{4}$ [2.5A, Example 1]   **12.** $3\frac{3}{5}$ [2.2B, Example 3]

13. $2\frac{2}{19}$ [2.7B, Example 4]   **14.** $\frac{45}{72}$ [2.3A, Example 1]   **15.** $1\frac{61}{90}$ [2.4B, Example 4]   **16.** $13\frac{81}{88}$ [2.5C, Example 5]

17. $\frac{7}{48}$ [2.5B, Example 2]   **18.** $\frac{1}{6}$ [2.8B, You Try It 2]   **19.** $1\frac{11}{12}$ [2.4A, Example 1]   **20.** $22\frac{4}{15}$ [2.4C, Example 7]

21. $\frac{11}{4}$ [2.2A, Example 2]   **22.** The electrician earns \$840.   [2.6C, Example 7]   **23.** 11 lots were available for sale.
[2.7C, Example 8]   **24.** The actual length of wall a is $12\frac{1}{2}$ feet. The actual length of wall b is 18 feet. The actual length of wall c is $15\frac{3}{4}$ feet.   [2.7C, Example 8]   **25.** The total rainfall for the 3-month period was $21\frac{11}{24}$ inches.   [2.4D, Example 9]

CUMULATIVE REVIEW EXERCISES

1. 290,000   [1.1D]   **2.** 291,278   [1.3B]   **3.** 73,154   [1.4B]   **4.** 540 r12   [1.5C]   **5.** 1   [1.6B]   **6.** $2 \cdot 2 \cdot 11$ [1.7B]

7. 210   [2.1A]   **8.** 20   [2.1B]   **9.** $\frac{23}{3}$ [2.2B]   **10.** $6\frac{1}{4}$ [2.2B]   **11.** $\frac{15}{48}$ [2.3A]   **12.** $\frac{2}{5}$ [2.3B]   **13.** $1\frac{7}{48}$ [2.4B]

14. $14\frac{11}{48}$ [2.4C]   **15.** $\frac{13}{24}$ [2.5B]   **16.** $1\frac{7}{9}$ [2.5C]   **17.** $\frac{7}{20}$ [2.6A]   **18.** $7\frac{1}{2}$ [2.6B]   **19.** $1\frac{1}{20}$ [2.7A]

20. $2\frac{5}{8}$ [2.7B]   **21.** $\frac{1}{9}$ [2.8B]   **22.** $5\frac{5}{24}$ [2.8C]   **23.** The amount in the checking account at the end of the week was \$862.   [1.3C]   **24.** The total income from the tickets was \$1410.   [1.4C]   **25.** The total weight is $12\frac{1}{24}$ pounds.   [2.4D]

26. The length of the remaining piece is $4\frac{17}{24}$ feet.   [2.5D]   **27.** The car travels 225 miles on $8\frac{1}{3}$ gallons of gas.   [2.6C]

28. 25 parcels can be sold from the remaining land.   [2.7C]

ANSWERS TO CHAPTER 3 SELECTED EXERCISES

PREP TEST

1. $\frac{3}{10}$ [2.2A]   **2.** 36,900   [1.1D]   **3.** Four thousand seven hundred ninety-one   [1.1B]   **4.** 6842   [1.1B]   **5.** 9394   [1.2A]

6. 1638   [1.3B]   **7.** 76,804   [1.4B]   **8.** 278 r18   [1.5C]

SECTION 3.1

1. Thousandths   **3.** Ten-thousandths   **5.** Hundredths   **7.** 0.3   **9.** 0.21   **11.** 0.461   **13.** $\frac{1}{10}$ **15.** $\frac{47}{100}$ **17.** $\frac{289}{1000}$

19. Thirty-seven hundredths   **21.** Nine and four tenths   **23.** Fifty-three ten-thousandths   **25.** Forty-five thousandths

27. Twenty-six and four hundredths   **29.** 3.0806   **31.** 407.03   **33.** 246.024   **35.** 73.02684   **37.** 6.2   **39.** 21.0

41. 18.41   **43.** 72.50   **45.** 936.291   **47.** 47   **49.** 7015   **51.** 2.97527   **53.** An entrant who completes the Boston Marathon runs 26.2 miles.   **55.** For example, 0.2701   **57.** For example, **a.** 0.15   **b.** 1.05   **c.** 0.001

SECTION 3.2

1. 150.1065   **3.** 95.8446   **5.** 69.644   **7.** 92.883   **9.** 113.205   **11.** 0.69   **13.** 16.305   **15.** 110.7666

17. 104.4959   **19.** Cal.: 234.192   **21.** Cal.: 781.943   **23.** Yes   **25.** The length of the shaft is 4.35 feet.
      Est.: 234   Est.: 782

27. The perimeter is 18.5 meters.   **29.** The total number of people who watched the three news programs is 26.3 million.

31. No, a 4-foot rope cannot be wrapped around the box.   **33.** Three possible answers are bread, butter, and mayonnaise; raisin bran, butter, and bread; and lunch meat, milk, and toothpaste.

SECTION 3.3

1. 5.627   **3.** 113.6427   **5.** 6.7098   **7.** 215.697   **9.** 53.8776   **11.** 72.7091   **13.** 0.3142   **15.** 1.023   **17.** 261.166   **19.** 655.32   **21.** 342.9268   **23.** 8.628   **25.** 7.01 − 2.325   **27.** 19.35 − 8.967   **29.** Cal.: 2.74506   **31.** Cal.: 7.14925
Est.: 3   Est.: 7

33. The missing dimension is 2.59 feet.   **35.** The difference in the average number of tickets sold is 320,000 tickets.
37. 33.5 million more people watched Super Bowl XLII than watched the Super Bowl post-game show.   **39. a.** 0.1
b. 0.01   **c.** 0.001

SECTION 3.4

1. 0.36   **3.** 0.25   **5.** 6.93   **7.** 1.84   **9.** 0.74   **11.** 39.5   **13.** 2.72   **15.** 0.603   **17.** 13.50   **19.** 4.316
21. 0.1323   **23.** 0.03568   **25.** 0.0784   **27.** 0.076   **29.** 34.48   **31.** 580.5   **33.** 20.148   **35.** 0.04255
37. 0.17686   **39.** 0.19803   **41.** 0.0006608   **43.** 0.536335   **45.** 0.429   **47.** 2.116   **49.** 0.476
51. 1.022   **53.** 37.96   **55.** 2.318   **57.** 3.2   **59.** 6.5   **61.** 6285.6   **63.** 3200   **65.** 35,700   **67.** 6.3
69. 3.9   **71.** 49,000   **73.** 6.7   **75.** 0.012075   **77.** 0.0117796   **79.** 0.31004   **81.** 0.082845   **83.** 5.175
85. Cal.: 91.2   **87.** Cal.: 1.0472   **89.** Cal.: 3.897   **91.** Cal.: 11.2406   **93.** Cal.: 0.371096   **95.** Cal.: 31.8528
Est.: 90   Est.: 0.8   Est.: 4.5   Est.: 12   Est.: 0.32   Est.: 30

97. A U.S. homeowner's average annual cost of electricity is $1147.92.   **99.** The amount received is $14.06.   **101.** You will
pay $2.40 in taxes.   **103.** The area is 23.625 square feet.   **105. a.** The nurse's overtime pay is $785.25.   **b.** The nurse's total
income is $2181.25.   **107.** It would cost the company $406.25.   **109.** The expression represents the cost of mailing 4 express
mail packages, each weighing $\frac{1}{2}$ pound or less, and 9 express mail packages, each weighing between $\frac{1}{2}$ pound and 1 pound, from
the post office to the addressee.   **111.** The added cost is $3,200,000.   **113. a.** The total cost for grade 1 is $56.32.
b. The total cost for grade 2 is $74.04.   **c.** The total cost for grade 3 is $409.56.   **d.** The total cost is $539.92.
115. $1\frac{3}{10} \times 2\frac{31}{100} = \frac{13}{10} \times \frac{231}{100} = \frac{3003}{1000} = 3\frac{3}{1000} = 3.003$

SECTION 3.5

1. 0.82   **3.** 4.8   **5.** 89   **7.** 60   **9.** 84.3   **11.** 32.3   **13.** 5.06   **15.** 1.3   **17.** 0.11   **19.** 3.8   **21.** 6.3   **23.** 0.6
25. 2.5   **27.** 1.1   **29.** 130.6   **31.** 0.81   **33.** 0.09   **35.** 40.70   **37.** 0.46   **39.** 0.019   **41.** 0.087   **43.** 0.360
45. 0.103   **47.** 0.009   **49.** 1   **51.** 3   **53.** 1   **55.** 57   **57.** 0.407   **59.** 4.267   **61.** 0.01037   **63.** 0.008295
65. 0.82537   **67.** 0.032   **69.** 0.23627   **71.** 0.000053   **73.** 0.0018932   **75.** 18.42   **77.** 16.07   **79.** 0.0135
81. 0.023678   **83.** 0.112   **85.** Cal.: 11.1632   **87.** Cal.: 884.0909   **89.** Cal.: 1.8269   **91.** Cal.: 58.8095   **93.** Cal.: 72.3053
Est.: 10   Est.: 1000   Est.: 1.5   Est.: 50   Est.: 100

95. a. Use division to find the cost.   **b.** Use multiplication to find the cost.   **97.** 6.23 yards are gained per carry.
99. The trucker must drive 35 miles.   **101.** Three complete shelves can be cut from a 12-foot board.   **103.** The dividend is
$1.72 per share.   **105.** The car travels 25.5 miles on 1 gallon of gasoline.   **107.** You will use 0.405 barrel of oil.
109. 2.57 million more women than men were attending institutions of higher learning.   **111.** The Army's advertising budget was
4.2 times greater than the Navy's advertising budget.   **113.** The population of this segment is expected to be 2.1 times greater in
2030 than in 2000.   **117.** ÷   **119.** −   **121.** +   **123.** 5.217   **125.** 0.025

SECTION 3.6

1. 0.625   **3.** 0.667   **5.** 0.167   **7.** 0.417   **9.** 1.750   **11.** 1.500   **13.** 4.000   **15.** 0.003   **17.** 7.080   **19.** 37.500
21. 0.160   **23.** 8.400   **25.** Less than 1   **27.** Greater than 1   **29.** $\frac{4}{5}$ **31.** $\frac{8}{25}$ **33.** $\frac{1}{8}$ **35.** $1\frac{1}{4}$ **37.** $16\frac{9}{10}$
39. $8\frac{2}{5}$ **41.** $8\frac{437}{1000}$ **43.** $2\frac{1}{4}$ **45.** $\frac{23}{150}$ **47.** $\frac{703}{800}$ **49.** $7\frac{19}{50}$ **51.** $\frac{57}{100}$ **53.** $\frac{2}{3}$ **55.** 0.15 < 0.5   **57.** 6.65 > 6.56
59. 2.504 > 2.054   **61.** $\frac{3}{8}$ > 0.365   **63.** $\frac{2}{3}$ > 0.65   **65.** $\frac{5}{9}$ > 0.55   **67.** 0.62 > $\frac{7}{15}$ **69.** 0.161 > $\frac{1}{7}$ **71.** 0.86 > 0.855
73. 1.005 > 0.5   **75.** 0.172 < 17.2   **77.** Cars 2 and 5 would fail the emissions test.

CHAPTER 3 CONCEPT REVIEW*

1. To round a decimal to the nearest tenth, look at the digit in the hundredths place. If the digit in the hundredths place is less than 5, that digit and all digits to the right are dropped. If the digit in the hundredths place is greater than or equal to 5, increase the digit in the tenths place by 1 and drop all digits to its right.   [3.1B]

2. The decimal 0.37 is read 37 hundredths. To write the decimal as a fraction, put 37 in the numerator and 100 in the denominator: $\frac{37}{100}$. [3.1A]

3. The fraction $\frac{173}{10,000}$ is read 173 ten-thousandths. To write the fraction as a decimal, insert one 0 after the decimal point so that the 3 is in the ten-thousandths place: 0.0173.   [3.1A]

4. When adding decimals of different place values, write the numbers so that the decimal points are on a vertical line.   [3.2A]

5. Write the decimal point in the product of two decimals so that the number of decimal places in the product is the sum of the numbers of decimal places in the factors.   [3.4A]

6. To estimate the product of two decimals, round each number so that it contains one nonzero digit. Then multiply. For example, to estimate the product of 0.068 and 0.0052, round the two numbers to 0.07 and 0.005; then multiply $0.07 \times 0.005 = 0.00035$. [3.4A]

7. When dividing decimals, move the decimal point in the divisor to the right to make the divisor a whole number. Move the decimal point in the dividend the same number of places to the right. Place the decimal point in the quotient directly over the decimal point in the dividend, and then divide as with whole numbers.   [3.5A]

8. First convert the fraction to a decimal: The fraction $\frac{5}{8}$ is equal to 0.625. Now compare the decimals: $0.63 > 0.625$. In the inequality $0.63 > 0.625$, replace the decimal 0.625 with the fraction $\frac{5}{8}$: $0.63 > \frac{5}{8}$. The answer is that the decimal 0.63 is greater than the fraction $\frac{5}{8}$.   [3.6C]

9. When dividing 0.763 by 0.6, the decimal points will be moved one place to the right: $7.63 \div 6$. The decimal 7.63 has digits in the tenths and hundredths places. We need to write a zero in the thousandths place in order to determine the digit in the thousandths place of the quotient so that we can then round the quotient to the nearest hundredth.   [3.5A]

10. To subtract a decimal from a whole number that has no decimal point, write a decimal point in the whole number to the right of the ones place. Then write as many zeros to the right of that decimal point as there are places in the decimal being subtracted from the whole number. For example, the subtraction $5 - 3.578$ would be written $5.000 - 3.578$.   [3.3A]

CHAPTER 3 REVIEW EXERCISES

1. 54.5   [3.5A]   **2.** 833.958   [3.2A]   **3.** $0.055 < 0.1$ [3.6C]   **4.** Twenty-two and ninety-two ten-thousandths   [3.1A]

5. 0.05678   [3.1B]   **6.** 2.33   [3.6A]   **7.** $\frac{3}{8}$ [3.6B]   **8.** 36.714   [3.2A]   **9.** 34.025   [3.1A]   **10.** $\frac{5}{8} > 0.62$ [3.6C]

11. 0.778   [3.6A]   **12.** $\frac{33}{50}$ [3.6B]   **13.** 22.8635   [3.3A]   **14.** 7.94   [3.1B]   **15.** 8.932   [3.4A]   **16.** Three hundred

forty-two and thirty-seven hundredths   [3.1A]   **17.** 3.06753   [3.1A]   **18.** 25.7446   [3.4A]   **19.** 6.594   [3.5A]

20. 4.8785   [3.3A]   **21.** The new balance in your account is \$661.51.   [3.3B]   **22.** The difference between the amount United

expects to pay per gallon of fuel and the amount Southwest expects to pay is \$.96.   [3.3B]   **23.** Northwest's cost per gallon of fuel

is \$3.34. Northwest's cost per gallon is more than United's cost per gallon.   [3.5B; 3.6C]   **24.** The number who drove is 6.4 times

greater than the number who flew.   [3.5B]   **25.** During a 5-day school week, 9.5 million gallons of milk are served.   [3.4B]

CHAPTER 3 TEST

1. $0.66 < 0.666$ [3.6C, Example 5]   **2.** 4.087   [3.3A, Example 1]   **3.** Forty-five and three hundred two ten-thousandths

[3.1A, Example 4]   **4.** 0.692   [3.6A, You Try It 1]   **5.** $\frac{33}{40}$ [3.6B, Example 3]   **6.** 0.0740   [3.1B, Example 6]

7. 1.583   [3.5A, Example 3]   **8.** 27.76626   [3.3A, Example 2]   **9.** 7.095   [3.1B, Example 6]   **10.** 232 [3.5A, Example 1]

11. 458.581   [3.2A, Example 2]   **12.** The missing dimension is 1.37 inches.   [3.3B, Example 4]   **13.** 0.00548

[3.4A, Example 2]   **14.** 255.957   [3.2A, Example 1]   **15.** 209.07086   [3.1A, Example 4]   **16.** Each payment is \$395.40.

[3.5B, Example 7]   **17.** Your total income is \$3087.14.   [3.2B, You Try It 4]   **18.** The cost of the call is \$4.63.   [3.4B, Example 8]

Note: The numbers in brackets following the answers in the Concept Review are a reference to the objective that corresponds to that problem. For example, the reference [1.2A] stands for Section 1.2, Objective A. This notation will be used for all Prep Tests, Concept Reviews, Chapter Reviews, Chapter Tests, and Cumulative Reviews throughout the text.

19. The yearly average computer use by a 10th-grade student is 348.4 hours. [3.4B, Example 7] **20.** On average, a 2nd-grade student uses a computer 36.4 hours more per year than a 5th-grade student. [3.4B, Example 8]

CUMULATIVE REVIEW EXERCISES

1. 235 r17 [1.5C] **2.** 128 [1.6A] **3.** 3 [1.6B] **4.** 72 [2.1A] **5.** $4\frac{2}{5}$ [2.2B] **6.** $\frac{37}{8}$ [2.2B] **7.** $\frac{25}{60}$ [2.3A]

8. $1\frac{17}{48}$ [2.4B] **9.** $8\frac{35}{36}$ [2.4C] **10.** $5\frac{23}{36}$ [2.5C] **11.** $\frac{1}{12}$ [2.6A] **12.** $9\frac{1}{8}$ [2.6B] **13.** $1\frac{2}{9}$ [2.7A] **14.** $\frac{19}{20}$ [2.7B]

15. $\frac{3}{16}$ [2.8B] **16.** $2\frac{5}{18}$ [2.8C] **17.** Sixty-five and three hundred nine ten-thousandths [3.1A] **18.** 504.6991 [3.2A]

19. 21.0764 [3.3A] **20.** 55.26066 [3.4A] **21.** 2.154 [3.5A] **22.** 0.733 [3.6A] **23.** $\frac{1}{6}$ [3.6B]

24. $\frac{8}{9} < 0.98$ [3.6C] **25.** Sweden mandates 14 more vacation days than Germany. [1.3C] **26.** The patient must lose $7\frac{3}{4}$ pounds the third month to achieve the goal. [2.5D] **27.** Your checking account balance is $617.38. [3.3B]

28. The resulting thickness is 1.395 inches. [3.3B] **29.** You paid $6008.80 in income tax last year. [3.4B]

30. The amount of each payment is $46.37. [3.5B]

ANSWERS TO CHAPTER 4 SELECTED EXERCISES

PREP TEST

1. $\frac{4}{5}$ [2.3B] **2.** $\frac{1}{2}$ [2.3B] **3.** 24.8 [3.6A] **4.** 4 × 33 [1.4A] **5.** 4 [1.5A]

SECTION 4.1

1. $\frac{1}{5}$ 1:5 1 to 5 **3.** $\frac{2}{1}$ 2:1 2 to 1 **5.** $\frac{3}{8}$ 3:8 3 to 8 **7.** $\frac{1}{1}$ 1:1 1 to 1 **9.** $\frac{7}{10}$ 7:10 7 to 10 **11.** $\frac{1}{2}$ 1:2 1 to 2

13. $\frac{2}{1}$ 2:1 2 to 1 **15.** $\frac{5}{2}$ 5:2 5 to 2 **17.** $\frac{5}{7}$ 5:7 5 to 7 **19.** days **21.** The ratio is $\frac{1}{3}$. **23.** The ratio is $\frac{3}{8}$.

25. The ratio is $\frac{1}{25,000}$. **27.** The ratio is $\frac{1}{5}$. **29.** The ratio is $\frac{1}{56}$. **31.** The ratio is 24 to 59.

SECTION 4.2

1. $\frac{3 \text{ pounds}}{4 \text{ people}}$ **3.** $\frac{\$20}{3 \text{ boards}}$ **5.** $\frac{20 \text{ miles}}{1 \text{ gallon}}$ **7.** $\frac{8 \text{ gallons}}{1 \text{ hour}}$ **9. a.** Dollars **b.** Seconds **11.** 1 **13.** 2.5 feet/second

15. $975/week **17.** 110 trees/acre **19.** $18.84/hour **21.** 35.6 miles/gallon **23.** The rate is 7.4 miles per dollar.

25. An average of 179.86 bushels of corn were harvested from each acre. **27.** The cost was $2.72 per disk. **29. a.** Australia has the least population density. **b.** There are 807 more people per square mile in India than in the United States.

31. 1.0179 × 2500 represents the value of 2500 American dollars in Canadian dollars.

SECTION 4.3

1. True **3.** Not true **5.** Not true **7.** True **9.** True **11.** True **13.** True **15.** Not true **17.** True

19. Yes **21.** Yes **23.** 3 **25.** 105 **27.** 2 **29.** 60 **31.** 2.22 **33.** 6.67 **35.** 21.33 **37.** 16.25

39. 2.44 **41.** 47.89 **43.** A 0.5-ounce serving contains 50 calories. **45.** The car can travel 329 miles. **47.** 12.5 gallons of water are required. **49.** The distance is 16 miles. **51.** 1.25 ounces are required. **53.** 160,000 people would vote.

55. The monthly payment is $176.75. **57.** 750 defective circuit boards can be expected in a run of 25,000. **59.** A bowling ball would weigh 2.67 pounds on the moon. **61.** The dividend would be $1071.

CHAPTER 4 CONCEPT REVIEW*

1. If the units in a comparison are different, then the comparison is a rate. For example, the comparison "50 miles in 2 hours" is a rate. [4.2A]

*Note: The numbers in brackets following the answers in the Concept Review are a reference to the objective that corresponds to that problem. For example, the reference [1.2A] stands for Section 1.2, Objective A. This notation will be used for all Prep Tests, Concept Reviews, Chapter Reviews, Chapter Tests, and Cumulative Reviews throughout the text.

2. To find a unit rate, divide the number in the numerator of the rate by the number in the denominator of the rate.   [4.2B]

3. To write the ratio $\frac{6}{7}$ using a colon, write the two numbers 6 and 7 separated by a colon: $6 : 7$.   [4.1A]

4 To write the ratio $12 : 15$ in simplest form, divide both numbers by the GCF of 3: $\frac{12}{3} : \frac{15}{3} = 4 : 5$.   [4.1A]

5. To write the rate $\frac{342 \text{ miles}}{9.5 \text{ gallons}}$ as a unit rate, divide the number in the numerator by the number in the denominator: $342 \div 9.5 = 36$. $\frac{342 \text{ miles}}{9.5 \text{ gallons}}$ is the rate. 36 miles/gallon is the unit rate.   [4.2B]

6. A proportion is true if the fractions are equal when written in lowest terms. Another way to describe a true proportion is to say that in a true proportion, the cross products are equal.   [4.3A]

7. When one of the numbers in a proportion is unknown, we can solve the proportion to find the unknown number. We do this by setting the cross products equal to each other and then solving for the unknown number.   [4.3B]

8. When setting up a proportion, keep the same units in the numerator and the same units in the denominator.   [4.3C]

9. To check the solution of a proportion, replace the unknown number in the proportion with the solution. Then find the cross products. If the cross products are equal, the solution is correct. If the cross products are not equal, the solution is not correct.   [4.3B]

10. To write the ratio $19 : 6$ as a fraction, write the first number as the numerator of the fraction and the second number as the denominator: $\frac{19}{6}$.   [4.1A]

CHAPTER 4 REVIEW EXERCISES

1. True   [4.3A]   **2.** $\frac{2}{5}$ 2 : 5 2 to 5   [4.1A]   **3.** 62.5 miles/hour   [4.2B]   **4.** True   [4.3A]   **5.** 68   [4.3B]

6. $12.50/hour   [4.2B]   **7.** $1.75/pound   [4.2B]   **8.** $\frac{2}{7}$ 2 : 7 2 to 7   [4.1A]   **9.** 36   [4.3B]   **10.** 19.44   [4.3B]

11. $\frac{2}{5}$ 2 : 5 2 to 5   [4.1A]   **12.** Not true   [4.3A]   **13.** $\frac{\$35}{4 \text{ hours}}$ [4.2A]   **14.** 27.2 miles/gallon   [4.2B]

15. $\frac{1}{1}$ 1 : 1 1 to 1   [4.1A]   **16.** True   [4.3A]   **17.** 65.45   [4.3B]   **18.** $\frac{100 \text{ miles}}{3 \text{ hours}}$ [4.2A]   **19.** The ratio is $\frac{2}{5}$.   [4.1B]

20. The property tax is $6400.   [4.3C]   **21.** The ratio is $\frac{3}{8}$.   [4.1B]   **22.** The cost per phone is $37.50.   [4.2C]

23. 1344 blocks would be needed.   [4.3C]   **24.** The ratio is $\frac{5}{2}$.   [4.1B]   **25.** The turkey costs $.93/pound.   [4.2C]

26. The average was 56.8 miles/hour.   [4.2C]   **27.** The cost is $493.50.   [4.3C]   **28.** The cost is $44.75/share.   [4.2C]

29. 22.5 pounds of fertilizer will be used.   [4.3C]   **30.** The ratio is $\frac{1}{2}$.   [4.1B]

CHAPTER 4 TEST

1. $3836.40/month   [4.2B, Example 2]   **2.** $\frac{1}{6}$ 1 : 6 1 to 6   [4.1A, You Try It 1]   **3.** $\frac{9 \text{ supports}}{4 \text{ feet}}$ [4.2A, Example 1]

4. Not true   [4.3A, Example 2]   **5.** $\frac{3}{2}$ 3 : 2 3 to 2   [4.1A, Example 2]   **6.** 144   [4.3B, Example 3]

7. 30.5 miles/gallon   [4.2B, Example 2]   **8.** $\frac{1}{3}$ 1 : 3 1 to 3   [4.1A, Example 2]   **9.** True   [4.3A, Example 1]

10. 40.5   [4.3B, Example 3]   **11.** $\frac{\$27}{2 \text{ boards}}$ [4.2A, Example 1]   **12.** $\frac{3}{5}$ 3 : 5 3 to 5   [4.1A, You Try It 1]

13. The dividend is $625.   [4.3C, Example 8]   **14.** The ratio is $\frac{1}{12}$.   [4.3C, Example 8]   **15.** The plane's speed is

538 miles/hour.   [4.2C, How To 1]   **16.** The college student's body contains 132 pounds of water.   [4.3C, Example 8]

17. The cost of the lumber is $1.73/foot.   [4.2C, How To 1]   **18.** The amount of medication required is 0.875 ounce.

[4.3C, Example 8]   **19.** The ratio is $\frac{4}{5}$.   [4.1B, Example 4]   **20.** 36 defective hard drives are expected to be found in the

production of 1200 hard drives.   [4.3C, Example 8]

CUMULATIVE REVIEW EXERCISES

1. 9158   [1.3B]   **2.** $2^4 \cdot 3^3$ [1.6A]   **3.** 3   [1.6B]   **4.** $2 \cdot 2 \cdot 2 \cdot 2 \cdot 2 \cdot 5$ [1.7B]   **5.** 36   [2.1A]   **6.** 14   [2.1B]

7. $\frac{5}{8}$ [2.3B]   **8.** $8\frac{3}{10}$ [2.4C]   **9.** $5\frac{11}{18}$ [2.5C]   **10.** $2\frac{5}{6}$ [2.6B]   **11.** $4\frac{2}{3}$ [2.7B]   **12.** $\frac{23}{30}$ [2.8C]

13. Four and seven hundred nine ten-thousandths   [3.1A]   **14.** 2.10   [3.1B]   **15.** 1.990   [3.5A]   **16.** $\frac{1}{15}$ [3.6B]

17. $\frac{1}{8}$ [4.1A]   **18.** $\frac{29¢}{2 \text{ pencils}}$ [4.2A]   **19.** 33.4 miles/gallon [4.2B]   **20.** 4.25 [4.3B]   **21.** The car's speed is 57.2 miles/hour.   [4.2C]   **22.** 36 [4.3B]   **23.** Your new balance is $744.   [1.3C]   **24.** The monthly payment is $570.   [1.5D]   **25.** 105 pages remain to be read.   [2.6C]   **26.** The cost per acre was $36,000.   [2.7C]   **27.** The change was $35.24.   [3.3B]   **28.** Your monthly salary is $3468.25.   [3.5B]   **29.** 25 inches will erode in 50 months.   [4.3C]   **30.** 1.6 ounces are required.   [4.3C]

ANSWERS TO CHAPTER 5 SELECTED EXERCISES

PREP TEST

1. $\frac{19}{100}$ [2.6B]   **2.** 0.23   [3.4A]   **3.** 47   [3.4A]   **4.** 2850   [3.4A]   **5.** 4000   [3.5A]   **6.** 32   [2.7B]   **7.** 62.5   [3.6A]   **8.** $66\frac{2}{3}$ [2.2B]   **9.** 1.75   [3.5A]

SECTION 5.1

1. $\frac{1}{4}$, 0.25   **3.** $1\frac{3}{10}$, 1.30   **5.** 1, 1.00   **7.** $\frac{73}{100}$, 0.73   **9.** $3\frac{83}{100}$, 3.83   **11.** $\frac{7}{10}$, 0.70   **13.** $\frac{22}{25}$, 0.88   **15.** $\frac{8}{25}$, 0.32   **17.** $\frac{2}{3}$ **19.** $\frac{5}{6}$ **21.** $\frac{1}{9}$ **23.** $\frac{5}{11}$ **25.** $\frac{3}{70}$ **27.** $\frac{1}{15}$ **29.** 0.065   **31.** 0.123   **33.** 0.0055   **35.** 0.0825   **37.** 0.0505   **39.** 0.02   **41.** Greater than   **43.** 73%   **45.** 1%   **47.** 294%   **49.** 0.6%   **51.** 310.6%   **53.** 70%   **55.** 37%   **57.** 40%   **59.** 12.5%   **61.** 150%   **63.** 166.7%   **65.** 87.5%   **67.** 48%   **69.** $33\frac{1}{3}$%   **71.** $166\frac{2}{3}$%   **73.** $87\frac{1}{2}$%   **75.** Less than   **77.** 6% of those surveyed named something other than corn on the cob, cole slaw, corn bread, or fries.   **79.** This represents $\frac{1}{2}$ off the regular price.

SECTION 5.2

1. 8   **3.** 10.8   **5.** 0.075   **7.** 80   **9.** 51.895   **11.** 7.5   **13.** 13   **15.** 3.75   **17.** 20   **19.** 5% of 95   **21.** 79% of 16   **23.** Less than   **25.** The number of people in the United States aged 18 to 24 without life insurance is less than 44 million.   **27.** 58,747 new student pilots are flying single-engine planes this year.   **29.** The piece contains 29.25 grams of gold, 8.75 grams of silver, and 12 grams of copper.   **31.** 77 million returns were filed electronically.   **33.** 49.59 billion of the email messages sent per day are not spam.   **35.** 6232 respondents did not answer yes to the question.

SECTION 5.3

1. 32%   **3.** $16\frac{2}{3}$%   **5.** 200%   **7.** 37.5%   **9.** 18%   **11.** 0.25%   **13.** 20%   **15.** 400%   **17.** 2.5%   **19.** 37.5%   **21.** 0.25%   **23.** False   **25.** 70% of couples disagree about financial matters.   **27.** Approximately 25.4% of the vegetables were wasted.   **29.** 29.8% of Americans with diabetes have not been diagnosed.   **31.** 98.5% of the slabs did meet safety requirements.   **33.** 26.7% of the total is spent on veterinary care.

SECTION 5.4

1. 75   **3.** 50   **5.** 100   **7.** 85   **9.** 1200   **11.** 19.2   **13.** 7.5   **15.** 32   **17.** 200   **19.** 9   **21.** 504   **23.** Less than   **25.** There were 15.8 million travelers who allowed their children to miss school to go along on a trip.   **27.** 22,366 runners started the Boston Marathon in 2008.   **29.** The cargo ship's daily fuel bill is $8000.   **31. a.** 3000 boards were tested.   **b.** 2976 of the boards tested were not defective.   **33.** The recommended daily amount of thiamin for an adult is 1.5 milligrams.

SECTION 5.5

1. 65   **3.** 25%   **5.** 75   **7.** 12.5%   **9.** 400   **11.** 19.5   **13.** 14.8%   **15.** 62.62   **17.** 15   **19. a.** (ii) and (iii)   **b.** (i) and (iv)   **21.** The drug will be effective for 4.8 hours.   **23. a.** $175 million is generated annually from sales of Thin Mints.   **b.** $63 million is generated annually from sales of Trefoil shortbread cookies.   **25.** The U.S. total turkey production was 7 billion pounds.   **27.** 57.7% of baby boomers have some college experience but have not earned a degree.   **29.** 46.8% of the deaths were due to traffic accidents.

CHAPTER 5 CONCEPT REVIEW*

1. To write 197% as a fraction, remove the percent sign and multiply by $\frac{1}{100}$: $197 \times \frac{1}{100} = \frac{197}{100}$. [5.1A]

2. To write 6.7% as a decimal, remove the percent sign and multiply by 0.01: $6.7 \times 0.01 = 0.067$. [5.1A]

3. To write $\frac{9}{5}$ as a percent, multiply by 100%: $\frac{9}{5} \times 100\% = 180\%$. [5.1B]

4. To write 56.3 as a percent, multiply by 100% : $56.3 \times 100\% = 5630\%$. [5.1B]

5. The basic percent equation is Percent $\times$ base = amount. [5.2A]

6. To find what percent of 40 is 30, use the basic percent equation: $n \times 40 = 30$. To solve for n, we *divide* 30 by 40: $30 \div 40 = 0.75 = 75\%$. [5.3A]

7. To find 11.7% of 532, use the basic percent equation and write the percent as a decimal: $0.117 \times 532 = n$. To solve for n, we *multiply* 0.117 by 532: $0.117 \times 532 = 62.244$. [5.2A]

8. To answer the question "36 is 240% of what number?", use the basic percent equation and write the percent as a decimal: $2.4 \times n = 36$. To solve for n, we *divide* 36 by 2.4: $36 \div 2.4 = 15$. [5.4A]

9. To use the proportion method to solve a percent problem, identify the percent, the amount, and the base. Then use the proportion $\frac{\text{percent}}{100} = \frac{\text{amount}}{\text{base}}$. Substitute the known values into this proportion and solve for the unknown. [5.5A]

10. To answer the question "What percent of 1400 is 763?" by using the proportion method, first identify the base (1400) and the amount (763). The percent is unknown. Substitute 1400 for the base and 763 for the amount in the proportion $\frac{\text{percent}}{100} = \frac{\text{amount}}{\text{base}}$. Then solve the proportion $\frac{n}{100} = \frac{763}{1400}$ for n. $n = 54.5$, so 54.5% of 1400 is 763. [5.5A]

CHAPTER 5 REVIEW EXERCISES

1. 60 [5.2A]   **2.** 20% [5.3A]   **3.** 175% [5.1B]   **4.** 75 [5.4A]   **5.** $\frac{3}{25}$ [5.1A]   **6.** 19.36 [5.2A]

7. 150% [5.3A]   **8.** 504 [5.4A]   **9.** 0.42 [5.1A]   **10.** 5.4 [5.2A]   **11.** 157.5 [5.4A]   **12.** 0.076 [5.1A]

13. 77.5 [5.2A]   **14.** $\frac{1}{6}$ [5.1A]   **15.** 160% [5.5A]   **16.** 75 [5.5A]   **17.** 38% [5.1B]   **18.** 10.9 [5.4A]

19. 7.3% [5.3A]   **20.** 613.3% [5.3A]   **21.** The student answered 85% of the questions correctly. [5.5B]

22. The company spent $4500 for newspaper advertising. [5.2B]   **23.** 31.7% of the cost is for electricity. [5.3B]

24. The total cost of the camcorder was $1041.25. [5.2B]   **25.** Approximately 78.6% of the women wore sunscreen often. [5.3B]

26. The world's population in 2000 was approximately 6,100,000,000 people. [5.4B]   **27.** The cost of the computer 4 years ago was $3000. [5.5B]   **28.** The total cranberry crop in that year was 572 million pounds. [5.3B/5.5B]

CHAPTER 5 TEST

1. 0.973 [5.1A, Example 3]   **2.** $\frac{5}{6}$ [5.1A, Example 2]   **3.** 30% [5.1B, Example 4]   **4.** 163% [5.1B, Example 4]

5. 150% [5.1B, HOW TO 1]   **6.** $66\frac{2}{3}\%$ [5.1B, Example 5]   **7.** 50.05 [5.2A, Example 1]   **8.** 61.36 [5.2A, Example 2]

9. 76% of 13 [5.2A, Example 1]   **10.** 212% of 12 [5.2A, Example 1]   **11.** The amount spent for advertising is $45,000. [5.2B, Example 3]   **12.** 1170 pounds of vegetables were not spoiled. [5.2B, Example 4]   **13.** 14.7% of the daily recommended amount of potassium is provided. [5.3B, Example 4]   **14.** 9.1% of the daily recommended number of calories is provided. [5.3B, Example 4]   **15.** The number of temporary employees is 16% of the number of permanent employees. [5.3B, Example 4]

16. The student answered approximately 91.3% of the questions correctly. [5.3B, Example 5]   **17.** 80 [5.4A, Example 2]

18. 28.3 [5.4A, Example 2]   **19.** 32,000 PDAs were tested. [5.4B, Example 4]   **20.** The increase was 60% of the original price. [5.3B, Example 5]   **21.** 143.0 [5.5A, Example 1]   **22.** 1000% [5.5A, Example 1]   **23.** The dollar increase is $1.74. [5.5B, Example 3]   **24.** The population now is 220% of the population 10 years ago. [5.5B, Example 4]

25. The value of the car is $25,000. [5.5B, Example 3]

Note: The numbers in brackets following the answers in the Concept Review are a reference to the objective that corresponds to that problem. For example, the reference [1.2A] stands for Section 1.2, Objective A. This notation will be used for all Prep Tests, Concept Reviews, Chapter Reviews, Chapter Tests, and Cumulative Reviews throughout the text.

CUMULATIVE REVIEW EXERCISES

1. 4　[1.6B]　　**2.** 240　[2.1A]　　**3.** $10\frac{11}{24}$　[2.4C]　　**4.** $12\frac{41}{48}$　[2.5C]　　**5.** $12\frac{4}{7}$　[2.6B]　　**6.** $\frac{7}{24}$　[2.7B]　　**7.** $\frac{1}{3}$　[2.8B]

8. $\frac{13}{36}$　[2.8C]　　**9.** 3.08　[3.1B]　　**10.** 1.1196　[3.3A]　　**11.** 34.2813　[3.5A]　　**12.** 3.625　[3.6A]　　**13.** $1\frac{3}{4}$　[3.6B]

14. $\frac{3}{8} < 0.87$　[3.6C]　　**15.** 53.3　[4.3B]　　**16.** \$19.20/hour　[4.2B]　　**17.** $\frac{11}{60}$　[5.1A]　　**18.** $83\frac{1}{3}\%$　[5.1B]

19. 19.56　[5.2A/5.5A]　　**20.** $133\frac{1}{3}\%$　[5.3A/5.5A]　　**21.** 9.92　[5.4A/5.5A]　　**22.** 342.9%　[5.3A/5.5A]

23. Sergio's take-home pay is \$592.　[2.6C]　　**24.** Each monthly payment is \$292.50.　[3.5B]　　**25.** 420 gallons were used during the month.　[3.5B]　　**26.** The real estate tax is \$10,000.　[4.3C]　　**27.** 22,577 hotels in the United States are located along highways.　[5.2B/5.5B]　　**28.** 45% of the people did not favor the candidate.　[5.3B/5.5B]　　**29.** The approximate average number of hours spent watching TV in a week is 61.3 hours.　[5.2B/5.5B]　　**30.** 18% of the children tested had levels of lead that exceeded federal standards.　[5.3B/5.5B]

ANSWERS TO CHAPTER 6 SELECTED EXERCISES

PREP TEST

1. 0.75　[3.5A]　　**2.** 52.05　[3.4A]　　**3.** 504.51　[3.3A]　　**4.** 9750　[3.4A]　　**5.** 45　[3.4A]　　**6.** 1417.24　[3.2A]

7. 3.33　[3.5A]　　**8.** 0.605　[3.5A]　　**9.** 0.379 < 0.397　[3.6C]

SECTION 6.1

1. The unit cost is \$.055 per ounce.　　**3.** The unit cost is \$.374 per ounce.　　**5.** The unit cost is \$.080 per tablet.

7. The unit cost is \$6.975 per clamp.　　**9.** The unit cost is \$.199 per ounce.　　**11.** Divide the price of one pint by 2.

13. The Kraft mayonnaise is the more economical purchase.　　**15.** The Cortexx shampoo is the more economical purchse.

17. The Ultra Mr. Clean is the more economical purchase.　　**19.** The Bertolli olive oil is the more economical purchase.

21. The Wagner's vanilla extract is the more economical purchase.　　**23.** Tea A is the more economical purchase.

25. The total cost is \$73.50.　　**27.** The total cost is \$16.88.　　**29.** The total cost is \$3.89.　　**31.** The total cost is \$26.57.

SECTION 6.2

1. The percent increase is 182.9%.　　**3.** The percent increase is 117.6%.　　**5.** The percent increase is 500%.

7. The percent increase is 111.2%.　　**9.** Yes　　**11.** Use equation (3) and then equation (2).　　**13.** The markup is \$17.10.

15. The markup rate is 60%.　　**17.** The selling price is \$304.50.　　**19.** The selling price is \$74.　　**21.** The percent decrease is 40%.　　**23. a.** The percent decrease in the population of Detroit is 13.7%.　　**b.** The percent decrease in the population of Philadelphia is 7.7%.　　**c.** The percent decrease in the population of Chicago is 1.8%.　　**25.** The car loses \$8460 in value.

27. a. The amount of the decrease was \$35.20.　　**b.** The new average monthly gasoline bill is \$140.80.　　**29.** The percent decrease is 31.9%.　　**31.** Use equation (3) and then (2).　　**33.** The discount rate is $33\frac{1}{3}\%$.　　**35.** The discount is \$80.　　**37.** The discount rate is 30%.　　**39. a.** The discount is \$.25 per pound.　　**b.** The sale price is \$1.00 per pound.　　**41.** The discount rate is 20%.

SECTION 6.3

1. a. \$10,000　　**b.** \$850　　**c.** 4.25%　　**d.** 2 years　　**3.** The simple interest owed is \$960.　　**5.** The simple interest due is \$3375.　　**7.** The simple interest due is \$1320.　　**9.** The simple interest due is \$84.76.　　**11.** The maturity value is \$5120.

13. The total amount due on the loan is \$12,875.　　**15.** The maturity value is \$14,543.70.　　**17.** The monthly payment is \$6187.50.

19. a. The interest charged is \$1080.　　**b.** The monthly payment is \$545.　　**21.** The monthly payment is \$3039.02.

23. a. Student A's principal is equal to student B's principal.　　**b.** Student A's maturity value is greater than student B's maturity value.　　**c.** Student A's monthly payment is less than student B's monthly payment.　　**25.** The finance charge is \$6.85.

27. You owe the company \$11.94.　　**29.** The difference between the finance charges is \$3.44.　　**31.** The finance charges for the first and second months are the same. No, you will not be able to pay off the balance.　　**33.** The value of the investment after 20 years is \$12,380.43.　　**35.** The value of the investment after 5 years is \$28,352.50.　　**37. a.** The value of the investment after 10 years will be \$6040.86.　　**b.** The amount of interest earned will be \$3040.86.　　**39.** The amount of interest earned is \$505.94.

SECTION 6.4

1. The mortgage is $172,450.   **3.** The down payment is $212,500.   **5.** The loan origination fee is $3750.   **7.** The mortgage is $315,000.   **9.** The mortgage is $189,000.   **11.** (iii)   **13.** The monthly mortgage payment is $644.79.   **15.** Yes, the lawyer can afford the monthly mortgage payment.   **17.** The monthly property tax is $166.   **19. a.** The monthly mortgage payment is $898.16.   **b.** The interest payment is $505.69.   **21.** The total monthly payment for the mortgage and property tax is $842.40.   **23.** The monthly mortgage payment is $2295.29.   **25.** The couple can save $63,408 in interest.

SECTION 6.5

1. No, Amanda does not have enough money for the down payment.   **3.** The sales tax is $1192.50.   **5.** The license fee is $650.   **7. a.** The sales tax is $1120.   **b.** The total cost of the sales tax and the license fee is $1395.   **9.** The amount financed is $12,150.   **11.** The amount financed is $36,000.   **13.** The expression represents the total cost of buying the car.   **15.** Use multiplication to find the cost.   **17.** The monthly car payment is $531.43.   **19.** The cost is $252.   **21.** Your cost per mile for gasoline was $.16.   **23. a.** The amount financed is $12,000.   **b.** The monthly car payment is $359.65.   **25.** The monthly payment is $810.23.   **27.** The amount of interest paid is $4665.

SECTION 6.6

1. Lewis earns $460.   **3.** The real estate agent's commission is $3930.   **5.** The stockbroker's commission is $84.   **7.** The teacher's monthly salary is $3244.   **9.** Carlos earned a commission of $540.   **11.** Steven receives $920 for installing the carpet.   **13.** The chemist's hourly wage is $125.   **15. a.** Gil's hourly wage for working overtime is $21.56.   **b.** Gil earns $344.96 for working 16 hours of overtime.   **17. a.** The increase in hourly pay is $1.23.   **b.** The clerk's hourly wage for working the night shift is $9.43.   **19.** Nicole's earnings were $475.   **21.** The starting salary for an accountant in the previous year was $40,312.   **23.** The starting salary for a computer science major in the previous year was $50,364.

SECTION 6.7

1. Your current checking account balance is $486.32.   **3.** The nutritionist's current balance is $825.27.   **5.** The current checking account balance is $3000.82.   **7.** Yes, there is enough money in the carpenter's account to purchase the refrigerator.   **9.** Yes, there is enough money in the account to make the two purchases.   **11.** The account's ending balance might be less than its starting balance on that day.   **13.** The bank statement and the checkbook balance.   **15.** The bank statement and the checkbook balance.   **17.** The ending balance on the monthly bank statement is greater than the ending balance on the check register.

CHAPTER 6 CONCEPT REVIEW*

1. To find the unit cost, divide the total cost by the number of units. The cost is $2.96. The number of units is 4. 2.96 ÷ 4 = 0.74, so the unit cost is $.74 per can.   [6.1A]

2. To find the total cost, multiply the unit cost by the number of units purchased. The unit cost is $.85. The number of units purchased is 3.4 pounds. 0.85 × 3.4 = 2.89, so the total cost is $2.89.   [6.1C]

3. To find the selling price when you know the cost and the markup, add the cost and the markup.   [6.2B]

4. To find the markup when you know the markup rate, multiply the markup rate by the cost.   [6.2B]

5. If you know the percent decrease, you can find the amount of decrease by multiplying the percent decrease by the original value.   [6.2C]

6. If you know the regular price and the sale price, you can find the discount by subtracting the sale price from the regular price.   [6.2D]

7. To find the discount rate, first subtract the sale price from the regular price to find the discount. Then divide the discount by the regular price.   [6.2D]

8. To find simple interest, multiply the principal by the annual interest rate by the time (in years).   [6.3A]

9. To find the maturity value for a simple interest loan, add the principal and the interest.   [6.3A]

10. Principal is the original amount deposited in an account or the original amount borrowed for a loan.   [6.3A]

11. If you know the maturity value of an 18-month loan, you can find the monthly payment by dividing the maturity value by 18 (the length of the loan in months).   [6.3A]

Note: The numbers in brackets following the answers in the Concept Review are a reference to the objective that corresponds to that problem. For example, the reference [1.2A] stands for Section 1.2, Objective A. This notation will be used for all Prep Tests, Concept Reviews, Chapter Reviews, Chapter Tests, and Cumulative Reviews throughout the text.

12. Compound interest is computed not only on the original principal but also on interest already earned. [6.3C]

13. For a fixed-rate mortgage, the monthly payment remains the same throughout the life of the loan. [6.4B]

14. The following expenses are involved in owning a car: car insurance, gas, oil, general maintenance, and the monthly car payment. [6.5B]

15. To balance a checkbook:

  1. In the checkbook register, put a check mark by each check paid by the bank and by each deposit recorded by the bank.

  2. Add to the current checkbook balance all checks that have been written but have not yet been paid by the bank and any interest paid on the account.

  3. Subtract any service charges and any deposits not yet recorded by the bank. This is the checkbook balance.

  4. Compare the balance with the bank balance listed on the bank statement. If the two numbers are equal, the bank statement and the checkbook "balance." [6.7B]

CHAPTER 6 REVIEW EXERCISES

1. The unit cost is $.195 per ounce or 19.5¢ per ounce. [6.1A]   **2.** The cost is $.279 or 27.9¢ per mile. [6.5B]

3. The percent increase is 30.4%. [6.2A]   **4.** The markup is $72. [6.2B]   **5.** The simple interest due is $3000. [6.3A]

6. The value of the investment after 10 years is $45,550.75. [6.3C]   **7.** The percent increase is 15%. [6.2A]   **8.** The total monthly payment for the mortgage and property tax is $1138.90. [6.4B]   **9.** The monthly payment is $518.02. [6.5B]

10. The value of the investment will be $53,593. [6.3C]   **11.** The down payment is $29,250. [6.4A]   **12.** The total cost of the sales tax and license fee is $2096.25. [6.5A]   **13.** The selling price is $2079. [6.2B]   **14.** The interest paid is $157.33. [6.5B]

15. The commission was $3240. [6.6A]   **16.** The sale price is $141. [6.2D]   **17.** The current checkbook balance is $943.68. [6.7A]   **18.** The maturity value is $31,200. [6.3A]   **19.** The origination fee is $1875. [6.4A]   **20.** The more economical purchase is 33 ounces for $6.99. [6.1B]   **21.** The monthly mortgage payment is $2131.62. [6.4B]   **22.** The total income was $655.20. [6.6A]   **23.** The donut shop's checkbook balance is $8866.58. [6.7A]   **24.** The monthly payment is $14,093.75. [6.3A]   **25.** The finance charge is $7.20. [6.3B]

CHAPTER 6 TEST

1. The cost per foot is $6.92. [6.1A, Example 1]   **2.** The more economical purchase is 3 pounds for $7.49. [6.1B, Example 2]

3. The total cost is $14.53. [6.1C, Example 3]   **4.** The percent increase in the cost of the exercise bicycle is 20%. [6.2A, Example 1]   **5.** The selling price of the blu-ray disc player is $441. [6.2B, Example 4]   **6.** The percent increase is 8.9%. [6.2A, Example 1]   **7.** The percent decrease is 20%. [6.2C, HOW TO 3]   **8.** The sale price of the corner hutch is $209.30. [6.2D, Example 8]   **9.** The discount rate is 40%. [6.2D, Example 7]   **10.** The simple interest due is $2000. [6.3A, You Try It 1]

11. The maturity value is $26,725. [6.3A, Example 2]   **12.** The finance charge is $4.50. [6.3B, Example 4]   **13.** The amount of interest earned in 10 years was $24,420.60. [6.3C, You Try It 5]   **14.** The loan origination fee is $3350. [6.4A, Example 2]

15. The monthly mortgage payment is $1713.44. [6.4B, HOW TO 2]   **16.** The amout financed is $19,000. [6.5A, Example 1]

17. The monthly truck payment is $686.22. [6.5B, Example 5]   **18.** Shaney earnes $1596. [6.6A, Example 1]

19. The current checkbook balance is $6612.25. [6.7A, Example 1]   **20.** The bank statement and the checkbook balance. [6.7B, Example 2]

CUMULATIVE REVIEW EXERCISES

1. 13 [1.6B]   **2.** $8\frac{13}{24}$ [2.4C]   **3.** $2\frac{37}{48}$ [2.5C]   **4.** 9 [2.6B]   **5.** 2 [2.7B]   **6.** 5 [2.8C]   **7.** 52.2 [3.5A]

8. 1.417 [3.6A]   **9.** $51.25/hour [4.2B]   **10.** 10.94 [4.3B]   **11.** 62.5% [5.1B]   **12.** 27.3 [5.2A]

13. 0.182 [5.1A]   **14.** 42% [5.3A]   **15.** 250 [5.4A]   **16.** 154.76 [5.4A/5.5A]   **17.** The total rainfall is $13\frac{11}{12}$ inches. [2.4D]   **18.** The amount paid in taxes is $970. [2.6C]   **19.** The ratio is $\frac{3}{5}$. [4.1B]   **20.** 33.4 miles are driven per gallon. [4.2C]   **21.** The unit cost is $1.10 per pound. [4.2C]   **22.** The dividend is $280. [4.3C]   **23.** The sale price is $720. [6.2D]   **24.** The selling price of the grinding rail is $119. [6.2B]   **25.** The percent increase in Sook Kim's salary is 8%. [6.2A]   **26.** The simple interest due is $2700. [6.3A]   **27.** The monthly car payment is $791.81. [6.5B]   **28.** The family's new checking account balance is $2243.77. [6.7A]   **29.** The cost per mile is $.33. [6.5B]   **30.** The monthly mortgage payment is $1232.26. [6.4B]

Glossary

addend In addition, one of the numbers added. [1.2]

addition The process of finding the total of two numbers. [1.2]

Addition Property of Zero Zero added to a number does not change the number. [1.2]

approximation An estimated value obtained by rounding an exact value. [1.1]

Associative Property of Addition Numbers to be added can be grouped (with parentheses, for example) in any order; the sum will be the same. [1.2]

Associative Property of Multiplication Numbers to be multiplied can be grouped (with parentheses, for example) in any order; the product will be the same. [1.4]

average The sum of all the numbers divided by the number of those numbers. [1.5]

balancing a checkbook Determining whether the checking account balance is accurate. [6.7]

bank statement A document showing all the transactions in a bank account during the month. [6.7]

basic percent equation Percent times base equals amount. [5.2]

borrowing In subtraction, taking a unit from the next larger place value in the minuend and adding it to the number in the given place value in order to make that number larger than the number to be subtracted from it. [1.3]

carrying In addition, transferring a number to another column. [1.2]

check A printed form that, when filled out and signed, instructs a bank to pay a specified sum of money to the person named on it. [6.7]

checking account A bank account that enables you to withdraw money or make payments to other people, using checks. [6.7]

commission That part of the pay earned by a salesperson that is calculated as a percent of the salesperson's sales. [6.6]

common factor A number that is a factor of two or more numbers is a common factor of those numbers. [2.1]

common multiple A number that is a multiple of two or more numbers is a common multiple of those numbers. [2.1]

Commutative Property of Addition Two numbers can be added in either order; the sum will be the same. [1.2]

Commutative Property of Multiplication Two numbers can be multiplied in either order; the product will be the same. [1.4]

composite number A number that has whole-number factors besides 1 and itself. For instance, 18 is a composite number. [1.7]

compound interest Interest computed not only on the original principal but also on interest already earned. [6.3]

cost The price that a business pays for a product. [6.2]

cross product In a proportion, the product of the numerator on the left side of the proportion times the denominator on the right, and the product of the denominator on the left side of the proportion times the numerator on the right. [4.3]

decimal A number written in decimal notation. [3.1]

decimal notation Notation in which a number consists of a whole-number part, a decimal point, and a decimal part. [3.1]

decimal part In decimal notation, that part of the number that appears to the right of the decimal point. [3.1]

decimal point In decimal notation, the point that separates the whole-number part from the decimal part. [3.1]

deposit slip A form for depositing money in a checking account. [6.7]

difference In subtraction, the result of subtracting two numbers. [1.3]

discount The difference between the regular price and the sale price. [6.2]

discount rate The percent of a product's regular price that is represented by the discount. [6.2]

dividend In division, the number into which the divisor is divided to yield the quotient. [1.5]

division The process of finding the quotient of two numbers. [1.5]

divisor In division, the number that is divided into the dividend to yield the quotient. [1.5]

down payment The percent of a home's purchase price that the bank, when issuing a mortgage, requires the borrower to provide. [6.4]

equivalent fractions Equal fractions with different denominators. [2.3]

expanded form The number 46,208 can be written in expanded form as $40,000 + 6000 + 200 + 0 + 8$. [1.1]

exponent In exponential notation, the raised number that indicates how many times the number to which it is attached is taken as a factor. [1.6]

exponential notation The expression of a number to some power, indicated by an exponent. [1.6]

factors In multiplication, the numbers that are multiplied. [1.4]

factors of a number The whole-number factors of a number divide that number evenly. [1.7]

finance charges Interest charges on purchases made with a credit card. [6.3]

fixed-rate mortgage A mortgage in which the monthly payment remains the same for the life of the loan. [6.4]

fraction The notation used to represent the number of equal parts of a whole. [2.2]

fraction bar The bar that separates the numerator of a fraction from the denominator. [2.2]

graph of a whole number A heavy dot placed directly above that number on the number line. [1.1]

greater than A number that appears to the right of a given number on the number line is greater than the given number. [1.1]

greatest common factor (GCF) The largest common factor of two or more numbers. [2.1]

hourly wage Pay calculated on the basis of a certain amount for each hour worked. [6.6]

improper fraction A fraction greater than or equal to 1. [2.2]

interest Money paid for the privilege of using someone else's money. [6.3]

interest rate The percent used to determine the amount of interest. [6.3]

inverting a fraction Interchanging the numerator and denominator. [2.7]

least common denominator (LCD) The least common multiple of denominators. [2.4]

least common multiple (LCM) The smallest common multiple of two or more numbers. [2.1]

less than A number that appears to the left of a given number on the number line is less than the given number. [1.1]

license fees Fees charged for authorization to operate a vehicle. [6.5]

loan origination fee The fee a bank charges for processing mortgage papers. [6.4]

markup The difference between selling price and cost. [6.2]

markup rate The percent of a product's cost that is represented by the markup. [6.2]

maturity value of a loan The principal of a loan plus the interest owed on it. [6.3]

minuend In subtraction, the number from which another number (the subtrahend) is subtracted. [1.3]

mixed number A number greater than 1 that has a whole-number part and a fractional part. [2.2]

monthly mortgage payment One of 12 payments due each year to the lender of money to buy real estate. [6.4]

mortgage The amount borrowed to buy real estate. [6.4]

multiples of a number The products of that number and the numbers 1, 2, 3, ... [2.1]

multiplication The process of finding the product of two numbers. [1.4]

Multiplication Property of One The product of a number and 1 is the number. [1.4]

number line A line on which a number can be graphed. [1.1]

numerator The part of a fraction that appears above the fraction bar. [2.2]

Order of Operations Agreement A set of rules that tells us in what order to perform the operations that occur in a numerical expression. [1.6]

percent Parts per hundred. [5.1]

percent decrease A decrease of a quantity, expressed as a percent of its original value. [6.2]

percent increase An increase of a quantity, expressed as a percent of its original value. [6.2]

period In a number written in standard form, each group of digits separated from other digits by a comma or commas. [1.1]

place value The position of each digit in a number written in standard form determines that digit's place value. [1.1]

place-value chart A chart that indicates the place value of every digit in a number. [1.1]

points A term banks use to mean percent of a mortgage; used to express the loan origination fee. [6.4]

prime factorization The expression of a number as the product of its prime factors. [1.7]

prime number A number whose only whole-number factors are 1 and itself. For instance, 13 is a prime number. [1.7]

principal The amount of money originally deposited or borrowed. [6.3]

product In multiplication, the result of multiplying two numbers. [1.4]

proper fraction A fraction less than 1. [2.2]

property tax A tax based on the value of real estate. [6.4]

proportion An expression of the equality of two ratios or rates. [4.3]

quotient In division, the result of dividing the divisor into the dividend. [1.5]

rate A comparison of two quantities that have different units. [4.2]

ratio A comparison of two quantities that have the same units. [4.1]

reciprocal of a fraction The fraction with the numerator and denominator interchanged. [2.7]

remainder In division, the quantity left over when it is not possible to separate objects or numbers into a whole number of equal groups. [1.5]

rounding Giving an approximate value of an exact number. [1.1]

salary Pay based on a weekly, biweekly, monthly, or annual time schedule. [6.6]

sale price The reduced price. [6.2]

sales tax A tax levied by a state or municipality on purchases. [6.5]

selling price The price for which a business sells a product to a customer. [6.2]

service charge An amount of money charged by a bank for handling a transaction. [6.7]

simple interest Interest computed on the original principal. [6.3]

simplest form of a fraction A fraction is in simplest form when there are no common factors in the numerator and denominator. [2.3]

simplest form of a rate A rate is in simplest form when the numbers that make up the rate have no common factor. [4.2]

simplest form of a ratio A ratio is in simplest form when the two numbers do not have a common factor. [4.1]

standard form A whole number is in standard form when it is written using the digits 0, 1, 2, . . . , 9. An example is 46,208. [1.1]

subtraction The process of finding the difference between two numbers. [1.3]

subtrahend In subtraction, the number that is subtracted from another number (the minuend). [1.3]

sum In addition, the total of the numbers added. [1.2]

total cost The unit cost multiplied by the number of units purchased. [6.1]

true proportion A proportion in which the fractions are equal. [4.3]

unit cost The cost of one item. [6.1]

unit rate A rate in which the number in the denominator is 1. [4.2]

whole numbers The whole numbers are 0, 1, 2, 3, ... [1.1]

whole-number part In decimal notation, that part of the number that appears to the left of the decimal point. [3.1]

Index

Index of Applications